A to Z GIS

An illustrated dictionary of geographic information systems

Edited by Iasha Wade and Shelly Sommer

ESRI Press
REDLANDS, CALIFORNIA

ESRI Press, 380 New York Street, Redlands, California 92373-8100

 First edition 2001
Second edition 2006
10 09 08 07 2 3 4 5 6 7 8 9 10
Originally published as *The ESRI Press Dictionary of GIS Terminology.* Copyright © 2001 ESRI

Printed in the United States of America

Library of Congress Cataloging-in-Publication Data
A to Z GIS : an illustrated dictionary of geographic information systems / edited by Tasha Wade and Shelly Sommer.—2nd ed.
p. cm.
ISBN-13: 978-1-58948-140-4
ISBN-10: 1-58948-140-2
1. Geographic information systems—Dictionaries. I. Wade, Tasha, 1973– II. Sommer, Shelly, 1969–
G70.212A86 2006
910'.285—dc22 2006017908

Ask for ESRI Press titles at your local bookstore or order by calling 1-800-447-9778. You can also shop online at www.esri.com/esripress. Outside the United States, contact your local ESRI distributor.

ESRI Press titles are distributed to the trade by the following:

In North America, South America, Asia, and Australia:
Independent Publishers Group (IPG)
Telephone (United States): 1-800-888-4741
Telephone (international): 312-337-0747
E-mail: frontdesk@ipgbook.com

In the United Kingdom, Europe, and the Middle East:
Transatlantic Publishers Group Ltd.
Telephone: 44 20 7373 2515
Fax: 44 20 7244 1018
E-mail: richard@tpgltd.co.uk

Book and cover design, and illustration by Jennifer Jennings

A to Z **GIS**

Preface

Geographic information systems (GIS) are now ubiquitous. They have spread far beyond the government, utility, and natural resource management organizations where they found their first homes. Because of this technological expansion, GIS users need a comprehensive dictionary of GIS language. Due to its interdisciplinary nature, GIS has borrowed terms from related fields. In many cases, the meanings of these words have evolved and shifted over time.

With more than 1,800 terms, *A to Z GIS: An Illustrated Dictionary of Geographic Information Systems* was designed to be a comprehensive technical dictionary for GIS students and professionals alike. Our goal for this dictionary was to create definitions that are technically accurate yet not intimidating to the GIS novice.

A to Z GIS is a dictionary of general GIS terminology, not software-specific terminology. Those who would like access to ESRI software-specific definitions can find them in the GIS Dictionary at support.esri.com/gisdictionary.

Choosing terms

The first step in creating a dictionary is choosing which terms to include and which to leave out. This process was a challenge, since there are surprisingly few "pure" GIS terms. Most terms used in GIS have been adopted from such related fields as cartography, computing, geodesy, mathematics, remote sensing, statistics, and surveying.

Our solution was to identify three classes of terms within the lexicon:

- Core GIS terms describe GIS concepts, processes, and operations. While many of these terms may have origins in other fields, they are more strongly correlated with GIS than with any related field. We included as many of these terms as possible.

- Terms from related fields such as cartography, computing, geodesy, geography, GPS, and remote sensing. These can be thought of as source fields that have lent methods, data, theory, and metaphors to GIS. We included the terms from related fields that a GIS practitioner or student is mostly likely to come across in the course of their activities, but left out terms that lay outside the GIS context.
- Terms from application areas of GIS that can be thought of as sink fields. Terms from these fields are less likely to appear in a broad range of GIS contexts. Terms were rarely selected from application areas.

Writing definitions

Once we decided on the fields from which to select terms, we chose terms from a database of terms and definitions used to construct glossaries and dictionaries at ESRI. The definitions in this database were written by more than a hundred subject-matter experts at ESRI and reviewed by others for accuracy and consistency. Five academic reviewers were kind enough to comment on all or part of the dictionary draft, and their insights into how words are really used and taught proved invaluable.

Creating the design and illustrations

The dictionary's design and illustrations were created by award-winning designer and artist Jennifer Jennings. Her challenge was to create a dictionary size and style that would be handy, usable, and approachable. Because the illustration size was limited, she had to combine conceptual clarity and technical accuracy with extreme simplicity. With that in mind, she created the illustrations to look like sophisticated sketches with hand-written notes. For the maps in this dictionary, the goal of conveying the main idea of a concept in a tight space often meant a departure from traditional cartographic methods.

The future

This dictionary has truly been a collaborative effort, and we are grateful to all the subject-matter experts, reviewers, and other staff and contributors who have been involved in its compilation. The expertise that has gone into this dictionary is broad, but we welcome feedback and alternative interpretations. Those who wish to provide feedback may do so via the online GIS Dictionary at support.esri.com/gisdictionary. We will consider all viewpoints and submissions for future printings and editions. GIS has a strong future, and as applications for GIS multiply, GIS terminology will continue to evolve and require definition.

—*The Editors*

Acknowledgments

We are indebted to Jack Dangermond, president of ESRI, for his support of this book, and for providing the kind of environment that fosters creativity and collaboration. We would also like to thank Nick Frunzi, director of ESRI's Educational Services Division, for his ongoing support and encouragement, and our managers, Christian Harder, Judy Hawkins, and Randy Worch, for all the advice and assistance they have provided to us over the past few years as we have worked on this book.

Special thanks to our associate editor, Tim Ormsby, who clarified many of the most complicated concepts in GIS. His thoughtful approach improved many of our most fundamental definitions.

More than a hundred people at ESRI contributed definitions, edits, or comments to *A to Z GIS*. Without their collaboration and expertise, this book would never have been produced. Our thanks to all for taking time out of their busy schedules to write and review terms and definitions, and answer our questions. We are especially thankful to the authors of our appendix articles, who provided additional information about some of the more complex topics in GIS.

Contributors include

Eric Akin, Jamil Alvi, Peter Aniello, David Arctur, Kim Avery, Jonathan Bailey, David Barnes, Mark Berry, Suzanne Boden, Bob Booth, Hal Bowman, Judy Boyd, Steve Bratt, Joe Breman, Patrick Brennan, Pat Breslin, Evan Brinton, Clint Brown, Aileen Buckley, Garry Burgess, Anthony Burgon, Rob Burke, Tarun Chandrasekhar, Colin Childs, Kristin Clark, Amy Collins, Clayton Crawford, Scott Crosier, Brian Cross, Matt Crowder, Greg Cunningham, Jason Cupp, Katy Dalton, David Danko, Eleanor Davies, David Davis, Tom DePuyt, Mara Dolan, Thomas Dunn, Cory Eicher, Gregory Emmanuel, Rupert Essinger, Witold Fraczek, Steve Frizzell, Charlie Frye, Tracy Gannon, Peng Gao, Joshua Giese, Shelly Gill, Craig Gillgrass, Rhonda Glennon, Brian Goldin,

John Grammer, Craig Greenwald, Michael Grossman, Paul Hardy, Melanie Harlow, Alan Hatakeyama, Harlan Heimgartner, Wayne Hewitt, Catherine Hill, Vicki Hill, Tim Hodson, Mike Hogan, Jennifer Jackson, Jennifer Jennings, Ann Johnson, Kevin Johnston, Catherine Jones, Rob Jordan, Gary Kabot, Kimberly Kearns, Timothy Kearns, Kevin M. Kelly, Heather Kennedy, Melita Kennedy, Steve Kopp, Kyle Krattiger, Konstantin Krivoruchko, Juan Laguna, Marcy LaViollette, Derek Law, Christine Leslie, Adrien Litton, Mike Livingston, Clint Loveman, Steve Lynch, Andy MacDonald, Gary MacDougall, Keith Mann, Michael Mannion, Frank Martin, Sean McCarron, Heather McCracken, Matt McGrath, Ginger McKay, Dan Meeks, Bill Moreland, Doug Morgenthaler, Makram Murad Al-Shaikh, Scott Murray, Jonathan Murphy, Nathan Noble, Nawajish Noman, Serene Ong, Tim Ormsby, Sarah Osborne, Krista Page, Jason Pardy, Brian Parr, Jamie Parrish, Andrew Perencsik, Kim Peter, Morakot Pilouk, Christie Pleiss, Greg Pleiss, Ghislain Prince, Edie Punt, Jeff Reinhart, Amir Razavi, Anne Reuland, Jaynya Richards, Mike Ridland, Rick Rossi, Phil Sanchez, Frederic Schettini, Lauren Scott, Charles Serafy, Jeff Shaner, Shannon Shields, Gillian Silvertand, Damian Spangrud, Marc St. Onge, Bjorn Svensson, Sally Swenson, Clark Swinehart, Agatha Tang, Emily Tessar, Corey Tucker, Francois Valois, Aleta Vienneau, Nathan Warmerdam, David Watkins, Kyle Watson, Craig Williams, Jason Willison, Jill Willison, Simon Woo, Randy Worch, Molly Zurn, Hong Xu.

We are deeply indebted to Pat Breslin, Mike Livingston, and the Content Studio team for their ongoing technical support of our database, schema, and database publishers. Without them, this book would not have been possible.

We are grateful to the ESRI Press staff who supported *A to Z GIS*. Jennifer Jennings created a book cover, design, and illustrations that are not only instructive, but beautiful. Michael Law provided cartographic support and contributed illustrations. Copy editor Tiffany Wilkerson gave us the full benefit of her attention to detail and immense knowledge of *The Chicago Manual of Style*, and David Boyles provided careful reviews and feedback on all our drafts. Thanks also to our production coordinator, Cliff Crabbe.

We greatly appreciate the assistance of librarians Patty Turner and Colleen Conner, who gave us access to hundreds of books and articles in the ESRI Library collection. Thanks also to our intern, Will Lewis, who spent his summer vacation reviewing definitions for accuracy with ESRI subject-matter experts.

We are grateful to Roland Viger for helping us sort out modeling terms.

We are deeply indebted to our academic reviewers, who helped us untangle important concepts and suggested innumerable ways to improve our drafts:

Dr. Barbara Buttenfield, Professor and Director of the Meridian Lab, Department of Geography, University of Colorado, Boulder

Dr. Michael N. DeMers, Associate Professor, Department of Geography, New Mexico State University

Mr. Jeffrey D. Hamerlinck, Director, Wyoming Geographic Information Science Center, University of Wyoming, Laramie

Dr. Duane F. Marble, Professor Emeritus of Geography and Natural Resources, The Ohio State University

Dr. Nicholas Nagle, Assistant Professor, Department of Geography, University of Colorado, Boulder

Dr. Marble, in particular, took the time to review the entire manuscript and meet with us about key concepts.

How to use this dictionary

Parts of an entry

- *Headword*—The term being defined is in bold type at the beginning of the entry.
- *Cross-references*—Listed after the last definition in an entry with "See also," these terms are related in some way to the headword. They might be synonyms, antonyms, broader terms, or narrower terms.
- *Definition*—The meaning of the headword is explained in the definition. For terms that have more than one definition (sense), the definitions are numbered and placed in order. GIS senses appear first. Senses from related fields follow and are given *field labels.*
- *Field label*—These are used to classify definitions by subject area. Only terms that come from outside GIS are given field labels. Field labels are set in brackets.
- *Illustration*—Illustrations appear after the definition they refer to and are labeled with the headword.

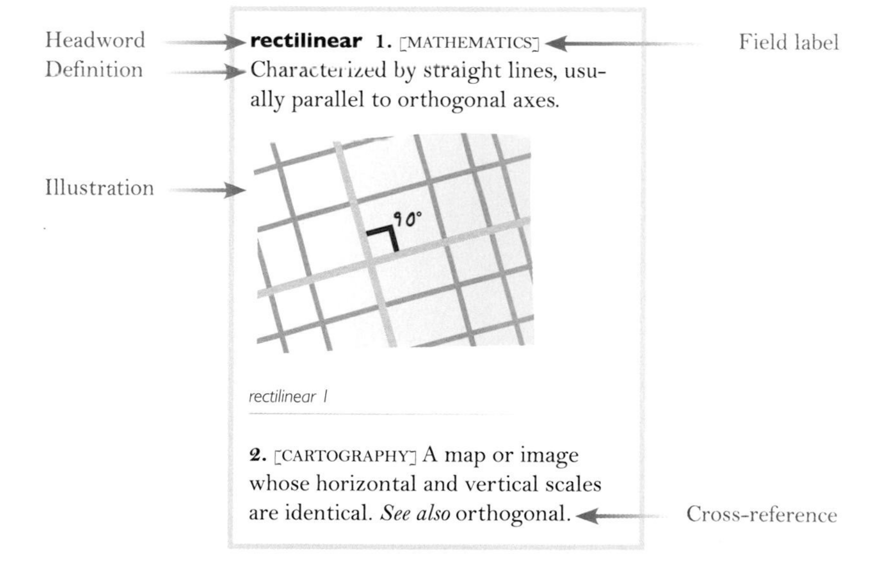

A note about the order of terms and selection of headwords

The terms in *A to Z GIS* are alphabetized letter by letter, according to lexicographical tradition. Spaces and hyphens between words are ignored. For example, baseline, basemap, and base station appear in that order. (A word-by-word alphabetization would place them in the following order: base station, baseline, basemap). The letter-by-letter system is advantageous for readers who may not know whether a compound is considered one word or two. Headwords for definitions have been selected based on usage, especially in cases where more than one spelling or form of a word exists, or with acronyms and abbreviations.

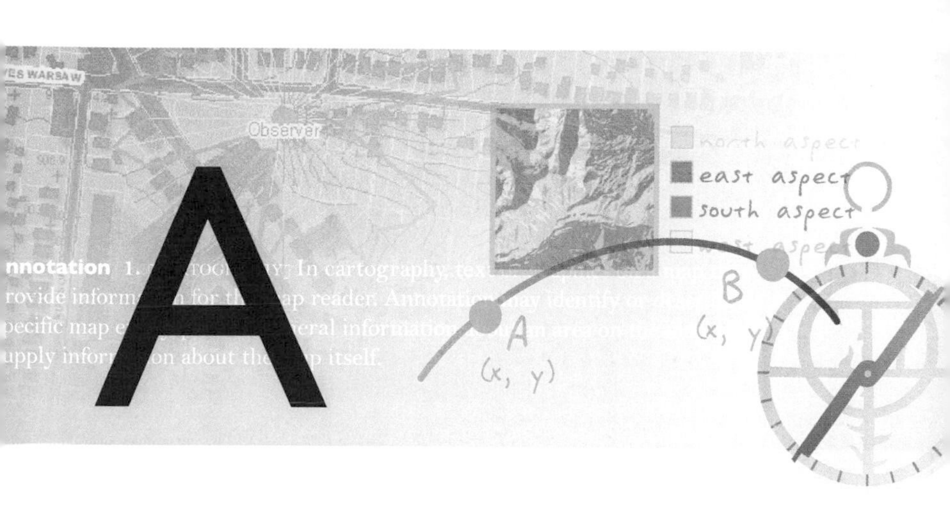

abbreviation A shortened form of a word or phrase that represents the whole. Abbreviations are commonly a letter or group of letters taken from the complete form of the word, such as the usage of *St.* in place of *Street.*

abscissa [CARTOGRAPHY] In a rectangular coordinate system, the distance of the x-coordinate along a horizontal axis from the vertical or y-axis. For example, a point with the coordinates (7,3) has an abscissa of 7. *See also* ordinate.

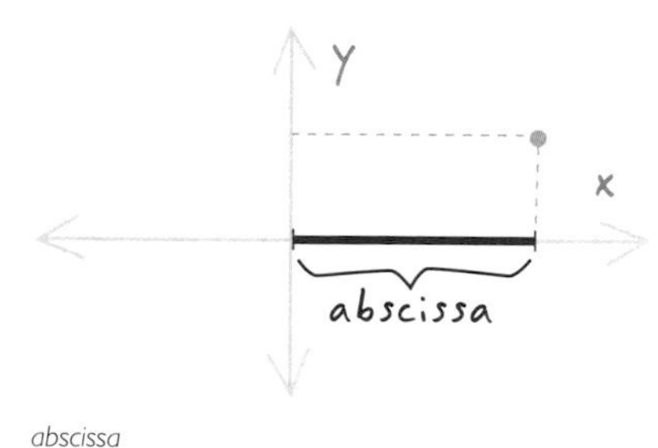

abscissa

absolute accuracy [CARTOGRAPHY] The degree to which the position of an object on a map conforms to its correct location on the earth according to an accepted coordinate system.

absolute coordinates [CARTOGRAPHY] Coordinates that are referenced to the origin of a given coordinate system. *See also* coordinates.

absolute mode *See* digitizing mode.

absorption [REMOTE SENSING] The amount of electromagnetic energy lost through interactions with gas molecules and matter during its passage through the atmosphere.

abstraction A simplified idea of a real-world object or system.

accessibility 1. An aggregate measure of the degree of ease with which a place, person, or thing can be reached, depending on factors such as slope, traffic, distance, and so on. 2. The degree to which Web sites, software, or computers provide equivalent information and functionality to a variety of people, including those with disabilities or visual impairment. *See also* impedance.

A

accuracy [MATHEMATICS] The degree to which a measured value conforms to true or accepted values. Accuracy is a measure of correctness. It is distinguished from precision, which measures exactness. *See also* precision, uncertainty.

across-track scanner [REMOTE SENSING] A remote-sensing tool with an oscillating mirror that moves back and forth across a satellite's direction of travel, creating scan line strips that are contiguous or that overlap slightly, thereby producing an image. *See also* along-track scanner.

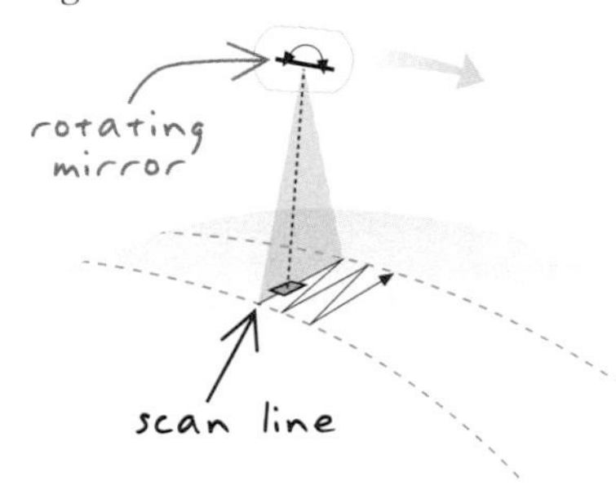

across-track scanner

active remote sensing [REMOTE SENSING] A remote-sensing system, such as radar, that produces electromagnetic radiation and measures its reflection back from a surface. *See also* passive remote sensing, remote sensing.

acutance [REMOTE SENSING] A measure, using a microdensitometer or other instrument, of how well a photographic system shows sharp edges between contiguous bright and dark areas.

address [SURVEYING] A designation of the location of a person's residence or workplace, an organization, or a building, consisting of numerical and text elements such as a street number, street name, and city arranged in a particular format.

address data Data that contains address information used for geocoding. Address data may consist of one individual address or a table containing many addresses. *See also* geocoding.

address element One of the components that comprise an address. House numbers, street names, street types, and street directions are examples of address elements.

address geocoding *See* geocoding.

address locator [ESRI SOFTWARE] A dataset in ArcGIS that stores the address attributes, associated indexes, and rules that define the process for translating nonspatial descriptions of places, such as street addresses, into spatial data that can be displayed as features on a map. An address locator contains a snapshot of the reference data used for geocoding, and parameters for standardizing addresses, searching for match locations, and creating output. Address locator files have a .loc file extension. In ArcGIS 8.3 and previous versions, an address locator was called a geocoding service.

address matching A process that compares an address or a table of addresses to the address attributes of a reference dataset to determine whether a particular address falls within an address range associated with a feature in the reference dataset. If an address falls within a feature's address range, it is considered a match and a location can be returned.

adjacency 1. [GEOGRAPHY] A type of spatial relationship in which two or more polygons share a side or boundary.

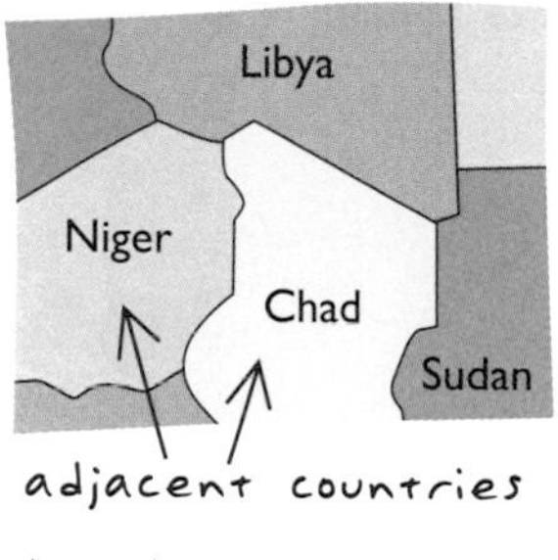

adjacency 1

2. [MATHEMATICS] The state or quality of lying close or contiguous.

adoption The process of appropriating a technology and putting it into use for one's own purposes; the act by an individual, organization, or community of choosing a technology and putting it into effect. *See also* diffusion.

Advanced Very High Resolution Radiometer *See* AVHRR.

aerial photograph [REMOTE SENSING] A photograph of the earth's surface taken from a platform flying above the surface but not in orbit, usually an aircraft. Aerial photography is often used as a cartographic data source for basemapping, locating geographic features, and interpreting environmental conditions. *See also* orthophotograph, oblique photograph, vertical photograph.

affine transformation [MATHEMATICS] A geometric transformation that scales, rotates, skews, and/or translates images or coordinates between any two Euclidean spaces. It is commonly used in GIS to transform maps between coordinate systems. In an affine transformation, parallel lines remain parallel, the midpoint of a line segment remains a midpoint, and all points on a straight line remain on a straight line. *See also* transformation.

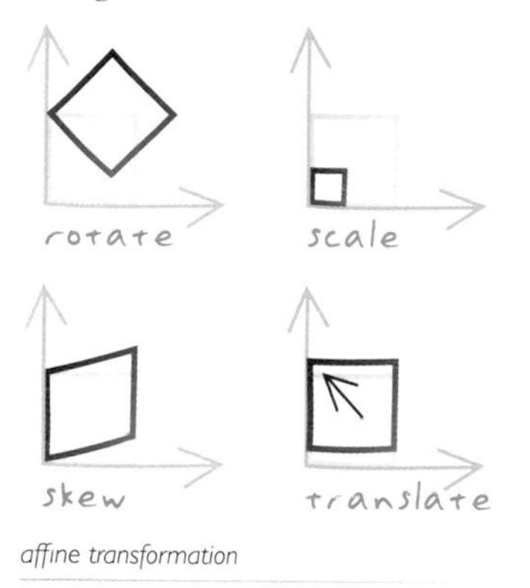

affine transformation

agent In modeling, an entity within a model that conducts transactions to simulate the actions of a human, group of humans, animal, or other actor. *See also* agent-based model.

agent-based model A simulation of the large-scale consequences of the decisions and interactions of individual members of a population. An agent-based model consists of an environment or framework that defines the scope and rules of actions, along with a number of agents representing one or more actors whose parameters and behaviors are defined. When the model is run, the characteristics of each agent are tracked through time and space. *See also* agent.

aggregation The process of collecting a set of similar, usually adjacent, polygons (with their associated attributes) to form a single, larger entity.

air station *See* exposure station.

albedo [PHYSICS] A measure of the reflectivity of an object or surface; the ratio of the amount of radiation reflected by a body to the amount of energy striking it.

algorithm [COMPUTING] A mathematical procedure used to solve problems with a series of steps. Algorithms are usually encoded as a sequence of computer commands.

alias [COMPUTING] An alternative name specified for fields, tables, files, or datasets that is more descriptive and user-friendly than the actual name. On computer networks, a single e-mail alias may refer to a group of e-mail addresses.

aliasing The jagged appearance of curves and diagonal lines in a raster image. Aliasing becomes more apparent as the size of the raster pixels is increased or the resolution of the image is decreased.

alidade 1. [SURVEYING] A peep sight mounted on a straightedge and used to measure direction. 2. [SURVEYING] The part of a theodolite containing the telescope and attachments. *See also* theodolite.

aligned dimension [SURVEYING] A drafting symbol that runs parallel to the baseline and indicates the true distance between beginning and ending dimension points. *See also* linear dimension.

allocation In network analysis, the process of assigning entities or edges and junctions to features until the feature's capacity or limit of impedance is reached. For example, streets may be assigned to the most accessible fire station within a six-minute radius, or students may be assigned to the nearest school until it is full.

almanac 1. [GPS] A file transmitted from a satellite to a receiver that contains information about the orbits of all satellites included in the satellite network. Receivers refer to the almanac to determine which satellite to track. 2. [ASTRONOMY] An annual publication containing weather forecasts, information on astronomical events, and miscellaneous facts, arranged according to the calendar of a given year.

along-track scanner [REMOTE SENSING] A remote-sensing tool with a line of many fixed sensors that record reflected radiation from the terrain along a satellite's direction of movement, creating scan-line strips that are contiguous or that overlap slightly, thereby producing an image. *See also* across-track scanner.

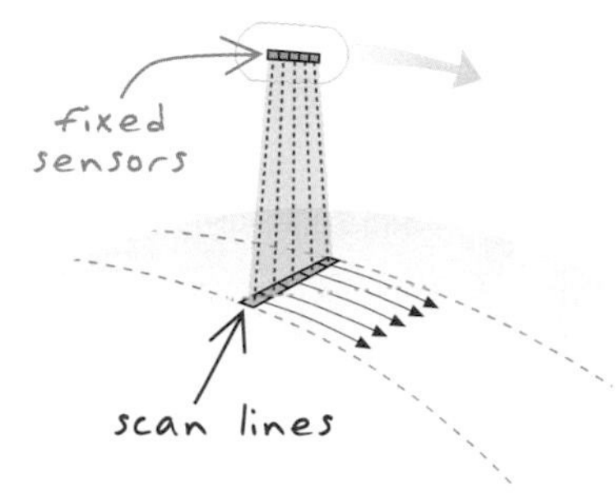

along-track scanner

alphanumeric grid [CARTOGRAPHY] A grid of numbered rows and lettered columns (or vice versa) superimposed on a map, used to find and identify features. Alphanumeric grids are commonly used as a reference system on local street maps.

alternate key An attribute or set of attributes in a relational database that provides a unique identifier for each record and could be used as an alternative to the primary key. *See also* key.

altitude **1.** [CARTOGRAPHY] The height or vertical elevation of a point above a reference surface. Altitude measurements are usually based on a given reference datum, such as mean sea level. **2.** The height above the horizon, measured in degrees, from which a light source illuminates a surface. Altitude is used when calculating a hillshade, or for controlling the position of a light source in a scene. *See also* elevation.

AM/FM *Acronym for automated mapping/facilities management.* GIS or CAD-based systems used by utilities and public works organizations for storing, manipulating, and mapping facility information such as the location of geographically dispersed assets.

ambiguity A state of uncertainty in data classification that exists when an object may appropriately be assigned two or more values for a given attribute. For example, coastal areas experiencing tidal fluctuations may be dry land at some times and under water at other times. Ambiguity may be caused by changeable conditions in reality, by incomplete or conflicting definitions of attributes, or by subjective differences in the evaluation of data. It may also be caused by disputes, as when two parties claim ownership of the same tract of land. *See also* vagueness, uncertainty.

American National Standards Institute *See* ANSI.

American Standard Code for Information Interchange *See* ASCII.

AML [ESRI SOFTWARE] *Acronym for ARC Macro Language.* A proprietary, high-level programming language created by ESRI for generating end-user applications in ArcInfo Workstation. *See also* high-level language.

amoeba *See* complex market area.

anaglyph A stereo image made by superimposing two images of the same area. The images are displayed in complementary colors, usually red and blue or green. When viewed through filters of corresponding colors, the images appear as one three-dimensional image.

anaglyph

analog 1. Represented continuously rather than in discrete steps; having value at any degree of precision. 2. In electronics, having a continuously variable signal, or a circuit or device that carries such signals. *See also* digital.

analog image An image represented by continuous variation in tone, such as a photograph. *See also* image, digital image, aerial photograph.

analog image

analysis A systematic examination of a problem or complex entity in order to provide new information from what is already known.

analysis mask *See* mask.

analysis of variance [STATISTICS] A procedure used to evaluate the variance of the mean values for two or more datasets in order to assess the probability that the data comes from the same sample or statistical population.

ancillary data 1. [DIGITAL IMAGE PROCESSING] Data from sources other than remote sensing, used to assist in analysis and classification or to populate metadata. 2. Supplementary data.

angular unit [GEODESY] The unit of measurement on a sphere or a spheroid, usually degrees. Some map projection parameters, such as the central meridian and standard parallel, are defined in angular units. *See also* linear unit.

anisotropic Having nonuniform spatial distribution of movement or properties, usually across a surface. *See also* anisotropy, isotropic.

anisotropy [STATISTICS] A property of a spatial process or data in which spatial dependence (autocorrelation) changes with both the distance and the direction between two locations. *See also* anisotropic, isotropy, autocorrelation.

annotation 1. [CARTOGRAPHY] Text or graphics on a map that provide information for the map reader. Annotation may identify or describe a specific map entity, provide general information about an area on the map, or supply information about the map itself.

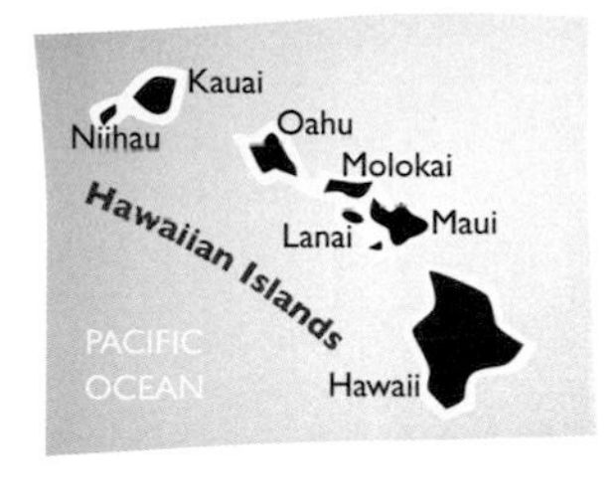

annotation 1

2. [ESRI SOFTWARE] In ArcGIS, text or graphics that can be individually selected, positioned, and modified. Annotation may be manually entered or generated from labels. Annotation can be stored as features in a geodatabase or as map annotation in a data frame. *See also* label. **For more information about annotation, see* Cartographic text in ArcGIS software: Annotation, labeling, and graphic text *on page 245.*

ANOVA *See* analysis of variance.

ANSI *Acronym for American National Standards Institute.* The private, nonprofit organization that develops U.S. industry standards through consensus and public review.

antipode [GEODESY] Any point on the surface of a sphere that lies 180 degrees (opposite) from a given point on the same surface, so that a line drawn between the two points through the center of the sphere forms a true diameter.

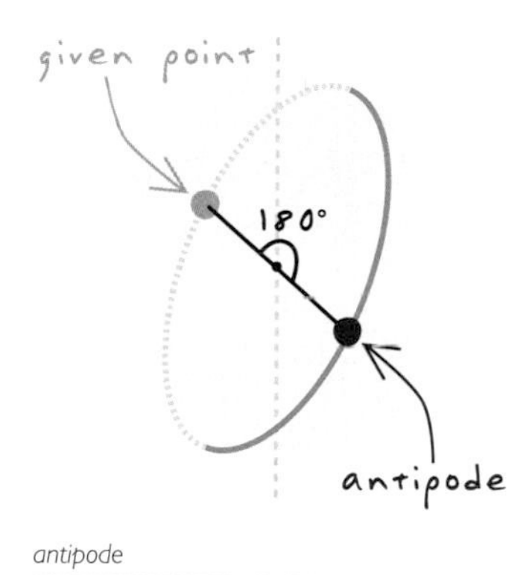

antipode

anywhere fix [GPS] A position that a GPS receiver can calculate without knowing its own location or the local time.

AOI *See* area of interest.

aphylactic projection *See* compromise projection.

API [PROGRAMMING] *Acronym for application programming interface.* A set of interfaces, methods, protocols, and tools that application developers use to build or customize a software program. APIs make it easier to develop a program by providing building blocks of prewritten, tested, and documented code that are incorporated into the new program. APIs can be built for any programming language.

apogee [ASTRONOMY] In an orbit path, the point at which the object in orbit is farthest from the center of the body being orbited. *See also* perigee.

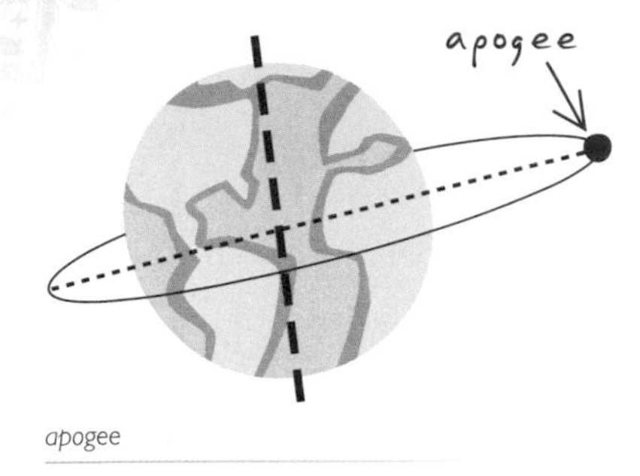

apogee

appending Adding features from multiple data sources of the same data type into an existing dataset. *See also* merging.

applet A small program that usually executes from within a Web browser. Applets are compatible with most platforms, and can also be used within applications or devices that support applets.

application 1. The use of a GIS to solve problems, automate tasks, or generate information within a specific field of interest. For example, a common agricultural application of GIS is determining fertilization requirements based on field maps of soil chemistry and previous crop yields. 2. [COMPUTING] A computer program used for a specific task or purpose, such as accounting or GIS.

application programming interface *See* API.

application server [COMPUTING] A computer program that receives user requests through a client application and returns results to the client. *See also* three-tier configuration.

arbitrary symbol [CARTOGRAPHY] A symbol that has no visual similarity to the feature it represents—for example, a circle used to represent a city, or a triangle used to represent a school. *See also* mimetic symbol.

arbitrary symbol

arc 1. On a map, a shape defined by a connected series of unique x,y coordinate pairs. An arc may be straight or curved.

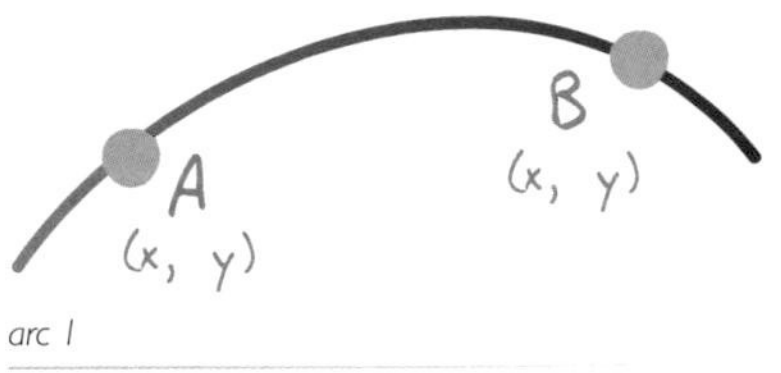

arc 1

2. A coverage feature class that represents lines and polygon boundaries. One line feature can contain many arcs. Arcs are topologically linked to nodes and to polygons. Their attributes are stored in an arc attribute table (AAT). Nodes indicate the endpoints and intersections of arcs; they do not exist as independent features. Together, the from-node and the to-node define the direction of the arc. *See also* path.

architecture [COMPUTING] The internal design of an application or software package; the way software or hardware components are organized into a functioning unit.

archive [COMPUTING] A collection of information or data that is stored on a permanent medium such as CDs, discs, or tapes. Information is archived to ensure its security or persistence.

ARC Macro Language *See* AML.

ArcObjects [ESRI SOFTWARE] A library of software components that make up the foundation of ArcGIS. ArcGIS Desktop, ArcGIS Engine, and ArcGIS Server are all built using the ArcObjects libraries.

arc second *See* second.

area 1. [MATHEMATICS] A closed, two-dimensional shape defined by its boundary or by a contiguous set of raster cells. 2. [MATHEMATICS] A calculation of the size of a two-dimensional feature, measured in square units. *See also* polygon.

areal scale *See* scale.

area of interest [CARTOGRAPHY] The extent used to define a focus area for either a map or database production.

argument 1. [COMPUTING] A value or expression passed to a function, command, or program. 2. [MATHEMATICS] An independent variable of a function.

arithmetic expression [MATHEMATICS] A number, variable, function, or combination of these, with operators or parentheses, or both, that can be evaluated to produce a single number. *See also* expression.

arithmetic operator *See* operator.

array 1. [GPS] A set of objects that are connected to function as a unit. An array of satellites is used to pinpoint locations on the earth. 2. [MATHEMATICS] A fundamental data structure consisting of a variable with multiple, sequentially indexed, cells that can each store a value of the same type. Each cell of the array acts as a variable, and the cells are referenced by an index value for each array dimension. One-dimensional arrays, called vectors, and two-dimensional arrays, called matrices, are most common, but arrays may have more dimensions. *See also* matrix, variable.

artificial neural network *See* neural network.

ascending node [ASTRONOMY] The point at which a satellite traveling south to north crosses the equator. *See also* descending node.

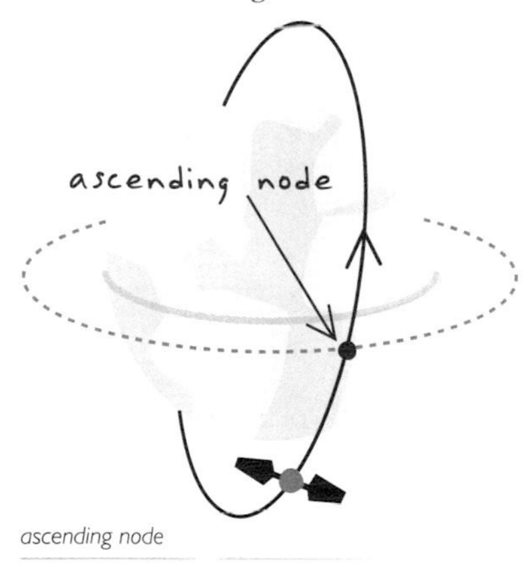

ascending node

ASCII *Acronym for American Standard Code for Information Interchange.* The de facto standard for the format of text files in computers and on the Internet that assigns a 7-bit binary number to each alphanumeric or special character. ASCII defines 128 possible characters.

aspatial data *See* nonspatial data.

aspatial query *See* attribute query.

aspect 1. The compass direction that a topographic slope faces, usually measured in degrees from north. Aspect can be generated from continuous elevation surfaces. For example, the aspect recorded for a TIN face is the steepest downslope direction of the face, and the aspect of a cell in a raster is the steepest downslope direction of a plane defined by the cell and its eight surrounding neighbors.

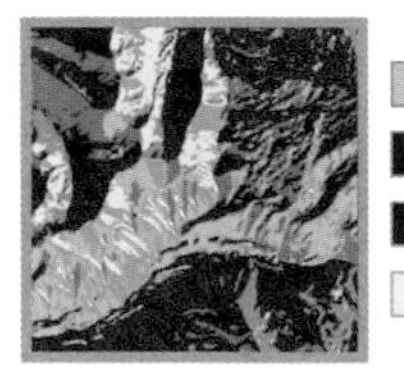

aspect 1

2. [CARTOGRAPHY] The conceptual center of a projection system. *See also* slope.

aspect ratio [COMPUTING] The ratio of the width of an image to its height. The aspect ratio of a standard computer monitor is 4:3 (rectangular).

association [COMPUTING] In UML, the relationship between two classes. In an association, instances of the classes in question usually exist together, but can exist on their own. *See also* class, UML.

assumed bearing [SURVEYING] A bearing measured from an arbitrarily chosen reference line called an assumed meridian. *See also* bearing.

astrolabe [ASTRONOMY] An instrument that measures the vertical angle between a celestial body and the horizontal plane at an observer's position. The astrolabe was replaced by the sextant in the fifteenth century for navigation, but modern versions are still used to determine local time and latitude. *See also* sextant.

astrolabe

asynchronous Not synchronous; that is, not occurring together or at the same time.

asynchronous request [COMPUTING] A request from a client application that does not require a response from the server for the client application to continue its process.

atlas [CARTOGRAPHY] A collection of maps usually related to a particular area or theme and presented together. Examples of atlases include world atlases, historical atlases, and biodiversity atlases.

atlas grid *See* alphanumeric grid.

atmospheric window [REMOTE SENSING] Parts of the electromagnetic spectrum that can be transmitted through the atmosphere with relatively little interference.

atomic clock [PHYSICS] A clock that keeps time by the radiation frequency associated with a particular atomic reaction. Atomic clocks are used in official timekeeping.

attenuation [REMOTE SENSING] The dimming and blurring effects in remotely sensed images caused by the absorption and scattering of light or other radiation that passes through the earth's atmosphere.

attribute **1.** Nonspatial information about a geographic feature in a GIS, usually stored in a table and linked to the feature by a unique identifier. For example, attributes of a river might include its name, length, and sediment load at a gauging station. **2.** In raster datasets, information associated with each unique value of a raster cell. **3.** [CARTOGRAPHY] Information that specifies how features are displayed and labeled on a map; for example, the graphic attributes of a river might include line thickness, line length, color, and font for labeling.

attribute data Tabular or textual data describing the geographic characteristics of features. *See also* nonspatial data.

attribute key *See* primary key.

attribute query A request for records of features in a table based on their attribute values. *See also* attribute.

attribute table A database or tabular file containing information about a set of geographic features, usually arranged so that each row represents a feature and each column represents one feature attribute. In raster datasets, each row of an attribute table corresponds to a certain zone of cells having the same value. In a GIS, attribute tables are often joined or related to spatial data layers, and the attribute values they contain can be used to find, query, and symbolize features or raster cells.

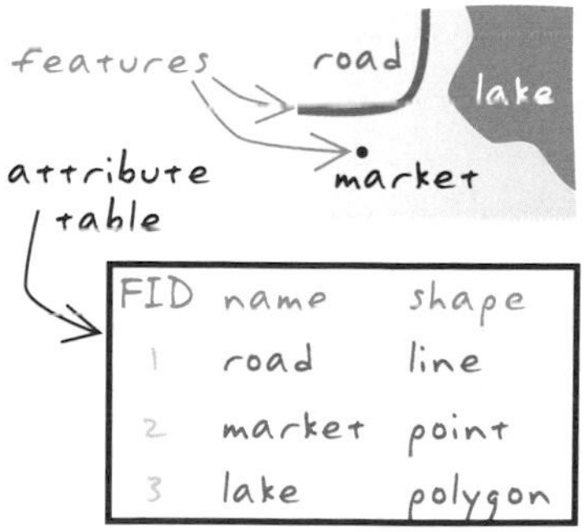

attribute table

authalic projection *See* equal-area projection.

authentication [COMPUTING] The process of validating the identity of a user who logs on to a computer system, network, or Web site.

A

autocorrelation [STATISTICS] The correlation or similarity of values, generally values that are nearby in a dataset. Temporal data is said to exhibit serial autocorrelation when values measured close together in time are more similar than values measured far apart in time. Spatial data is said to exhibit spatial autocorrelation when values measured nearby in space are more similar than values measured farther away from each other. *See also* correlation, spatial autocorrelation, Tobler's First Law of Geography.

automated cartography [CARTOGRAPHY] The process of making maps using computer systems that carry out many of the tasks associated with map production. *See also* GIS.

automated digitizing *See* autovectorization.

automated feature extraction The identification of geographic features and their outlines in remote-sensing imagery through postprocessing technology that enhances feature definition, often by increasing feature-to-background contrast or using pattern recognition software.

automated mapping/facilities management *See* AM/FM.

automated text placement An operation in which text is automatically placed on or next to features on a digital map by a software application according to rules set by the software user.

automation scale The scale at which nondigital data is made digital; for example, a map digitized at a scale of 1:24,000 has an automation scale of 1:24,000. The data can be rendered at different display scales. *See also* display scale.

autovectorization The creation of vector data from raster data through automated tracing of pixels that are in close proximity and of the same or similar value. *See also* vectorization, digitizing.

availability The degree of ease with which a dataset or other object may be found or obtained.

AVHRR [REMOTE SENSING] *Acronym for Advanced Very High Resolution Radiometer.* A scanner flown on National Oceanic and Atmospheric Administration (NOAA) polar-orbiting satellites for measuring visible and infrared radiation reflected from vegetation, cloud cover, shorelines, water, snow, and ice. AVHRR data is often used for weather prediction and vegetation mapping.

axis 1. [CARTOGRAPHY] A line along which measurements are made in order to determine the coordinates of a location.

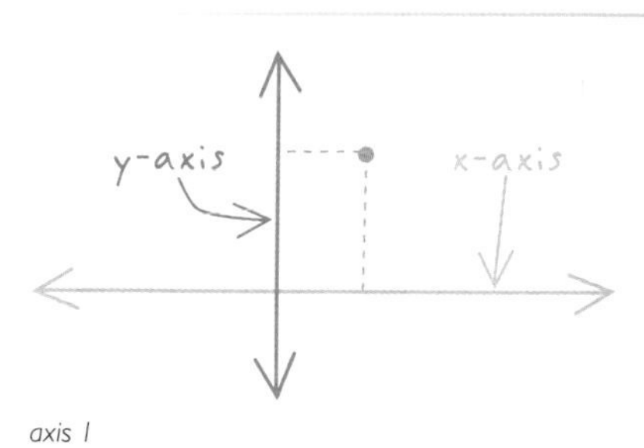

axis 1

2. [CARTOGRAPHY] In a spherical coordinate system, the line that directions are related to and from which angles are measured. **3.** [ASTRONOMY] The imaginary line through the poles about which a rotating body turns.

azimuth **1.** [CARTOGRAPHY] The horizontal angle, measured in degrees, between a baseline drawn from a center point and another line drawn from the same point. Normally, the baseline points true north and the angle is measured clockwise from the baseline.

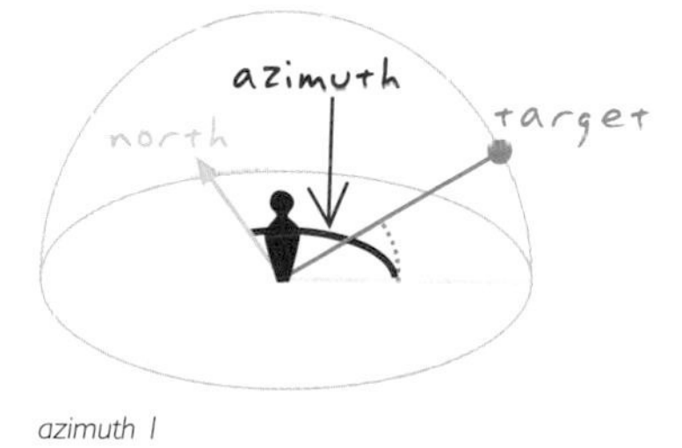

azimuth 1

2. A compass direction. For example, in some GIS software, the direction from which a light source illuminates a surface is called the azimuth. **3.** [NAVIGATION] The horizontal angle, measured in degrees, between a reference line drawn from a point and another line drawn from the same point to a point on the celestial sphere. Normally, the reference line points true north and the angle is measured clockwise from the reference line.

A

azimuthal projection [CARTOGRAPHY] A map projection that transforms points from a spheroid or sphere onto a tangent or secant plane. The azimuthal projection is also known as a planar or zenithal projection.

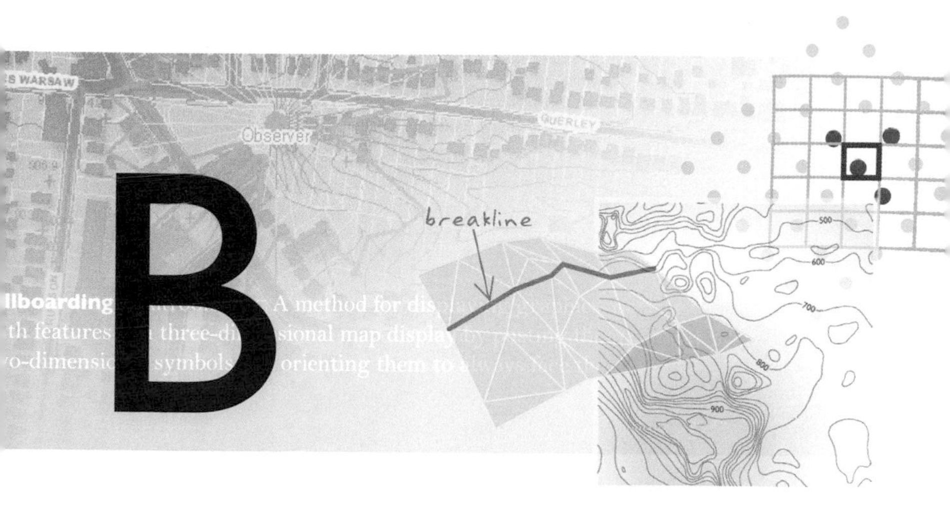

background image A satellite image, aerial photograph, or scanned map over which vector data is displayed. Although a background image can be used to align coordinates, it is not linked to attribute information and is not part of the spatial analysis in a GIS.

backscatter [REMOTE SENSING] Electromagnetic energy that is reflected back toward its source by terrain or particles in the atmosphere.

backup [COMPUTING] A copy of a file, a set of files, or a disk for safekeeping in case the original is lost or damaged.

band [REMOTE SENSING] A set of adjacent wavelengths or frequencies with a common characteristic. For example, visible light is one band of the electromagnetic spectrum, which also includes radio, gamma, and infrared waves. *See also* raster dataset band.

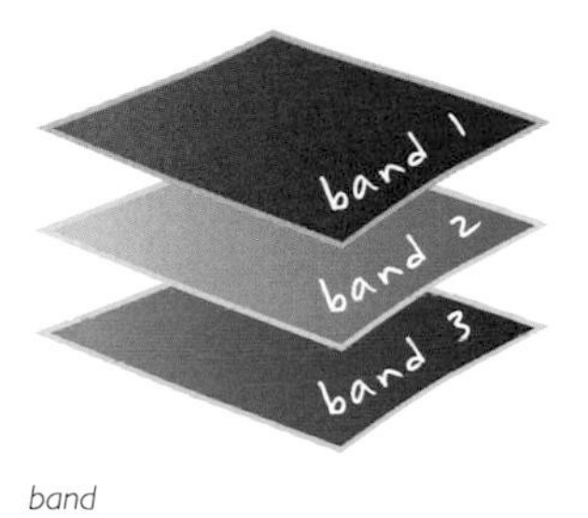

band

band-pass filter [REMOTE SENSING] A wave filter that allows signals in a certain frequency to pass through, while blocking or attenuating signals at other frequencies. *See also* low-pass filter, high-pass filter.

band ratio [DIGITAL IMAGE PROCESSING] A technique that enhances contrast between features by dividing a measure of reflectance for the pixels in one image band by the measure of reflectance for the pixels in the other image band.

B

band separate [REMOTE SENSING] An image format that stores each band of data in a separate file.

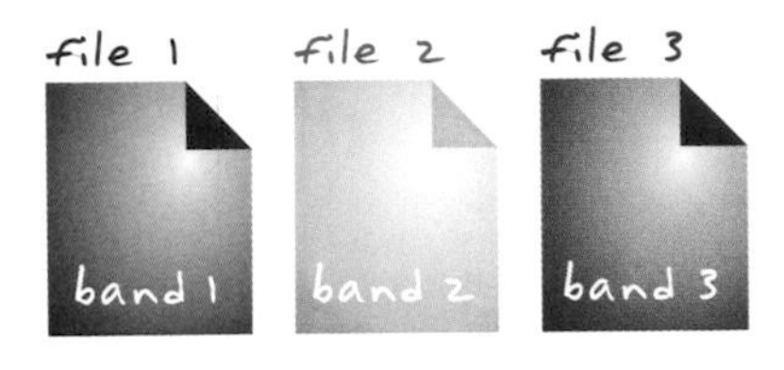

band separate

bandwidth [COMPUTING] The amount of digital data that can be transferred over a computer network within a specified time period, usually measured in bits per second (bps).

barrier 1. In network analysis, an entity that prevents flow from traversing a network edge or junction. 2. A line feature used to keep certain points from being used in the calculation of new values when a raster is interpolated. The line can represent a cliff, ridge, or some other interruption in the landscape. Only the sample points on the same side of the barrier as the current processing cell will be considered.

bar scale *See* scale bar.

base data Map data over which other, thematic information is placed.

base height [REMOTE SENSING] In aerial photography, the height or altitude from which a photograph is taken.

base height ratio [REMOTE SENSING] In aerial photography, the distance on the ground between the centers of overlapping photos, divided by aircraft altitude. In a stereomodel, base height ratio is used to determine vertical exaggeration. *See also* base height.

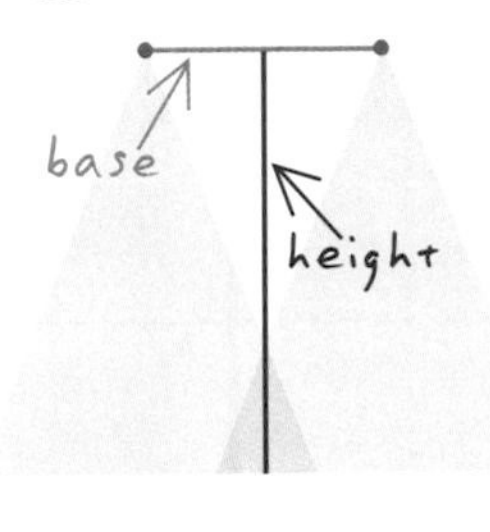

base height ratio

base layer A data layer in a GIS to which all other layers are geometrically referenced. *See also* layer.

baseline 1. [SURVEYING] An accurately surveyed line from which other lines or the angles between them are measured. 2. [SURVEYING] In a U.S. land survey system, a line passing east and west through the origin, used to establish township, section, and quarter-section corners.

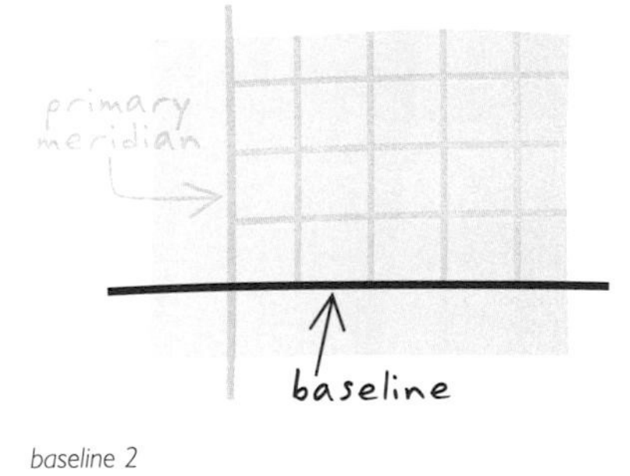

baseline 2

3. [GPS] The physical distance between a base station and a rover.

basemap 1. A map depicting background reference information such as landforms, roads, landmarks, and political boundaries, onto which other, thematic information is placed. A basemap is used for locational reference and often includes a geodetic control network as part of its structure. 2. A map to which GIS data layers are registered and rescaled.

base station [GPS] A GPS receiver at a known location that broadcasts and collects correction information for roving GPS receivers. *See also* differential correction.

batch file [COMPUTING] A text file containing commands that is sent to the CPU to be executed automatically. A batch file allows the central processing unit (CPU) to process the commands at off-peak times or at a regularly scheduled time, rather than on demand from the user. *See also* batch processing.

batch geocoding The process of geocoding many address records at the same time.

batch processing [COMPUTING] A method for processing data automatically in which the data is grouped into batches and executed by the computer at one time, without user interaction.

batch vectorization An automated process that converts raster data into vector features for an entire raster or a portion of it based on user-defined settings. *See also* vectorization.

bathymetric curve *See* depth contour.

bathymetric map [CARTOGRAPHY] A map representing the topography of a seafloor or lake bed, using contour lines to indicate depth.

bathymetric map

bathymetry [CARTOGRAPHY] The science of measuring and charting the depths of water bodies to determine the topography of a lake bed or seafloor.

battleships grid *See* alphanumeric grid.

baud rate In communications, the number of electrical cycles, or signals, transmitted per second. At lower transfer speeds the baud rate equals the data transfer rate measured in bps, or bits per second. Baud rate and bps are still sometimes used interchangeably, though inaccurately, since current standards allow for the encoding of multiple bits into a single cycle. *See also* bit.

Bayesian statistics [STATISTICS] A statistical approach to measuring likelihood. Bayesian estimates are based on the synthesis of a prior distribution and current sample data. Classical approaches to statistics estimate the probability of an event by averaging all possible data. The Bayesian approach, in contrast, weights probability according to actual data from a particular situation. It also factors in data from sources outside the statistical investigation, such as past experience, expert opinion, or prior belief. This outside information is described by a distribution that includes all possible values for the parameter.

bearing [GEODESY] The horizontal direction of a point in relation to another point, expressed as an angle from a known direction, usually north, and usually measured from 0 degrees at the reference direction clockwise through 360 degrees. Bearings are often referred to as true bearings, magnetic bearings, or assumed bearings, depending on whether the meridian is true, magnetic, or assumed. *See also* heading.

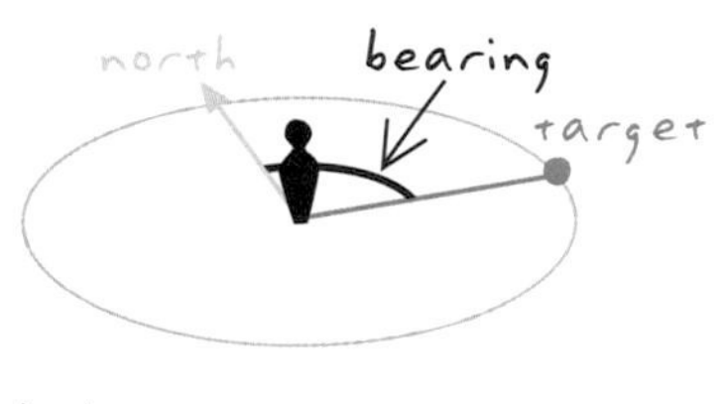

bearing

behavior [ESRI SOFTWARE] The actions or characteristics exhibited by an object in a database, as defined by a set of rules.

benchmark 1. [SURVEYING] A brass or bronze disk, set in a concrete base or similarly permanent structure, inscribed with a mark showing its elevation above or below an adopted vertical datum. 2. [COMPUTING] A performance test of hardware, software, or an application that sets a standard of quality using test data or operations typical of working conditions.

best route The route of least impedance between two or more locations, taking into account connectivity and travel restrictions such as one-way streets and rush-hour traffic.

Bézier curve [MATHEMATICS] A curved line whose shape is derived mathematically rather than by a series of connected vertices. In graphics programs, a Bézier curve usually has two endpoints and two handles that can be moved to change the direction and the steepness of the curve. Bézier curves are named for the French engineer Pierre Bézier (1910–1999).

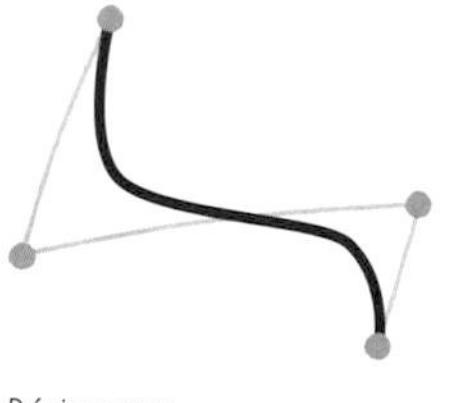

Bézier curve

Bhattacharyya distance [DIGITAL IMAGE PROCESSING] A measure of the theoretical distance between two normal distributions of spectral classes, which acts as an upper limit on the probability of error in a Bayesian estimate of correct classification. Bhattacharyya distance is named for the Indian mathematician Anil Kumar Bhattacharyya (1915–1996).

big endian [COMPUTING] A computer hardware architecture in which, within a multibyte numeric representation, the most significant byte has the lowest address and the remaining bytes are encoded in decreasing order of significance. *See also* little endian.

bilinear interpolation [MATHEMATICS] A resampling method that uses a weighted average of the four nearest cells to determine a new cell value.

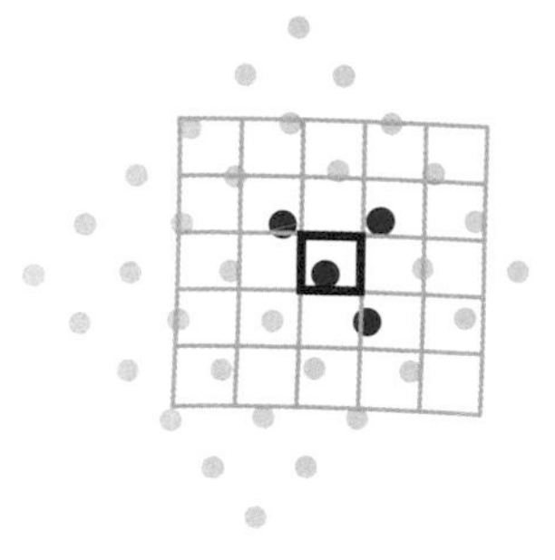

bilinear interpolation

billboarding [CARTOGRAPHY] A method for displaying graphics associated with features in a three-dimensional map display by posting them vertically as two-dimensional symbols and orienting them to always face the user.

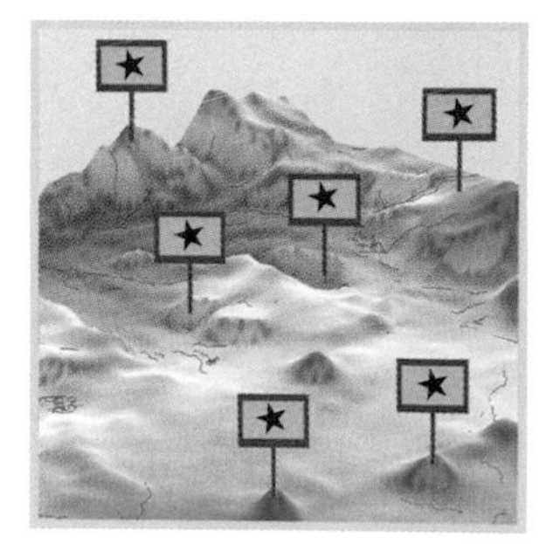

billboarding

bin 1. In a variogram map, each cell that groups lags with similar distance and direction. Bins are commonly formed by dividing the sample area into a grid of cells or sectors, and are used to calculate the empirical semivariogram for kriging. 2. In a histogram, user-defined size classes for a variable. *See also* lag, bin, kriging, histogram.

binary [COMPUTING] Having only two states, such as yes or no, on or off, true or false, or 0 or 1.

binary file [COMPUTING] A file that contains data encoded as a sequence of bits (ones and zeros) instead of plain text. A binary file, such as a DLL or an executable file, contains information that can be directly loaded or executed by a computer.

binary large object *See* BLOB.

binding [PROGRAMMING] The process by which a program discovers an object's methods and properties. *See also* early binding, late binding.

B

bingo grid *See* alphanumeric grid.

binomial distribution [STATISTICS] A distribution describing the probability of obtaining exactly K successes in N independent trials, where each trial results in either a success or a failure.

biogeography The study of the geographical distribution of living things.

biomass The total amount of organic matter in a defined area; usually refers to vegetation.

bit [COMPUTING] The smallest unit of information within a computer. A bit can have one of two values, 1 and 0, that can represent on and off, yes and no, or true and false.

bit depth The range of values that a particular raster format can store, based on the formula 2^n. An 8-bit depth dataset can store 256 unique values.

bitmap An image format in which one or more bits represent each pixel on the screen. The number of bits per pixel determines the shades of gray or number of colors that a bitmap can represent.

blind digitizing A method of manual digitizing in which the operator has no graphic display on hand with which to see the digitized coordinates as they are captured. *See also* digitizing.

BLOB *Acronym for binary large object.* A large block of data, such as an image, a sound file, or geometry, stored in a database. The database cannot read the BLOB's structure and only references it by its size and location.

block group A unit of U.S. census geography that is a combination of census blocks. A block group is the smallest unit for which the U.S. Census Bureau reports a full range of demographic statistics. There are about 700 residents per block group. A block group is a subdivision of a census tract.

block kriging [STATISTICS] A kriging method in which the average expected value in an area around an unsampled point is generated rather than the estimated exact value of an unsampled point. Block kriging is commonly used to provide better variance estimates and smooth interpolated results. *See also* kriging.

blunder [SURVEYING] A defective measurement that can be detected by a statistical test.

Boolean expression [MATHEMATICS] An expression, named for the English mathematician George Boole (1815–1864), that results in a true or false (logical) condition. For example, in the Boolean expression "HEIGHT > 70 AND DIAMETER = 100," all locations where the height is greater than 70 and the diameter is equal to 100 would be given a value of 1, or true, and all locations where this criteria is not met would be given a value of 0, or false. *See also* expression.

Boolean operation A GIS operation that uses Boolean operators to combine input datasets into a single output dataset. *See also* Boolean operator.

Boolean operator [MATHEMATICS] A logical operator used in the formulation of a Boolean expression. Common Boolean operators include AND, which specifies a combination of conditions (A and B must be true); OR, which specifies a list of alternative conditions (A or B must be true); NOT, which negates a condition (A but not B must be true); and XOR (exclusive or), which makes conditions mutually exclusive (A or B may be true but not both A and B). *See also* operator, Boolean expression.

boundary [SURVEYING] A line separating adjacent political entities, such as countries or districts; adjacent tracts of privately-owned land, such as parcels; or adjacent geographic zones, such as ecosystems. A boundary is a line that may or may not follow physical features, such as rivers, mountains, or walls. *See also* boundary line.

boundary effect A problem created during spatial analysis, caused by arbitrary or discrete boundaries being imposed on spatial data representing nondiscrete or unbounded spatial phenomena. Boundary problems include edge effects, in which patterns of interaction or interdependency across the borders of the bounded region are ignored or distorted, and shape effects, in which the shape imposed on the bounded area affects the perceived interactions between phenomena. *See also* MAUP.

boundary line [CARTOGRAPHY] A division between adjacent political entities, tracts of private land, or geographic zones. Boundary lines may be imaginary lines, physical features that follow those lines, or the graphical representation of those lines on a map. Boundary lines between privately owned land parcels are usually called property lines. *See also* boundary.

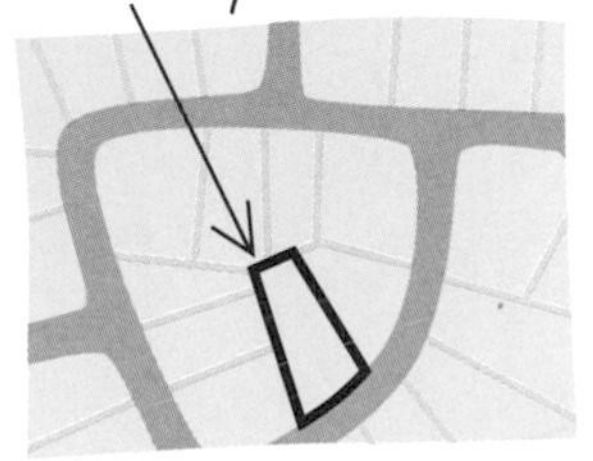

boundary line

boundary monument [SURVEYING] An object that marks an accurately surveyed position on or near a boundary.

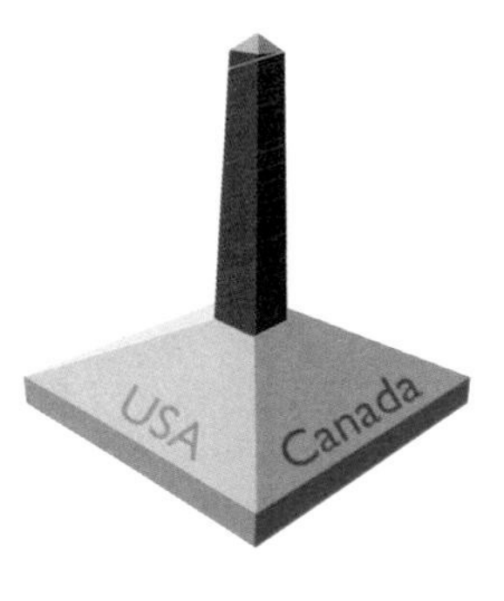

boundary monument

boundary survey 1. [SURVEYING] A map that shows property lines and corner monuments of a parcel of land. 2. [SURVEYING] The survey taken to gather data for a map that shows property lines and corner monuments of a parcel of land.

bounding rectangle The rectangle, aligned with the coordinate axes and placed on a map display, that encompasses a geographic feature or group of features or an area of interest. It is defined by minimum and maximum coordinates in the x and y directions and is used to represent, in a general way, the location of a geographic area. *See also* envelope.

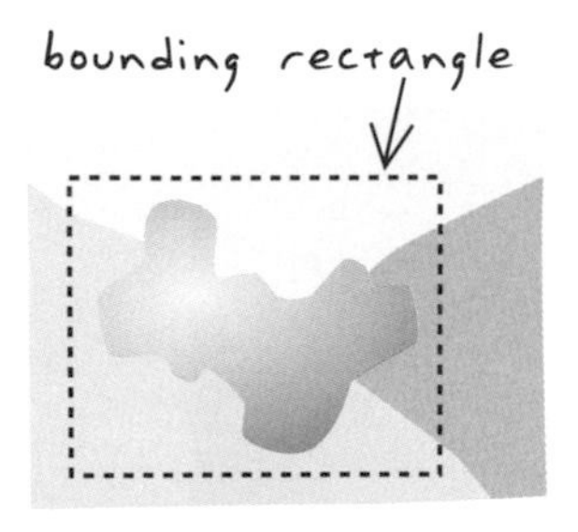

bounding rectangle

Bowditch rule *See* compass rule.

breakline A line in a TIN that represents a distinct interruption in the slope of a surface, such as a ridge, road, or stream. No triangle in a TIN may cross a breakline (in other words, breaklines are enforced as triangle edges). Z-values along a breakline can be constant or variable. *See also* structure line.

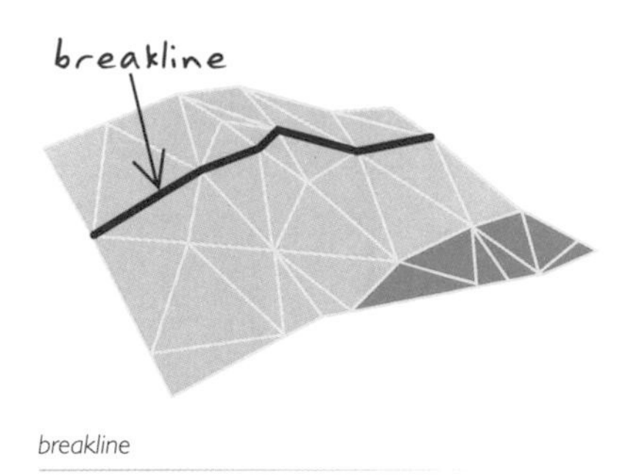

breakline

B-tree A tree data structure used for indexing data within a database or file system implementation. In a B-tree structure, data is sorted into a set of hierarchical nodes, usually using only three or four levels. The limited number of levels makes effective searches possible, because most of the nodes in the tree do not have to be accessed during a search. *See also* tree data structure, hierarchical database, R-tree.

buffer 1. A zone around a map feature measured in units of distance or time. A buffer is useful for proximity analysis.

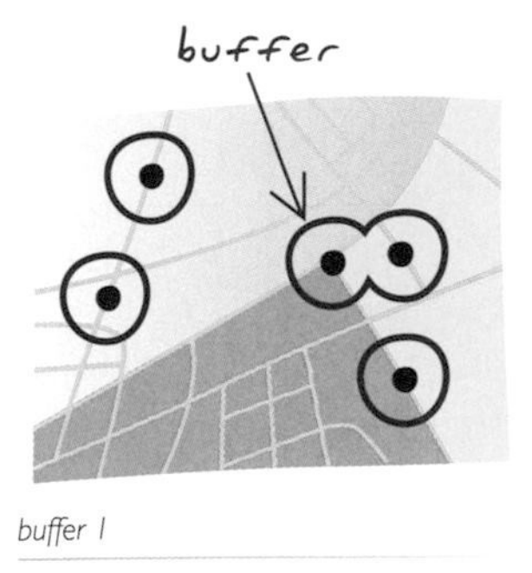

buffer 1

2. [COMPUTING] Space on a computer disk or RAM that has been allocated

for temporary storage. This temporary storage may also be called a spooler when it is used to hold data in memory before the data is sent to another machine, such as a printer.

bug [COMPUTING] A flaw or error in a software program or hardware component that prevents it from performing the way it should.

bus [COMPUTING] A set of conductors that provide communications links between the various functional components of a computer, such as memory and peripheral devices.

bust *See* closure error.

byte [COMPUTING] The smallest addressable unit of data storage within a computer, almost always equivalent to 8 bits and containing one character.

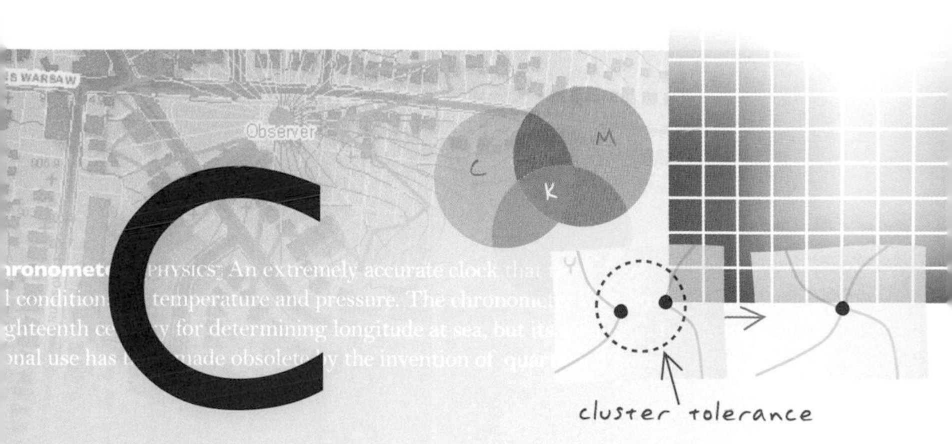

C/A code *See* civilian code.

CAD *Acronym for computer-aided design.* A computer-based system for the design, drafting, and display of graphical information. Also known as computer-aided drafting, such systems are most commonly used to support engineering, planning, and illustrating activities.

cadastral survey [SURVEYING] A boundary survey taken for the purposes of ownership and taxation. *See also* cadastre.

cadastre [SURVEYING] An official record of the dimensions and value of land parcels, used to record ownership and assist in calculating taxes.

cadastral map

cadastre

calibration 1. The comparison of the accuracy of an instrument's measurements to a known standard. 2. In spatial analysis, the selection of attribute values and computational parameters that will cause a model to properly represent the situation being analyzed. For example, in pathfinding and allocation, calibration generally refers to assigning or calculating impedance values.

callout line [CARTOGRAPHY] A line on a map extending between a feature's geographic position and its corresponding symbol or label, used in areas where there is not enough room to display a symbol or label in its correct location.

C

camera station *See* exposure station.

candidate key [COMPUTING] In a relational database, any key that can be used as the primary key in a table. *See also* primary key.

capacity In location-allocation, the maximum number of people or units that a center can service, contain, or have assigned to it. *See also* location-allocation.

cardinal direction *See* cardinal point.

cardinality [MATHEMATICS] The correspondence or equivalency between sets; how sets relate to each other. For example, if one row in a table is related to three rows in another table, the cardinality is one to many.

cardinal point [NAVIGATION] One of the four compass directions on the earth's surface: north, south, east, or west.

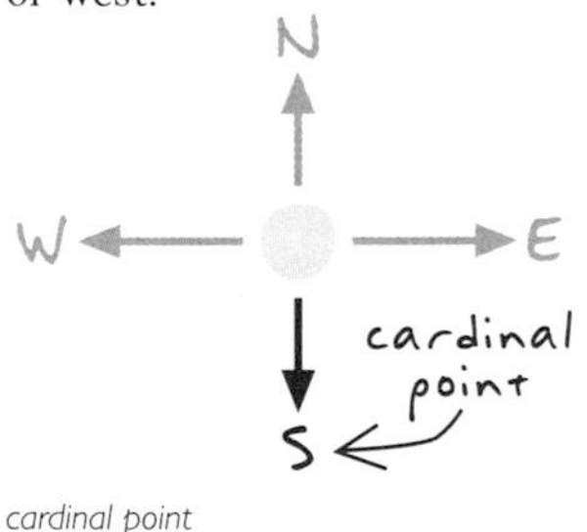

cardinal point

carrier [PHYSICS] An electromagnetic wave, such as radio, with modulations that are used as signals to transmit information.

carrier-aided tracking [GPS] Signal processing that uses the GPS carrier signal to lock onto the PRN code generated by the satellite. *See also* PRN code.

carrier-phase GPS [GPS] GPS measurements that are calculated using the carrier signal of a satellite. *See also* code-phase GPS.

carrying contour [CARTOGRAPHY] A single line representing multiple coincident contour lines, used to show vertical topographic features such as cliffs, cuts, and fills. *See also* contour line, contour interval.

Cartesian coordinate system [CARTOGRAPHY] A two-dimensional, planar coordinate system in which horizontal distance is measured along an x-axis and vertical distance is measured along a y-axis. Each point on the plane is defined by an x,y coordinate. Relative measures of distance, area, and direction are constant throughout the Cartesian coordinate plane. The Cartesian coordinate system is named for the French mathematician and philosopher René Descartes (1596–1650). *See also* coordinate system.

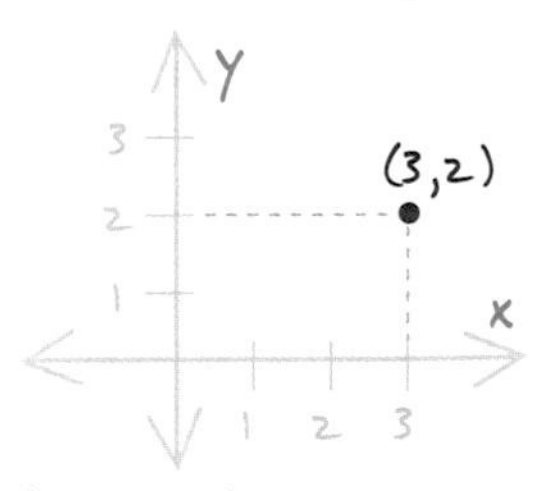

Cartesian coordinate system

cartogram [CARTOGRAPHY] A diagram or abstract map in which geographical areas are distorted proportionally to the value of an attribute.

cartogram

cartographer [CARTOGRAPHY] One who practices the art and science of expressing graphically, usually through maps, the natural and social features of the earth.

cartographic generalization [CARTOGRAPHY] The abstraction, reduction, and simplification of features so that a map is clear and uncluttered at a given scale. *See also* generalization.

cartography [CARTOGRAPHY] The art and science of expressing graphically, usually through maps, the natural and social features of the earth.

cartouche [CARTOGRAPHY] An ornamental frame on a map, usually around the map's title. Cartouches are rarely used on modern maps.

cartouche

catchment *See* watershed.

categorical raster *See* discrete raster.

celestial sphere [ASTRONOMY] The sky, considered as the inside of a sphere of infinitely large radius that surrounds the earth, on which all celestial bodies except the earth are imagined to be projected.

cell 1. The smallest unit of information in raster data, usually square in shape. In a map or GIS dataset, each cell represents a portion of the earth, such as a square meter or square mile, and usually has an attribute value associated with it, such as soil type or vegetation class.

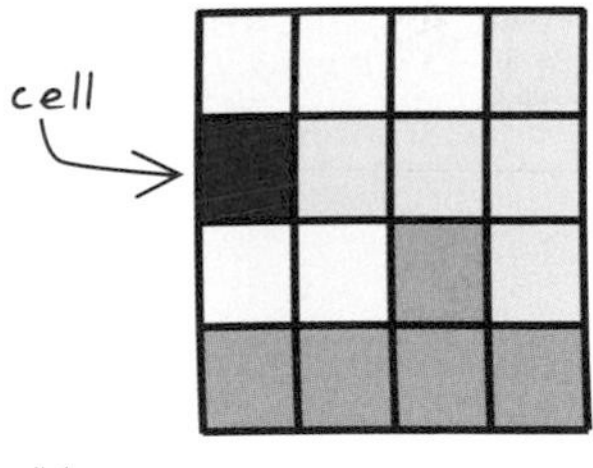

cell 1

2. A pixel. 3. A small drawing, usually of a frequently used or complex symbol, notation, or detail. Cells are similar to blocks in AutoCAD drawings.

cell size The dimensions on the ground of a single cell in a raster, measured in map units. Cell size is often used synonymously with pixel size.

C

cellular automaton A mathematical construction consisting of a row or grid of cells in which each cell has an initial value—from a known and limited number of possible values—and all cells are simultaneously evaluated and updated according to their internal states and the values of their neighbors. The simplest cellular automaton is a row in which each cell has one of two values, such as red or green. In this case, there are eight possible value combinations for a cell and its neighbors. (If a green cell with two red neighbors is notated RGR, then the eight combinations are RRR, RRG, RGR, GRR, RGG, GRG, GGR, GGG.) A set of rules determines whether or not a cell changes value when it is evaluated. A sample rule might be, "A green cell becomes red if it has a red neighbor on both sides." Successive updates, or generations, of a cellular automaton may produce complex patterns. Cellular automata are of interest in spatial modeling and are often used to model land-cover change.

census block The smallest geographic entity for which the U.S. Census Bureau tabulates decennial census data. Many blocks correspond to city blocks bounded by streets, but blocks in rural areas may include several square miles and have some boundaries that are not streets. The Census Bureau established blocks covering the entire nation for the first time in 1990. Previous censuses dating back to 1940 had blocks established only for part of the nation.

census geography Any one of various types of precisely defined geographic areas used by the U.S. Census Bureau to collect and aggregate data. The largest unit of area is the entire United States, while the smallest is a census block.

census tract A small, statistical subdivision of a county that usually includes approximately 4,000 inhabitants, but which may include from 2,500 to 8,000 inhabitants. A census tract is designed to encompass a population with relatively uniform economic status, living conditions, and some demographic characteristics. Tract boundaries normally follow physical features, but may also follow administrative boundaries or other nonphysical features.

center 1. [MATHEMATICS] The point in a circle or in a sphere equidistant from all other points on the object.

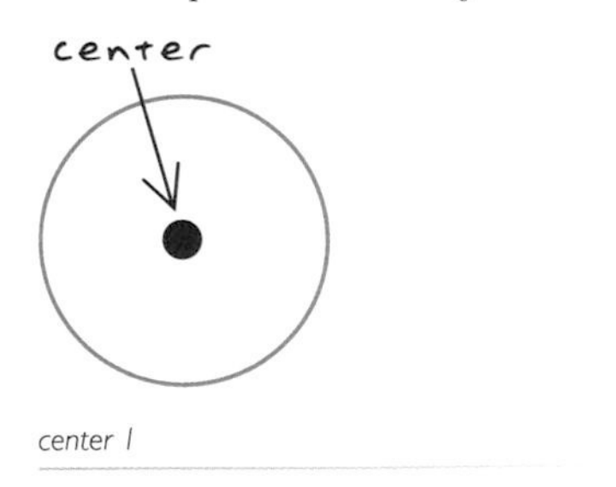

center 1

2. [MATHEMATICS] The point from which angles or distances are measured. **3.** In network allocation, a location from which resources are distributed or to which they are brought.

centerline A line digitized along the center of a linear geographic feature, such as a street or a river, that at a large enough scale would be represented by a polygon.

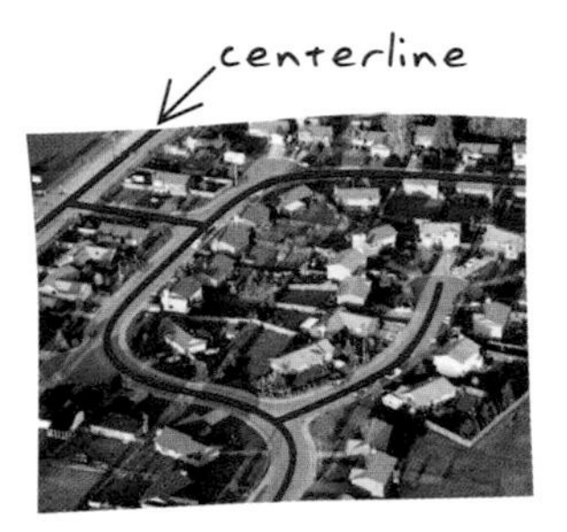

centerline

centerline vectorization The generation of vector features along the center of connected cells. It is typically used for vectorizing scanned parcel and survey maps. *See also* outline vectorization.

centerpoint [REMOTE SENSING] In aerial photography, the point at the exact center of an aerial photograph.

central meridian [CARTOGRAPHY] The line of longitude that defines the center and often the x-origin of a projected coordinate system. In planar rectangular coordinate systems of limited extent, such as state plane, grid north coincides with true north at the central meridian.

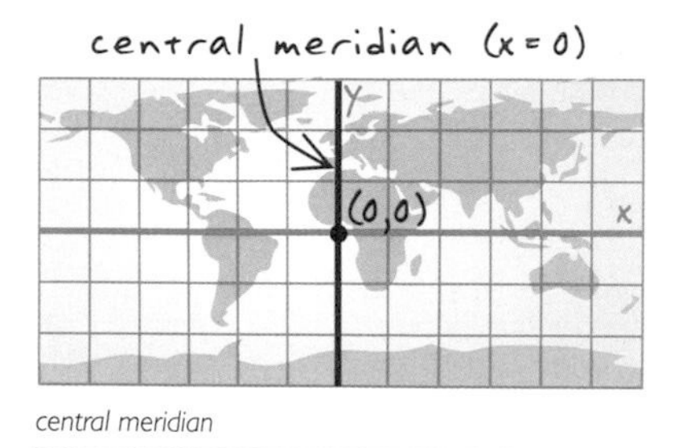

central meridian

centroid The geometric center of a feature. Of a line, it is the midpoint; of a polygon, the center of area; of a three-dimensional figure, the center of volume.

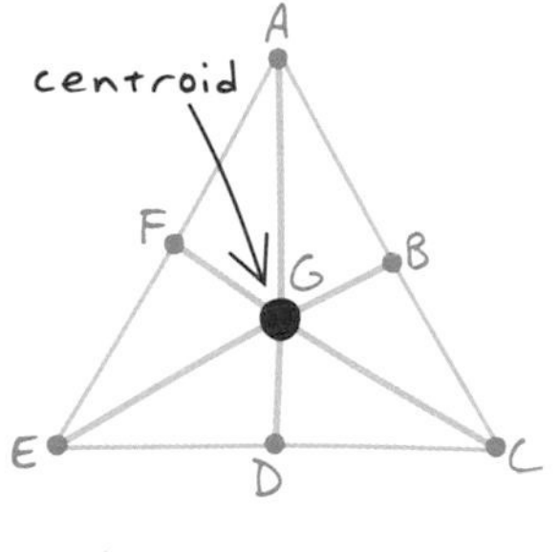

centroid

CGI [COMPUTING] *Acronym for Common Gateway Interface.* A standard for scripts that run external programs from a World Wide Web server. CGI typically specifies how to pass arguments to the program via HTTP requests; defines a set of environmental variables made available to the program; and generates output, usually in HTML format, that is passed back to the browser. CGI scripts are frequently designed to access information in a database and format the results as HTML, convert information retrieved from an interactive Web page into a database, send datasets, and so on.

C

chain [SURVEYING] A unit of length equal to 66 feet, used especially in U.S. public land surveys. Ten square chains equal 1 acre.

chain code A method of drawing a polygon as a series of straight line segments defined as a set of directional codes, with each code following the last like links in a chain.

change detection [REMOTE SENSING] A process that measures how the attributes of a particular area have changed between two or more time periods. Change detection often involves comparing aerial photographs or satellite imagery of the area taken at different times. The process is most frequently associated with environmental monitoring, natural resource management, or measuring urban development.

character [COMPUTING] A letter, digit, or special graphic symbol treated as a single unit of data and usually stored as one byte.

chart 1. [CARTOGRAPHY] A map used to plot a course for air or water navigation. 2. [MATHEMATICS] A graphic representation of tabular data; a diagram showing the relationship between two or more variable quantities, usually measured along two perpendicular axes. A chart may also be referred to as a graph.

chi-square statistic [STATISTICS] A statistic used to assess how well a model fits the data. It compares categorized data with a multinomial model that predicts the relative frequency of outcomes in each category to see to what extent they agree.

chord [MATHEMATICS] A straight line that joins two points on a curve.

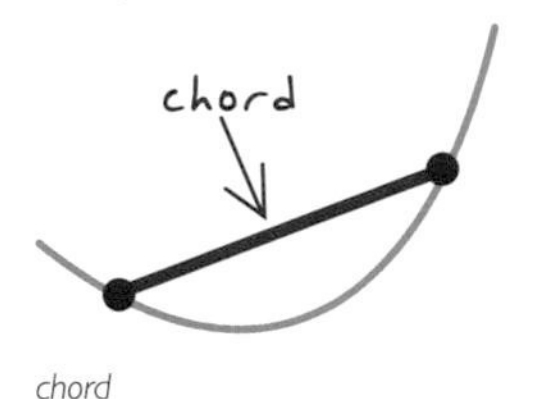

chord

choropleth map [CARTOGRAPHY] A thematic map in which areas are distinctly colored or shaded to represent classed values of a particular phenomenon. *See also* thematic map.

chroma The saturation, purity, or intensity of a color. *See also* saturation, value, intensity, hue.

chronometer [PHYSICS] An extremely accurate clock that remains accurate through all conditions of temperature and pressure. The chronometer was developed in the eighteenth century for determining longitude at sea, but its scientific and navigational use has been made obsolete by the invention of quartz and atomic clocks.

chronometer

circle [MATHEMATICS] A two-dimensional geometric shape for which the distance from the center to any point on the edge is equal; the closed curve defining such a shape.

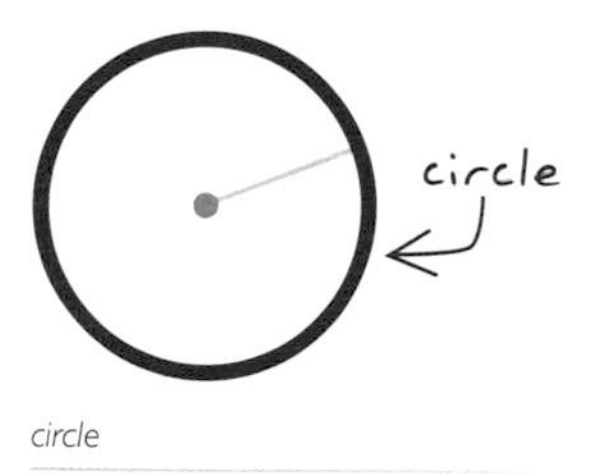

circle

circular arc [MATHEMATICS] A curved line that is a section of a circle, with two vertices, one situated at each endpoint.

circular arc

circular variance [STATISTICS] A measure of directional variation, on a scale from zero to one, among a set of line vectors. Circular variance approaches zero when all vectors point in roughly the same direction and approaches one when the vectors point in markedly different directions. *See also* vector.

civilian code [GPS] The standard PRN code used by most civilian GPS receivers. *See also* P-code, PRN code.

Clarke Belt [ASTRONOMY] An orbit 22,245 miles (35,800 kilometers) above the equator in which a satellite travels at the same speed that the earth rotates. The Clarke Belt was named after the writer and scientist Arthur C. Clarke. It is also referred to as a geostationary orbit. *See also* geostationary.

Clarke ellipsoid of 1866 [GEODESY] A reference ellipsoid having a semimajor axis of approximately 6,378,206.4 meters and a flattening of 1/294.9786982. It is the basis for the North American Datum of 1927 (NAD27) and other datums. The Clarke ellipsoid of 1866 is also known as the Clarke spheroid of 1866. *See also* datum.

Clarke spheroid of 1866 *See* Clarke ellipsoid of 1866.

class 1. A set of entities grouped together on the basis of shared attribute values. **2.** Pixels in a raster file that represent the same condition. **3.** [COMPUTING] A template for a type of object in an object-oriented programming language. A class is used to create objects that share the same structure and behavior.

C

classification [CARTOGRAPHY] The process of sorting or arranging entities into groups or categories; on a map, the process of representing members of a group by the same symbol, usually defined in a legend. *See also* equal-area classification, equal-interval classification, natural breaks classification, quantile classification, standard deviation classification, Jenks' optimization.

class intervals [CARTOGRAPHY] A set of categories for classification that divide the range of all values so that each piece of data is contained within a nonoverlapping category. *See also* classification.

cleaning Improving the appearance of scanned or digitized data by correcting overshoots and undershoots, closing polygons, performing coordinate editing, and so on.

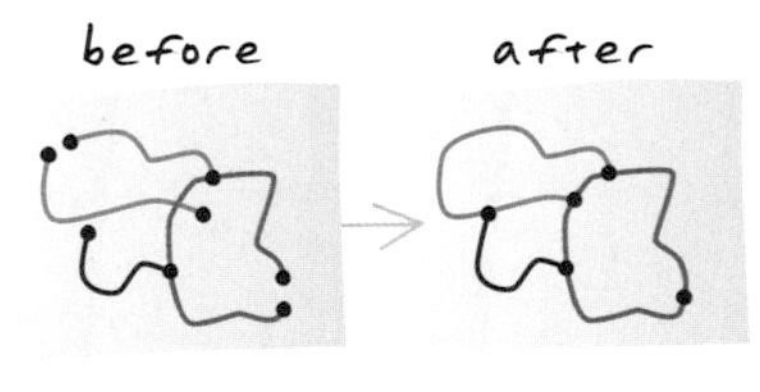

cleaning

clearinghouse A repository structure, physical or virtual, that collects, stores, and disseminates information, metadata, and data. A clearinghouse provides widespread access to information and is generally thought of as reaching or existing outside organizational boundaries.

client [COMPUTING] An application, computer, or device in a client/server model that makes requests to a server.

client/server architecture [COMPUTING] A software system with a central processor (server) that accepts requests from one or more user applications, computers, or devices (clients). Although client/server architecture can exist on one computer, it is more relevant to (and is typically thought of as relating to) network systems that distribute applications over computers in different locations. *See also* client, server.

clinometric map [CARTOGRAPHY] A map that represents slope with colors or shading.

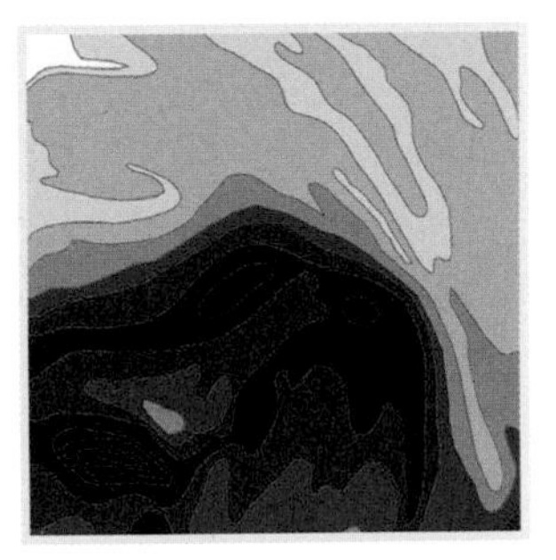

clinometric map

cloning [COMPUTING] In object-oriented programming, the process of creating a new instance of a class with the same state as an existing instance.

close coupling *See* tight coupling.

closed loop traverse [SURVEYING] A traverse that starts and ends with the same survey point. *See also* open traverse.

closure error [SURVEYING] A discrepancy between existing coordinates and computed coordinates that occurs when the final point of a closed traverse has known coordinates and the final course of a traverse computes different coordinates for the same survey point.

cluster analysis [STATISTICS] A statistical classification technique for dividing a population into relatively homogeneous groups. The similarities between members belonging to a class, or cluster, are high, while similarities between members belonging to different clusters are low. Cluster analysis is frequently used in market analysis for consumer segmentation and locating customers, but it is also applied to other fields.

clustering [ESRI SOFTWARE] A part of the topology validation process in which vertices that fall within a specified distance (cluster tolerance) of each other are snapped together. *See also* snapping, cluster tolerance.

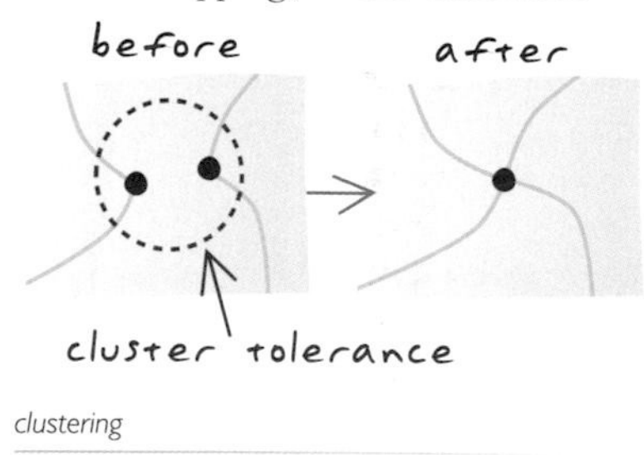

clustering

cluster tolerance [ESRI SOFTWARE] The minimum tolerated distance between vertices in a topology. Vertices that fall within the set cluster tolerance are snapped together during the topology validation process. *See also* snapping.

CMYK A color model that combines the printing inks cyan, magenta, yellow, and black to create a range of colors. Most commercial printing uses this color model. *See also* RGB.

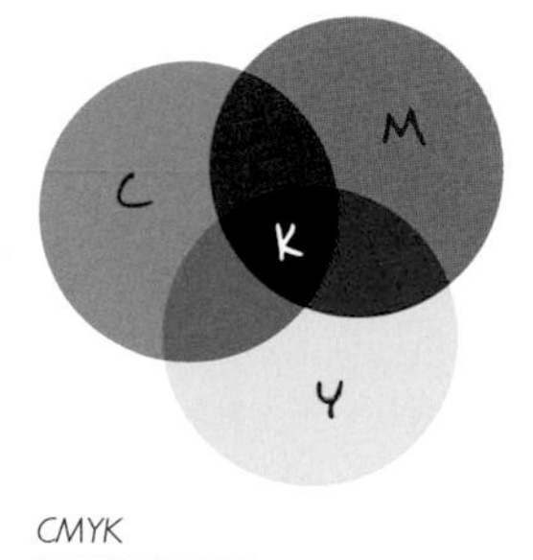

CMYK

Coarse/Acquisition code *See* civilian code.

code-phase GPS [GPS] GPS measurements calculated using the PRN code transmitted by a GPS satellite. *See also* carrier-phase GPS.

cognitive map *See* mental map.

COGO **1.** [SURVEYING] *Acronym for coordinate geometry.* A method for calculating coordinate points from surveyed bearings, distances, and angles. **2.** [SURVEYING] Automated mapping software used in land surveying that calculates locations using distances and bearings from known reference points.

C

coincident [MATHEMATICS] Occupying the same space. Coincident features or parts of features occupy the same space in the same plane.

cokriging [STATISTICS] A form of kriging in which the distribution of a second, highly correlated variable (covariate) is used along with the primary variable to provide interpolation estimates. Cokriging can improve estimates if the primary variable is difficult, impossible, or expensive to measure, and the second variable is sampled more intensely than the primary variable. *See also* kriging.

color composite [REMOTE SENSING] A color image made by assigning red, green, and blue colors to each of the separate monochrome bands of a multispectral image and then superimposing them.

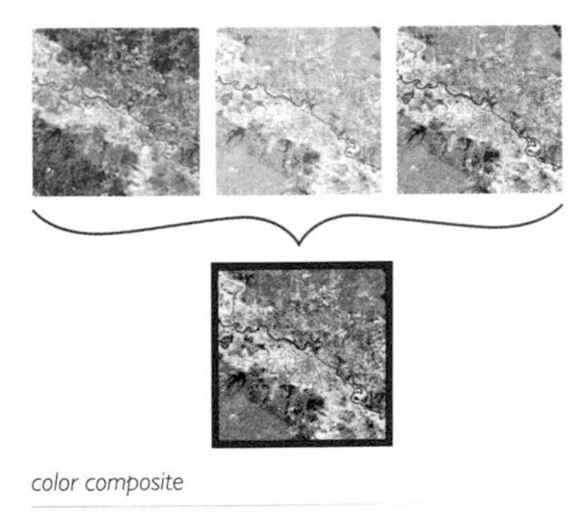

color composite

color map A set of values that are associated with specific colors. Color maps are most commonly used to display a raster dataset consistently on many different platforms.

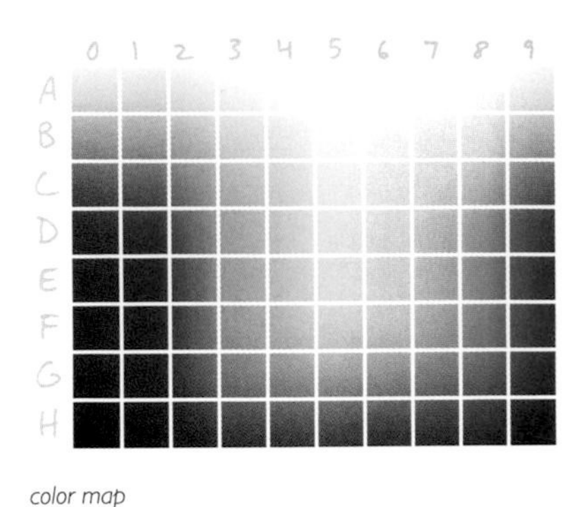

color map

color model Any system that organizes colors according to their properties for printing or display. Examples include RGB (red, green, blue), CMYK (cyan, magenta, yellow, black), HSB (hue, saturation, brightness), HSV (hue, saturation, value), HLS (hue, lightness, saturation), and CIE-L*a*b (Commission Internationale de l'Eclairage-luminance, a, b).

color ramp [CARTOGRAPHY] A range of colors used to show ranking or order among classes on a map.

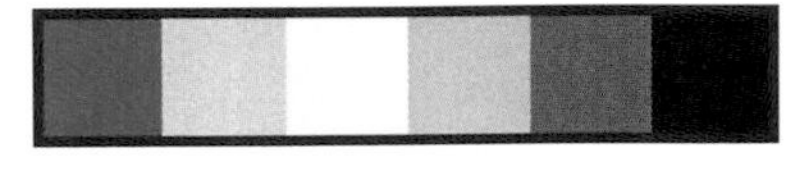

color ramp

color separation 1. In printing, the use of a separate printing plate for each ink color used. 2. The process of scanning with color filters to separate the original image into single-color negatives.

column 1. An item in an attribute table. 2. [COMPUTING] The vertical dimension of a table. Each column stores the values of one type of attribute for all the records, or rows, in the table. All the values in a given column are of the same data type; for example, number, string, BLOB, or date. **3.** A vertical group of cells in a raster, or pixels in an image.

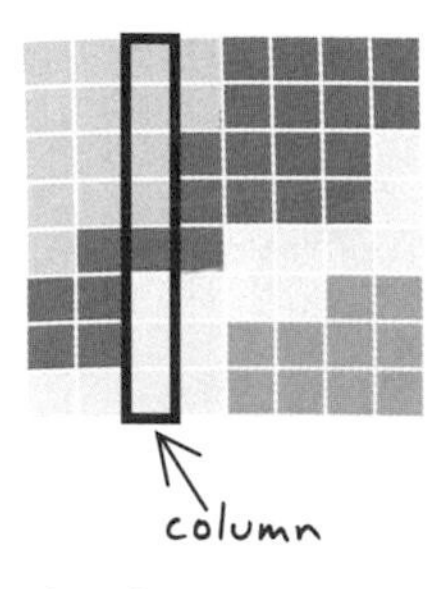

column 3

COM [COMPUTING] *Acronym for Component Object Model.* A binary standard that enables software components to interoperate in a networked environment regardless of the language in which they were developed. Developed by Microsoft, COM technology provides the underlying services of interface negotiation, life-cycle management (determining when an object can be removed from a system), licensing, and event handling. The ArcGIS system is created using COM objects.

command [COMPUTING] An instruction to a computer program, usually one word or concatenated words or letters, given by the user from a control device, such as a keyboard, or read from a file by a command interpreter.

command line [COMPUTING] A string of text that acts as a command, typed at an interface prompt.

command line interface [COMPUTING] A format of the input and output of a program in which the user enters commands by means of strings of text typed on a keyboard, as opposed to selecting commands from graphical prompts such as icons or dialog boxes.

Common Gateway Interface
See CGI.

comparison threshold [COMPUTING] The degree of uncertainty that can be tolerated in the spelling of a keyword used in a search, including phonetic errors and the random insertion, deletion, replacement, or transposition of characters.

compass **1.** [NAVIGATION] An instrument used to find the direction of north from one's current location, consisting of a case with compass points marked around its edge and a floating magnetic needle that pivots to point to magnetic north.

compass 1

2. [CARTOGRAPHY] An instrument with two legs connected by a joint, used to draw and measure circles. *See also* magnetic north, compass point, cardinal point.

C

compass north *See* magnetic north.

compass point [CARTOGRAPHY] An indication of direction. One of the 32 divisions into which the circle around the needle of a compass is divided, each equal to 11.25 degrees. *See also* compass.

compass rose [CARTOGRAPHY] A diagram of compass points drawn on a map or chart, subdivided clockwise from 0 to 360 degrees with 0 indicating true north. On older maps and charts a compass rose was a decorated diagram of cardinal points, divided into 16 or 32 points. *See also* compass.

compass rule [SURVEYING] A widely used rule for adjusting a traverse that assumes the precision in angles or directions is equivalent to the precision in distances. This rule distributes the closure error over the whole traverse by changing the northings and eastings of each traverse point in proportion to the distance from the beginning of the traverse. More specifically, a correction factor is computed for each point as the sum of the distances along the traverse from the first point to the point in question, divided by the total length of the traverse. The correction factor at each point is multiplied by the overall closure error to get the amount of error correction distributed to the point's coordinates. The compass rule is also known as the Bowditch rule, named for the American mathematician and navigator Nathaniel Bowditch (1773–1838). *See also* closure error.

compiler [PROGRAMMING] A program used in software development that translates the lines of a programmer's code from one programming language to another, usually from a high-level language to the ones and zeros of machine language. *See also* high-level language, low-level language.

complex market area An area calculated by finding the outermost customers of a store along several vectors and connecting them. Complex market areas are more accurate than simple market areas because they respond to physical and cultural barriers. They are sometimes called amoebas because of their irregular shapes. *See also* simple market area.

Component Object Model *See* COM.

composite relationship [COMPUTING] A link or association between objects where the lifetime of one object controls the lifetime of its related objects. For example, the association between a highway and its shield markers is a composite relationship, since the shield markers should not exist without the highway. *See also* simple relationship.

compound element Within metadata, a group of data elements (including other compound elements) that together describe a characteristic of a spatial dataset in more detail than can be described by an individual data element.

compound key A primary key that requires two or more fields to be unique. *See also* key.

compression [COMPUTING] The process of reducing the size of a file or database. Compression improves data handling, storage, and database performance. Examples of compression methods include quadtrees, run-length encoding, and wavelets.

compromise projection [CARTOGRAPHY] A projection that does not have equal area, conformal, or equidistant characteristics. The compromise projection is an attempt at balance between these characteristics, and is often used in thematic mapping. *See also* projection.

computational geometry [MATHEMATICS] A branch of mathematics that uses algorithms to solve geometry problems. Computational geometry is used in many GIS operations, including proximity analysis, feature generalization, and automated text placement.

computer-aided design *See* CAD.

computer-assisted learning Instruction or training that uses computer-based media instead of hard-copy materials. Computer-assisted learning is generally designed to use the strengths of computer-based media such as the ability to navigate in a nonlinear fashion through the use of hyperlinks.

concatenate [COMPUTING] To join two or more character strings together end to end; for example, to combine the two strings "spatial" and "analysis" into the single string "spatial analysis."

concatenated key [COMPUTING] In a relational database table, a primary key made by combining two or more keys that together form a unique identifier.

concurrency management A database management process for maintaining the consistency of data while supporting simultaneous editing by more than one user. A typical technique involves locking portions of the database to prevent data corruption caused by multiple users simultaneously editing data.

conditional operator [COMPUTING] A symbol or keyword that specifies the relationship between two values and is used to construct queries to a database. Examples include = (equal to), < (less than), and > (greater than).

conditional statement [COMPUTING] A programming language statement that executes one option if the statement is true, and another if it is false. The if-then-else statement is an example of a conditional statement.

confidence level [STATISTICS] In a statistical test, the risk, expressed as a percentage, that the null hypothesis will be incorrectly rejected because of sampling error when the null hypothesis is true. For example, a confidence level of 95 percent means that if the same test were performed 100 times on 100 different samples, the null hypothesis would be incorrectly rejected five times. *See also* null hypothesis, significance level.

conflation A set of procedures that aligns the features of two geographic data layers and then transfers the attributes of one to the other. *See also* rubber sheeting.

conflict In database editing, a state of incompatibility that occurs when multiple users simultaneously edit a version or reconcile two versions. Conflicts occur when the same feature or topologically related features are edited in two versions, and it is unclear which representation of the database is valid.

conformality [CARTOGRAPHY] The characteristic of a map projection that preserves the shape of any small geographic area. *See also* projection.

conformal projection [CARTOGRAPHY] A projection that preserves the correct shapes of small areas. In a conformal projection, graticule lines intersect at 90-degree angles, and at any point on the map the scale is the same in all directions. A conformal projection maintains all angles at each point, including those between the intersections of arcs; therefore, the size of areas enclosed by many arcs may be greatly distorted. No map projection can preserve the shapes of larger regions. *See also* projection.

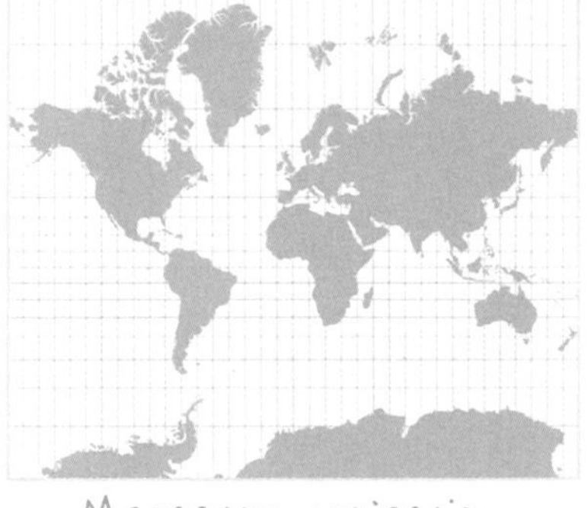

conformal projection

conic projection [CARTOGRAPHY] A projection that transforms points from a spheroid or sphere onto a tangent or secant cone that is wrapped around the globe in the manner of a party hat. The cone is then sliced from the apex (top) to the bottom, and flattened into a plane. *See also* projection.

conjoint boundary *See* shared boundary.

connectivity 1. The way in which features in GIS data are attached to one another functionally or spatially. **2.** [ESRI SOFTWARE] In a geodatabase, the state of association between edges and junctions in a network system for network data models. Connectivity helps define and control flow, tracing, and pathfinding in a network.

3. [ESRI SOFTWARE] In a coverage, topological identification of connected arcs by recording the from-node and to-node for each arc. Arcs that share a common node are connected.

connectivity analysis *See* network analysis.

connector A visual representation of the relationship between elements in a model. Connectors join elements together to create processes. Typical processes connect an input data element, a tool element, and a derived data element.

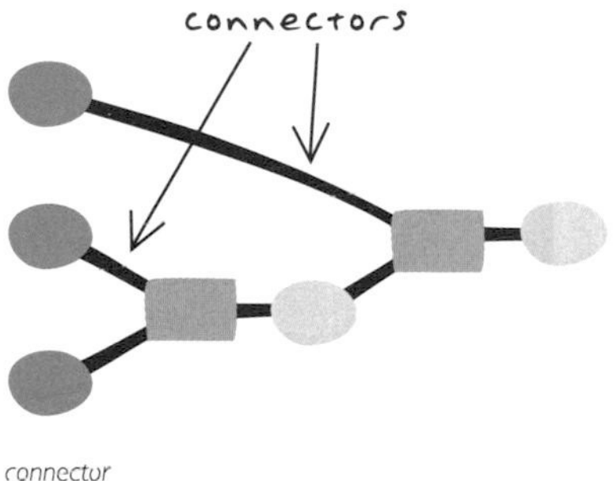

connector

constant azimuth *See* rhumb line.

containment A spatial relationship in which a point, line, or polygon feature or set of features is enclosed completely within a polygon.

Content Standard for Digital Geospatial Metadata A publication authored by the FGDC that specifies the information content of metadata for digital geospatial datasets. The purpose of the standard is to provide a common set of terminology and definitions for concepts related to the metadata. All U.S. government agencies (federal, state, and local) that receive federal funds to create metadata must follow this standard. *See also* FGDC.

conterminous [MATHEMATICS] Having the same or coincident boundaries.

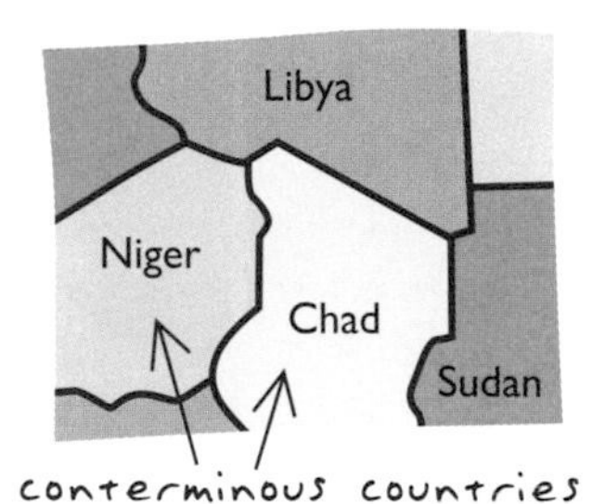

conterminous

contiguous **1.** [MATHEMATICS] Lying next to or close to one another.

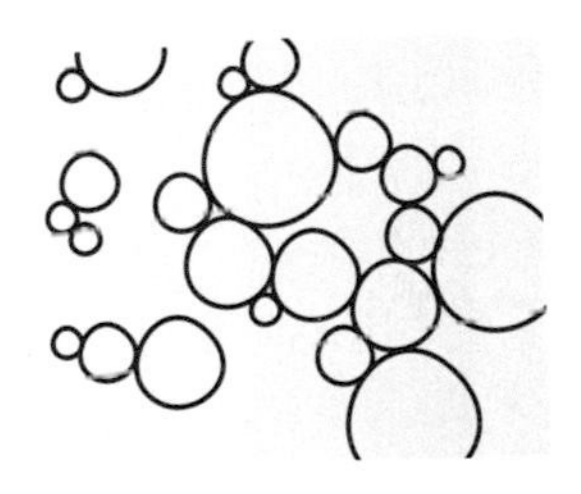

contiguous 1

2. Of polygons: adjacent; having a common boundary; sharing an edge. **3.** Of raster cells: connected orthogonally or diagonally; or, sometimes, connected strictly orthogonally. **4.** Of TIN edges: having no gaps or overlaps.

C

continuous data Data such as elevation or temperature that varies without discrete steps. Since computers store data discretely, continuous data is usually represented by TINs, rasters, or contour lines, so that any location has either a specified value or one that can be derived. *See also* discrete data.

continuous raster A raster in which cell values vary continuously to form a surface. In a continuous raster, the phenomena represented have no clear boundaries. Values exist on a scale relative to each other. It is assumed that the value assigned to each cell is what is found at the center of the cell. Rasters representing elevation, precipitation, chemical concentrations, suitability models, or distance from a road are examples of continuous rasters. *See also* discrete raster.

continuous raster

continuous tone image A photograph that has not been screened and so displays all the varying tones from dark to light. *See also* halftone image, dot screen.

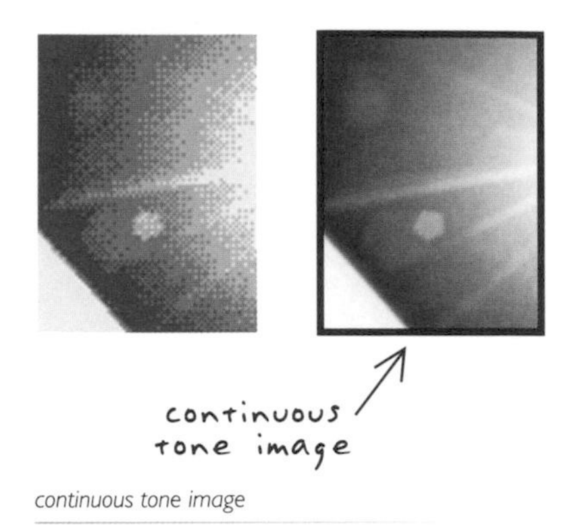

continuous tone image

contour interval [CARTOGRAPHY] The difference in elevation between adjacent contour lines. *See also* contour line.

contour line [CARTOGRAPHY] A line on a map that connects points of equal elevation based on a vertical datum, usually sea level. *See also* form lines, contour interval, supplemental contour.

contour line

contour tagging Assigning elevation values to contour lines. *See also* contour line.

contrast [REMOTE SENSING] In remote sensing and photogrammetry, the ratio between the energy emitted or reflected by an object and that emitted or reflected by its immediate surroundings.

contrast ratio The ratio between the maximum and minimum brightness values in an image.

contrast stretch Increasing the contrast in an image by expanding its grayscale range to the range of the display device.

control [COMPUTING] A basic element of a software application's GUI. Examples of controls include menus, buttons, tools, check boxes, slider bars, text input boxes, and combo boxes. *See also* GUI.

control point 1. [SURVEYING] An accurately surveyed coordinate location for a physical feature that can be identified on the ground. Control points are used in least-squares adjustments as the basis for improving the spatial accuracy of all other points to which they are connected. 2. [CARTOGRAPHY] One of various locations on a paper or digital map that has known coordinates and is used to transform another dataset—spatially coincident but in a different coordinate system—into the coordinate system of the control point. Control points are used in digitizing data from paper maps, in georeferencing both raster and vector data, and in performing spatial adjustment operations such as rubber sheeting.

convergence angle [CARTOGRAPHY] The angle between a vertical line (grid north) and true north on a map.

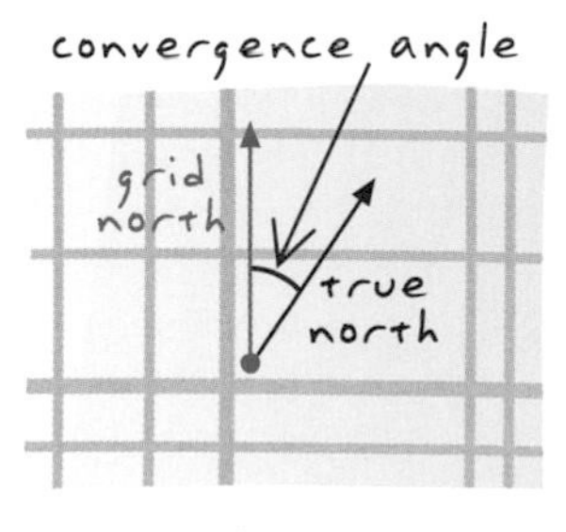

convergence angle

conversion The process of changing input data from one representation or format to another, such as from raster to vector, or from one file format to another, such as from x,y coordinate table to point shapefile.

convex hull [MATHEMATICS] The smallest convex polygon that encloses a group of objects, such as points. In ArcGIS, TIN boundaries are convex hulls by default. *See also* convex polygon.

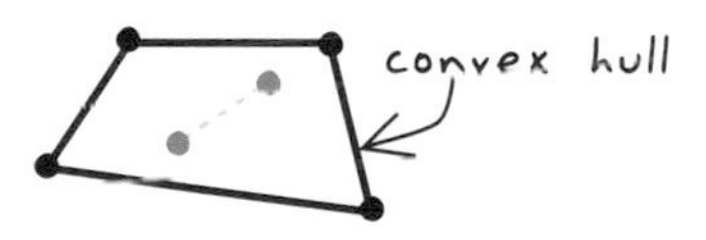

convex hull

convex polygon [MATHEMATICS] A polygon in which a straight line drawn between any two points inside the polygon is completely contained within the polygon. Visually, the boundary of a convex polygon is the shape a rubber band would take around a group of objects. *See also* convex hull.

coordinated universal time [ASTRONOMY] The official timekeeping system of the world's nations since 1972.

It refers local time throughout the world to time at the prime meridian, and is based on atomic clocks, but is periodically artificially adjusted so as to always remain within 0.9 seconds of universal time. The adjustment is made by the addition of leap seconds to the course of atomic time. Coordinated universal time is abbreviated "UTC." (The abbreviation UTC does not represent the word order of "coordinated universal time" in either English or French. It is an extension of the "UT*" pattern established for versions of universal time.) *See also* atomic clock, universal time, prime meridian.

coordinate geometry *See* COGO.

coordinates [CARTOGRAPHY] A set of values represented by the letters x, y, and optionally z or m (measure), that define a position within a spatial reference. Coordinates are used to represent locations in space relative to other locations. *See also* geographic coordinates, x,y coordinates, absolute coordinates.

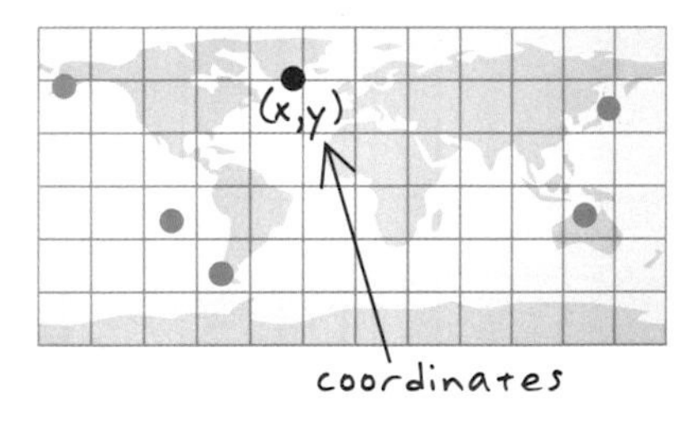

coordinates

coordinate system [CARTOGRAPHY] A reference framework consisting of a set of points, lines, and/or surfaces, and a set of rules, used to define the positions of points in space in either two or three dimensions. The Cartesian coordinate system and the geographic coordinate system used on the earth's surface are common examples of coordinate systems. *See also* coordinates, geographic coordinate system, projected coordinate system, vertical coordinate system. **For more information about coordinate systems, see* Projected and geographic coordinate systems: What is the difference? *on page 259.*

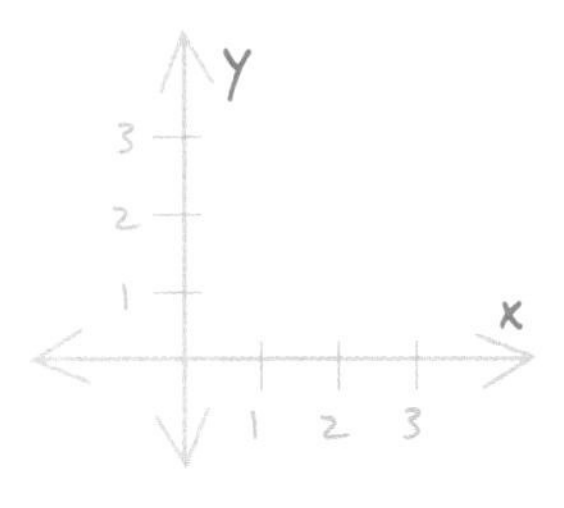

coordinate system

coordinate transformation [CARTOGRAPHY] The process of converting the coordinates in a map or image from one coordinate system to another, typically through rotation and scaling.

correlation [STATISTICS] An association between data or variables that change or occur together. For example, a positive correlation exists between housing costs and distance from the beach; generally, the closer a home is to the beach, the more it costs. Correlation does not imply causation. For example, there is a statistical correlation between ice cream sales and crime rates, but neither causes the other. The correlation coefficient is an index number between −1 and 1 indicating the strength of the association between two variables.

corridor A buffer drawn around a line. *See also* buffer.

corridor

corridor analysis A form of spatial analysis usually applied to environmental and land-use data in order to find the best locations for building roads, pipelines, and other linear transportation features. *See also* buffer.

cost A function of time, distance, or any other factor that incurs difficulty or an outlay of resources. *See also* impedance, restriction.

cost-benefit analysis An appraisal that attempts to compare the benefits (including social benefits) expected from a project with the costs (sometimes including social costs) incurred by the project over its lifetime. Generally cost-benefit analyses are used to compare alternative proposals, or to make a case for the implementation of a particular plan or system.

cost-distance analysis The calculation of the least cumulative cost from each cell to specified source locations over a cost raster. *See also* cost raster.

cost grid *See* cost raster.

cost raster A raster dataset that identifies the cost of traveling through each cell in the raster. A cost raster can be used to calculate the cumulative cost of traveling from every cell in the raster to a source or a set of sources. *See also* cost.

county The primary legal subdivision of all U.S. states except Alaska and Louisiana. The U.S. Census Bureau uses counties or equivalent entities (boroughs in Alaska, parishes in Louisiana, the District of Columbia in its entirety, and municipios in Puerto Rico) as statistical subdivisions.

county subdivision A statistical division of a county recognized by the U.S. Census Bureau for data presentation.

covariance [STATISTICS] A statistical measure of the linear relationship between two variables. Covariance measures the degree to which two variables move together relative to their individual mean returns. *See also* variogram, correlation.

coverage [ESRI SOFTWARE] A data model for storing geographic features. A coverage stores a set of thematically associated data considered to be a unit. It usually represents a single layer, such as soils, streams, roads, or land use. In a coverage, features are stored as both primary features (points, arcs, polygons) and secondary features (tics, links, annotation). Feature attributes are described and stored in feature attribute tables. Coverages cannot be edited in ArcGIS 8.3 and subsequent versions.

cracking [ESRI SOFTWARE] In ArcGIS, a part of the topology validation process in which vertices are created at the intersection of feature edges.

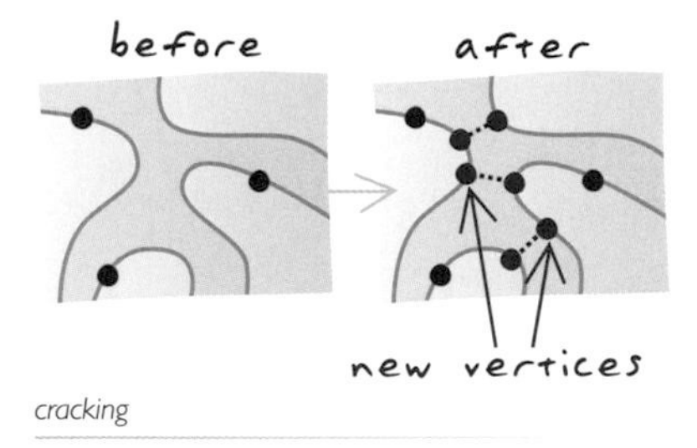

cracking

Crandall rule [SURVEYING] A special-case, least-squares-based method for adjusting the closure error in a traverse. The Crandall rule is most frequently used in a closed traverse that represents a parcel from a subdivision plan to ensure that tangency between courses remains intact as, for example, when applied to a tangent curve. It assumes that course directions and angles have no error and, therefore, all error corrections are applied only to the distances. This method uses a least-squares adjustment to distribute the closure error, and applies infinite weight to the angles or direction measurements to ensure that they are not adjusted. In some circumstances the results of this adjustment method may be unexpected, or the adjustment may not be possible, and an alternative method is required. The Crandall rule was developed by C.L. Crandall around 1901. *See also* closure error.

crop guide *See* crop marks.

crop marks Marks that indicate the edge of the page of a finished, printed map. Cropmarks are used as a reference for trimming excess paper after printing.

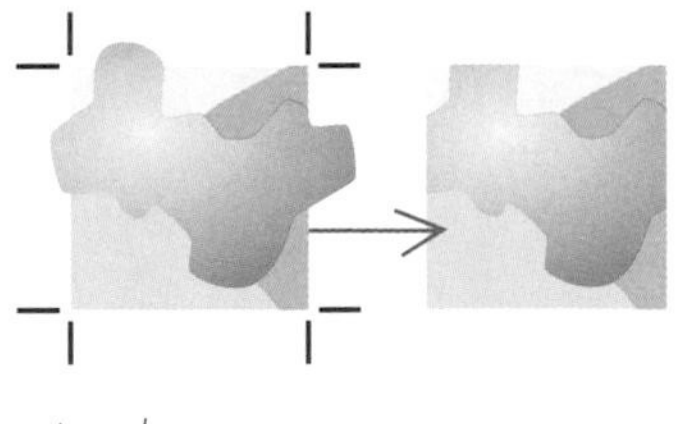

crop marks

cross correlation [STATISTICS] Statistical correlation between spatial random variables of different types, attributes, names, and so on, where the correlation depends on the distance or direction that separates the locations. *See also* autocorrelation.

cross covariance [STATISTICS] The statistical tendency of variables of different types, attributes, names, and so on, to vary in ways that are related to each other. Positive cross covariance occurs when both variables tend to be above their respective means together, and negative cross covariance occurs if one variable tends to be above its mean when the other variable is below its mean. *See also* covariance.

cross tabulation 1. In a GIS, comparing attributes in different coverages or map layers according to location. **2.** [STATISTICS] A method for showing the relationship between two or more data characteristics by repeating each of the categories of one variable for each category of the other variables. For example, a cross tabulation of census data might show households by number of occupants by income.

cross validation [STATISTICS] A procedure for testing the quality of a predicted data distribution. In cross validation, a piece of data whose value is known independently is removed from the dataset and the rest of the data is used to predict its value. Full cross validation is done by removing, in turn, each piece of data from the dataset and using the rest of the data to predict its value.

cross variogram [STATISTICS] A function of the distance and direction separating two locations, used to quantify cross correlation. The cross variogram is defined as the variance of the difference between two variables of different types or attributes at two locations. The cross variogram generally increases with distance, and is described by nugget, sill, and range parameters.

CSDGM *See* Content Standard for Digital Geospatial Metadata.

cubic convolution [MATHEMATICS] A technique for resampling raster data in which the average of the nearest 16 cells is used to calculate the new cell value. *See also* resampling.

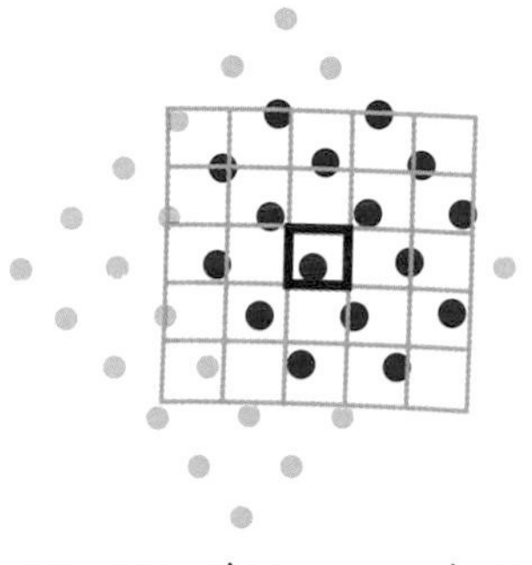

cubic convolution

cultural feature [CARTOGRAPHY] A human-made feature represented on a map, such as a building, road, tower, or bridge.

cultural geography The field of geography concerning the spatial distribution and patterns created by human cultures and their effects on the earth.

curb approach In network analysis, a network location property that models a path for approaching a stop from a specific side based on edge direction. For example, a school bus must approach a school from its door side so that students exiting the bus will not have to cross the street. There are three types of curb approaches: left, right, or both.

C

curb approach

curve fitting Converting short connected straight lines into smooth curves to represent features such as rivers, shorelines, and contour lines. The curves that result pass through or close to the existing points.

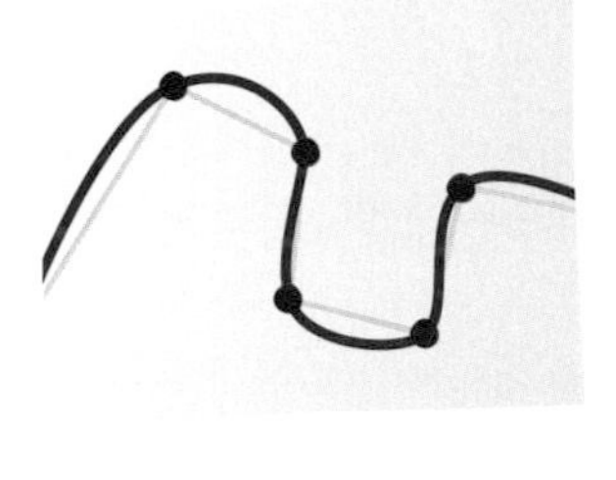

curve fitting

custom behavior [COMPUTING] A set of methods, functions or operations associated with a database object that has been specifically created or overridden by a programmer.

customer profiling A process that establishes common demographic characteristics for a set of customers within a geographic area.

customer prospecting A type of market analysis that locates regions with appropriate demographic characteristics for targeting new customers.

custom functionality [COMPUTING] A modification to or enhancement of standard software functionality to meet a specific user's needs.

custom object [COMPUTING] An object with custom behavior provided by a developer. *See also* custom behavior.

cycle 1. A set of lines forming a closed polygon. 2. In network analysis, a path or tour beginning and ending at the same location. 3. [PHYSICS] One oscillation of a wave.

cylindrical projection [CARTOGRAPHY] A projection that transforms points from a spheroid or sphere onto a tangent or secant cylinder. The cylinder is then sliced from top to bottom and flattened into a plane. *See also* projection.

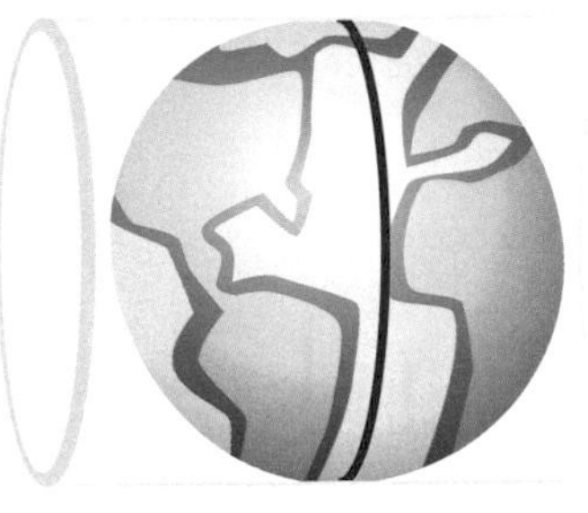

cylindrical projection

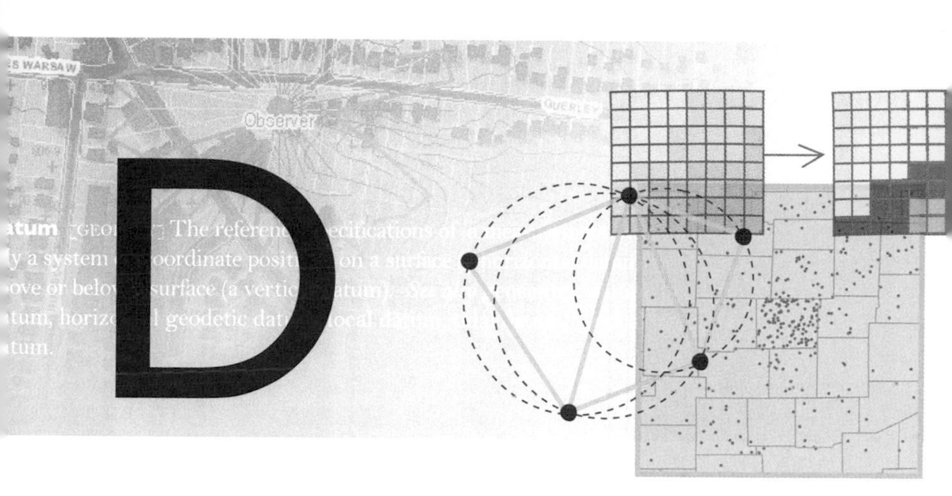

D

dangle The endpoint of a dangling arc. *See also* dangling arc.

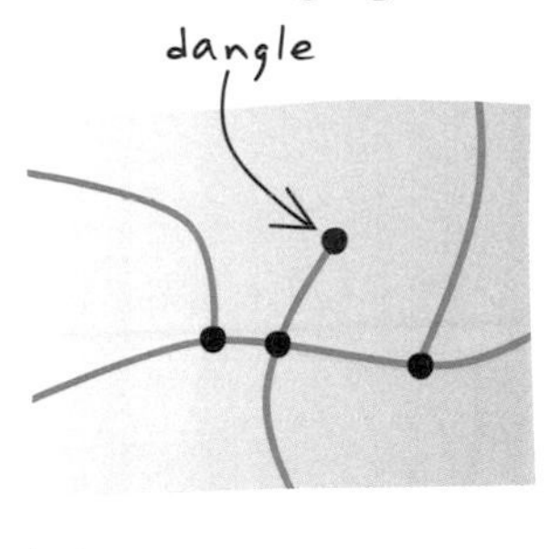

dangle

dangle length *See* dangle tolerance.

dangle tolerance [ESRI SOFTWARE] In ArcInfo coverages, the minimum length allowed for dangling arcs by the clean process, which removes dangling arcs shorter than the dangle tolerance.

dangling arc An arc having the same polygon on both its left and right sides and having at least one node that does not connect to any other arc. It often occurs where a polygon does not close properly, where arcs do not connect properly (an undershoot), or where an arc was digitized past its intersection with another arc (an overshoot). A dangling arc is not always an error; for example, it can represent a cul-de-sac in a street network. *See also* undershoot, overshoot.

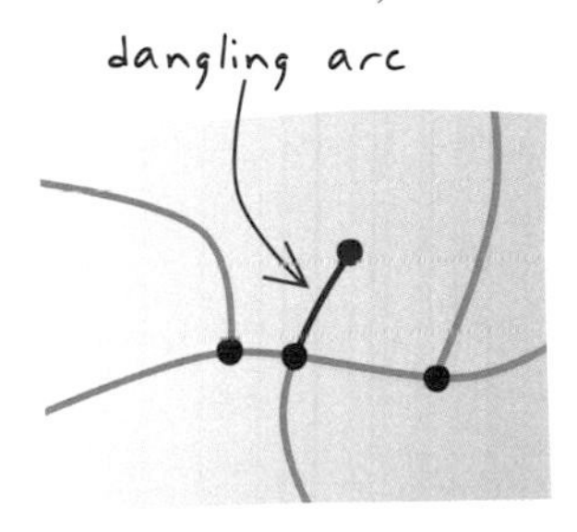

dangling arc

dangling node *See* dangle.

dasymetric mapping A technique in which attribute data that is organized by a large or arbitrary area unit is more accurately distributed within that unit by the overlay of geographic boundaries that exclude, restrict, or confine the attribute in question. For example, a population attribute organized by census tract might be more accurately distributed by the overlay of water bodies, vacant land, and other land-use boundaries within which it is reasonable to infer that people do not live.

D

data Any collection of related facts arranged in a particular format; often, the basic elements of information that are produced, stored, or processed by a computer.

database One or more structured sets of persistent data, managed and stored as a unit and generally associated with software to update and query the data. A simple database might be a single file with many records, each of which references the same set of fields. A GIS database includes data about the spatial locations and shapes of geographic features recorded as points, lines, areas, pixels, grid cells, or TINs, as well as their attributes. *See also* geodatabase.

database administrator The person who manages a database. Database administration includes user setup, security, backup and recovery procedures for data, and optimization of physical data storage for best performance. *See also* database.

database connection A link to a database from a software application. Database connections have two states: connected to or disconnected from the database. Deletion of a database connection only deletes the connection itself, not the database or its contents. Creation of a database connection requires selection of a data provider for data retrieval. *See also* database.

database design The development of the conceptual, logical, and physical structures of a database in order to meet user requirements. *See also* database.

database generalization The abstraction, reduction, and simplification of features and feature classes for deriving a simpler model of reality or decreasing stored data volumes. *See also* generalization.

database management system A set of software applications used to create and maintain databases according to a schema. Database management systems provide tools for adding, storing, changing, deleting, and retrieving data. *See also* database.

database support The proprietary database platforms supported by a program or component. *See also* database.

data capture Any operation that converts GIS data into computer-readable form. Geographic data can be captured by being downloaded directly into a GIS from sources such as remote-sensing or GPS data, or it can be digitized, scanned, or keyed in manually from paper maps or photographs.

data conversion The process of translating data from one format to another.

data definition language [PROGRAMMING] A set of SQL statements that can be used either interactively or within a programming language to create a new database, set permissions on it, and define its attributes.

data dictionary A catalog or table containing information about the datasets stored in a database. In a GIS, a data dictionary might contain the full names of attributes, meanings of codes, scale of source data, accuracy of locations, and map projections used. *See also* metadata.

data element The smallest unit of information used to describe a particular characteristic of a spatial dataset. A data element is a logically primitive description that cannot be further subdivided. *See also* compound element.

data file [COMPUTING] A file that holds text, graphics, or numbers. *See also* executable file.

data flow The route of data passage through a system. *See also* data.

data format The structure used to store a computer file or record.

data frame [ESRI SOFTWARE] A map element that defines a geographic extent, a page extent, a coordinate system, and other display properties for one or more layers in ArcMap. A dataset can be represented in one or more data frames. In data view, only one data frame is displayed at a time; in layout view, all a map's data frames are displayed at the same time. Many cartography texts use the term "map body" to refer to what ESRI calls a data frame.

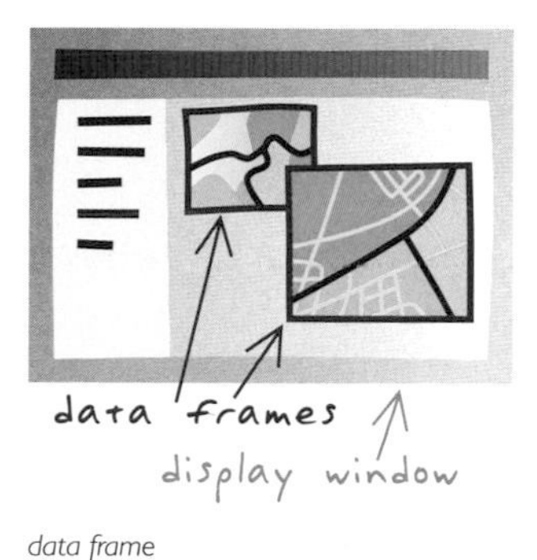

data frame

data integration The process of sharing and combining data between two organizations or systems.

data integrity The degree to which the data in a database is accurate and consistent according to data model and data type.

data logger *See* data recorder.

data message [GPS] Information in a satellite's GPS signal that reports its orbital position, operating health, and clock corrections.

data model 1. In GIS, a mathematical construct for representing geographic objects or surfaces as data. For example, the vector data model represents geography as collections of points, lines, and polygons; the raster data model represents geography as cell matrixes that store numeric values; and the TIN data model represents geography as sets of contiguous, nonoverlapping triangles. **2.** [ESRI SOFTWARE] In ArcGIS, a set of database design specifications for objects in a GIS application. A data model describes the thematic layers used in the application (for example, hamburger stands, roads, and counties); their spatial representation (for example, point, line, or polygon); their attributes; their integrity rules and relationships (for example, counties must nest within states); their cartographic portrayal; and their metadata requirements. **3.** In information theory, a description of the rules by which data is defined, organized, queried, and updated within an information system (usually a database management system). *See also* data structure.

data recorder [GPS] A lightweight, handheld field computer used to store data collected by a GPS receiver.

data recorder

data repository *See* clearinghouse.

dataset Any collection of related data, usually grouped or stored together. *See also* feature dataset.

dataset precision The mathematical exactness or detail with which a value is stored within a dataset, based on the number of significant digits that can be stored for each coordinate.

data sharing Making data available and accessible to organizations or individuals other than the creator of the data.

data structure The organization of data within a specific computer system that allows the data to be stored and manipulated effectively; a representation of a data model in computer form. *See also* data model.

data transfer The process of moving data from one system to another or from one point on a network to another.

data type The attribute of a variable, field, or column in a table that determines the kind of data it can store. Common data types include character, integer, decimal, single, double, and string.

datum [GEODESY] The reference specifications of a measurement system, usually a system of coordinate positions on a surface (a horizontal datum) or heights above or below a surface (a vertical datum). *See also* geocentric datum, geodetic datum, horizontal geodetic datum, local datum, tidal datum, vertical geodetic datum.

datum level [GEODESY] A surface to which heights, elevations, or depths are referenced. *See also* vertical geodetic datum.

datum plane *See* datum level.

datum shift *See* geographic transformation.

datum transformation *See* geographic transformation.

DBMS *See* database management system.

DDL *See* data definition language.

dead reckoning [NAVIGATION] A navigation method of last resort that uses the most recently recorded position of a ship or aircraft, along with its speed and drift, to calculate a new position.

debug [COMPUTING] To test for, detect, and correct errors in a computer program or component.

decimal degrees [CARTOGRAPHY] Values of latitude and longitude expressed in decimal format rather than in degrees, minutes, and seconds.

decision support system [COMPUTING] A computer program that includes data presentation and modeling tools that help people understand problems and find solutions.

declination 1. [CARTOGRAPHY] In a spherical coordinate system, the angle between the equatorial plane and a line to a point somewhere on the sphere.

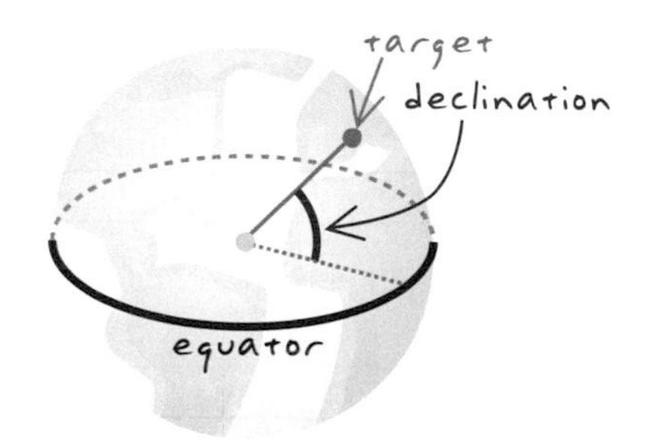

declination 1

2. [CARTOGRAPHY] The arc between the equator and a point on a great circle perpendicular to the equator. 3. [ASTRONOMY] The angular distance between a star or planet and the celestial equator. *See also* magnetic declination, celestial sphere.

degree 1. [GEODESY] A unit of angular measure represented by the symbol °. The earth is divided into 360 degrees of longitude and 180 degrees of latitude. 2. [MATHEMATICS] The angle equal to 1/360th of the circumference of a circle. A degree can be divided into 60 minutes of arc or 3600 seconds of arc. *See also* minute, second.

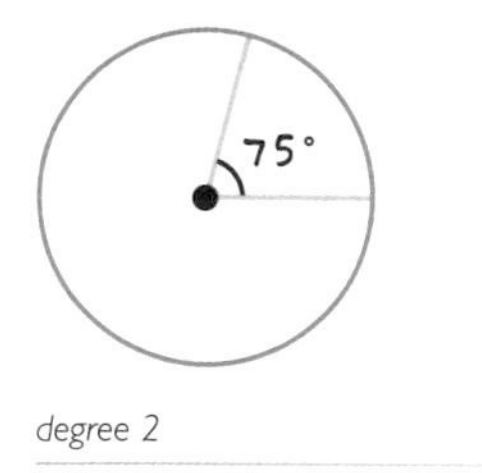

degree 2

degree slope One method for representing the measurement of an inclined surface. The steepness of

D

a slope may be measured from 0 to 90 degrees. *See also* percent slope.

degree slope

degrees/minutes/seconds [CARTOGRAPHY] The unit of measure for describing latitude and longitude. A degree is 1/360th of a circle. A degree is further divided into 60 minutes, and a minute is divided into 60 seconds.

Delaunay triangles The components of Delaunay triangulation. Delaunay triangles cannot exist alone; they must exist as part of a set or collection that is typically referred to as a triangulated irregular network (TIN). A circle circumscribed through the three nodes of a Delaunay triangle will not contain any other points from the collection in its interior. *See also* Delaunay triangulation, Voronoi diagram, Thiessen polygons.

Delaunay triangulation A technique for creating a mesh of contiguous, nonoverlapping triangles from a dataset of points. Each triangle's circumscribing circle contains no points from the dataset in its interior. Delaunay triangulation is named for the Russian mathematician Boris Nikolaevich Delaunay. *See also* Voronoi diagram, Delaunay triangles, Thiessen polygons.

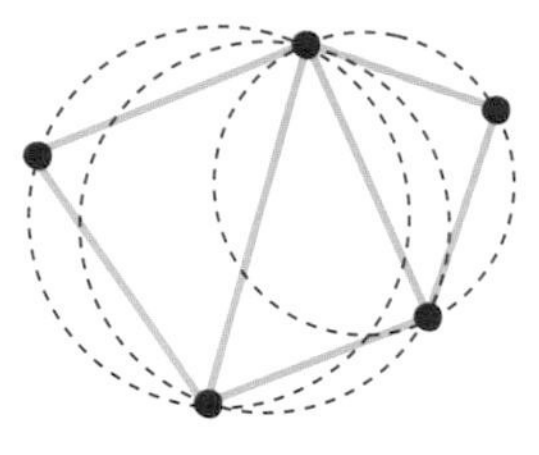

Delaunay triangulation

delimiter [COMPUTING] A character, such as a space or comma, that separates words or values.

DEM *Acronym for digital elevation model.* **1.** The representation of continuous elevation values over a topographic surface by a regular array of z-values, referenced to a common datum. DEMs are typically used to represent terrain relief.

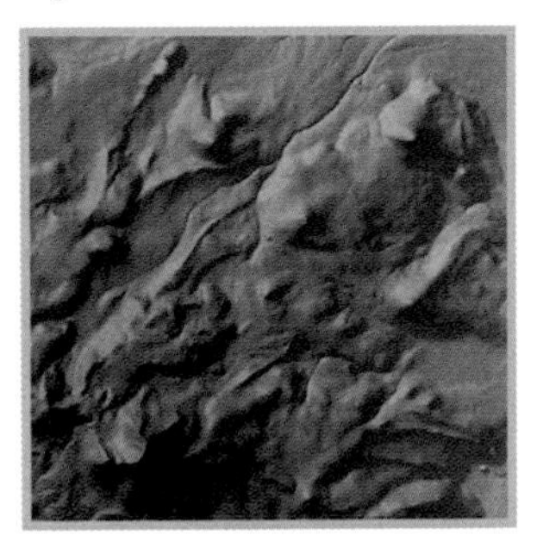

DEM 1

2. A format for elevation data, tiled by map sheet, produced by the National Mapping Division of the USGS. *See also* z-value, datum, DTED.

demographics [GEOGRAPHY] The statistical characteristics (such as age, birth rate, and income) of a human population.

demography [GEOGRAPHY] The statistical study of human populations, especially their locations, distribution, economic statistics, and vital statistics.

densify To add vertices to a line at specified distances without altering the line's shape. *See also* spline.

densitometer [CARTOGRAPHY] An instrument for measuring the opacity of translucent materials such as photographic negatives and optical filters. *See also* microdensitometer.

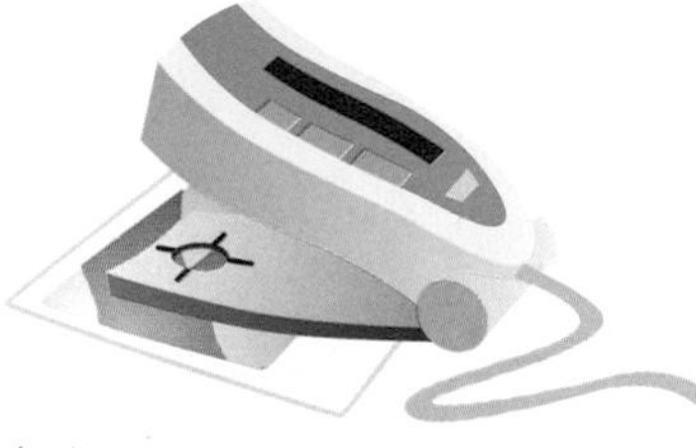

densitometer

density 1. [MATHEMATICS] In spatial measurements, the quantity per unit area or length. 2. [PHYSICS] In a substance such as a gas, solid, or liquid, a measurement of the ratio of mass to volume.

density slicing [REMOTE SENSING] A technique normally applied to a single-band monochrome image for highlighting areas that appear to be uniform in tone, but are not. Grayscale values (0-255) are converted into a series of intervals, or slices, and different colors are assigned to each slice. Density slicing is often used to highlight variations in vegetation.

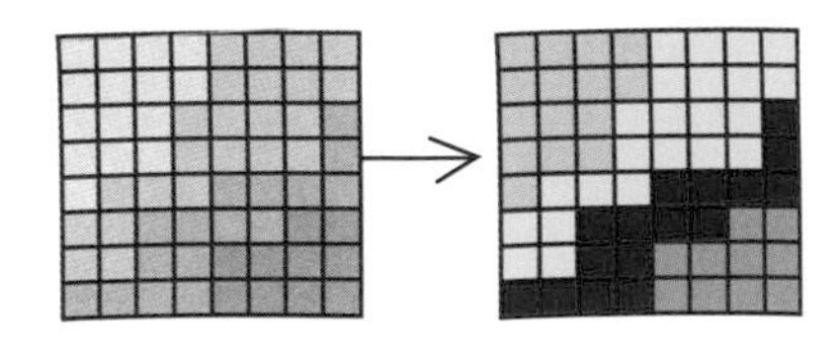

density slicing

depression contour *See* hachured contour.

depth contour [CARTOGRAPHY] A line on a map connecting points of equal depth below a hydrographic datum. *See also* contour line.

depth curve *See* depth contour.

descending node [ASTRONOMY] The point at which a satellite traveling north to south crosses the equator. *See also* ascending node.

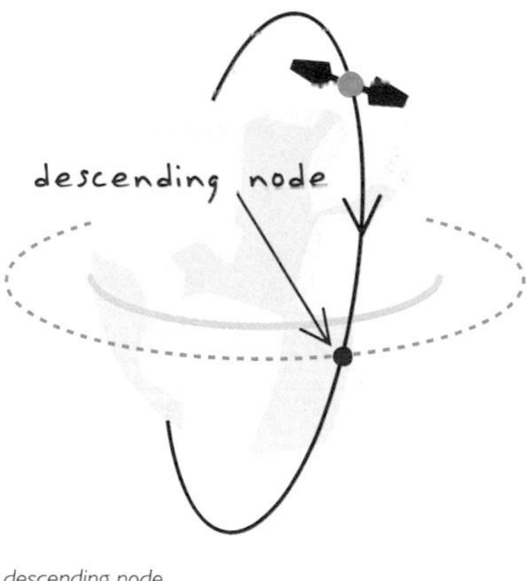

descending node

descriptive data *See* attribute data.

desire-line analysis A type of market analysis that draws lines from

a set of geocoded points (usually customers) to a single, central point (usually a store). Desire lines can be weighted. *See also* weight.

desktop GIS Mapping software that is installed onto and runs on a personal computer and allows users to display, query, update, and analyze data about geographic locations and the information linked to those locations.

destination [COMPUTING] The secondary object in a relationship class, such as a table containing attributes associated with features in a related table. *See also* origin.

determinate flow direction A conclusively definitive line or course in which something is issuing or moving in a stream. For an edge feature, this occurs when the flow direction can be ascertained from the connectivity of a network, the locations of sources and sinks, and the enabled or disabled states of features. *See also* indeterminate flow direction.

deterministic model [STATISTICS] In spatial modeling, a type of model or a part of a model in which the outcome is completely and exactly known based on known input; the fixed or nonrandom components of a spatial model. The spline and inverse distance weighted interpolation methods are deterministic since they have no random components. The kriging and cokriging interpolation methods may have a deterministic component, often called the trend. *See also* model, stochastic model, trend.

detrending [STATISTICS] The process of removing the trend from a spatial model by subtracting the trend surface (usually polynomial functions of the spatial x- and y-coordinates) from the original data values. The resulting detrended values are called residuals. *See also* trend.

developable surface [CARTOGRAPHY] A geometric shape such as a cone, cylinder, or plane that can be flattened without being distorted. Many map projections are classified in terms of these shapes.

development environment [PROGRAMMING] A software product used to write, compile, and debug components or applications.

device coordinates The coordinates shown on a digitizer or display, as opposed to those of a recognized datum or coordinate system.

DGIWG *Acronym for Digital Geographic Information Working Group.* A group established in 1983 to develop standards for spatial data exchange among nations participating in the North Atlantic Treaty Organization (NATO). The goals of the group are interoperability and burden sharing among nations, and its membership has recently expanded beyond NATO nations. While DGIWG is not an official NATO body, its work on standards has been recognized by the NATO Geographic Conference (NGC). *See also* interoperability, data sharing, DIGEST.

DGPS *See* differential correction.

DHTML *See* dynamic HTML.

diazo process A way of quickly and inexpensively copying maps using a diazo compound, ultraviolet light, and ammonia.

difference image [DIGITAL IMAGE PROCESSING] An image made by subtracting the pixel values of one image from those in another.

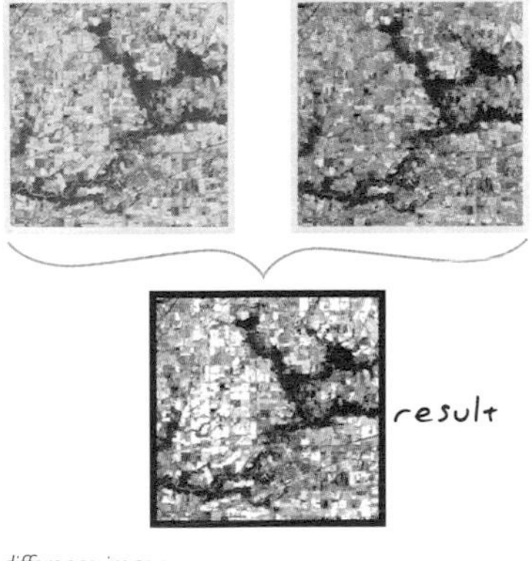

difference image

differential correction [GPS] A technique for increasing the accuracy of GPS measurements by comparing the readings to two receivers—one roving and the other a fixed base station—and a known location.

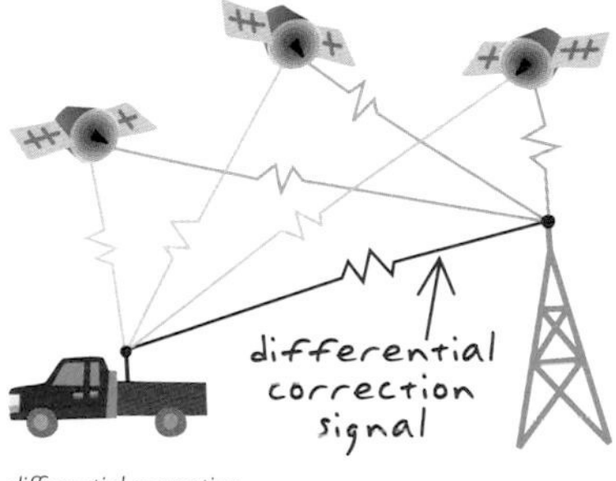

differential correction

differential Global Positioning System *See* differential correction.

diffusion The spread of an innovation or technology use among a group of people or organizations. *See also* adoption.

DIGEST *Acronym for Digital Geographic Information Exchange Standard.* A standard for spatial data transfer among nations, data producers, and data users. The Digital Geographic Information Working Group (DGIWG) developed the standard to support interoperability within and between nations and share the burden of digital data production. The standard addresses the exchange of raster, matrix, and vector data (and associated text) and a range of levels of topological structures. *See also* DGIWG.

digital Represented in discrete, quantified units rather than continuously. Computers process and store information in digital form. *See also* analog.

digital elevation model *See* DEM.

Digital Geographic Information Exchange Standard *See* DIGEST.

Digital Geographic Information Working Group *See* DGIWG.

digital image An image stored in binary form and divided into a matrix of pixels. Each pixel consists of a digital value of one or more bits, defined by the bit depth. The digital value may represent, but is not limited to, energy, brightness, color, intensity, sound, elevation, or a classified value derived through image processing. A digital image is stored as a raster and may contain one or more bands. *See also* image, analog image, raster.

D

digital image processing [REMOTE SENSING] Any technique that changes the digital values of an image for the sake of analysis or enhanced display, such as density slicing or low- and high-pass filtering.

digital line graph *See* DLG.

digital nautical chart *See* DNC.

digital number [REMOTE SENSING] In a digital image, a value assigned to a pixel.

digital orthophoto quadrangle *See* DOQ.

digital orthophoto quarter quadrangle *See* DOQQ.

digital raster graphic *See* DRG.

digital terrain elevation data *See* DTED.

digital terrain model *See* DEM.

digitizer A device connected to a computer, consisting of a tablet and a handheld puck, that converts positions on the tablet surface as they are traced by an operator to digital x,y coordinates, yielding vector data consisting of points, lines, and polygons. *See also* puck.

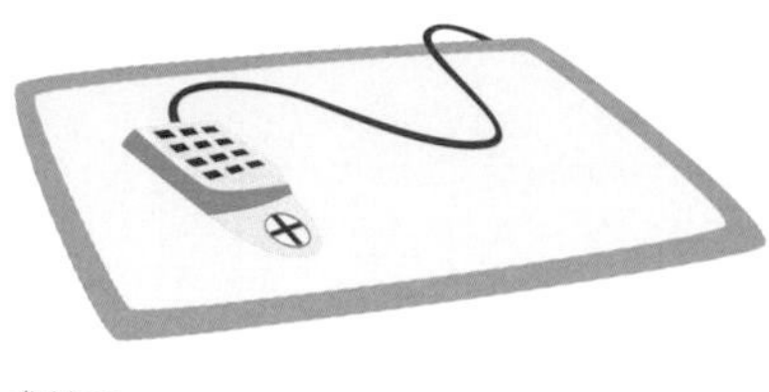

digitizer

digitizing The process of converting the geographic features on an analog map into digital format using a digitizing tablet, or digitizer, which is connected to a computer. Features on a paper map are traced with a digitizer puck, a device similar to a mouse, and the x,y coordinates of these features are automatically recorded and stored as spatial data. *See also* heads-up digitizing, blind digitizing.

digitizing mode A way of using a digitizing tablet in which locations on the tablet are mapped to specific locations on the screen. Moving the digitizer puck on the tablet surface causes the screen pointer to move to precisely the same position on the screen. *See also* mouse mode.

Dijkstra's algorithm An algorithm that examines the connectivity of a network to find the shortest path between two points. Dijkstra's algorithm is named after the Dutch computer scientist Edsger Dijkstra (1930–2002).

dilution of precision *See* DOP.

DIME *Acronym for Dual Independent Map Encoding.* A data storage format for geographic data developed by the U.S. Census Bureau in the 1960s. DIME-encoded data was stored in Geographic Base Files (GBF). The Census Bureau replaced the DIME format with Topologically Integrated Geocoding and Referencing (TIGER) in 1990. *See also* TIGER, GBF/DIME.

dimension 1. [PHYSICS] A length of a certain distance and bearing. 2. [PHYSICS] The area over which an entity extends. 3. [PHYSICS] The number of axes that are essential to the existence of an entity in space. For example, the identity of a location on a plane requires two axes; therefore, a plane exists in the second dimension, and an entity with two axes, or dimensions, may be uniquely identified as a plane.

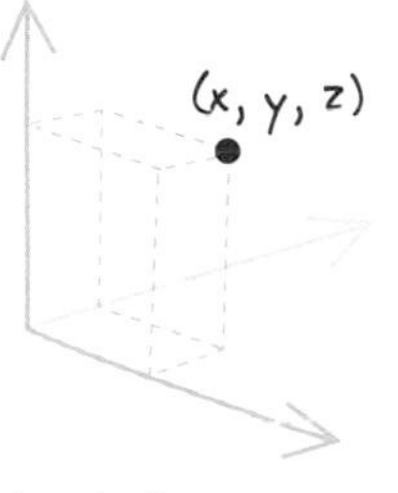

dimension 3

DIP *See* digital image processing.

directed network flow A network state in which edges have an associated direction of flow. In a directed network flow, the resource that traverses a network's components cannot choose a direction to take, as in hydrologic and utility systems. *See also* undirected network flow, downstream, upstream.

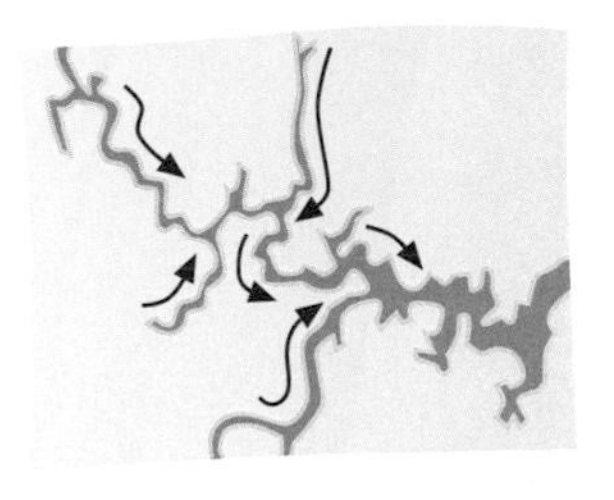

directed network flow

directional filter [DIGITAL IMAGE PROCESSING] An edge-detection filter that enhances those linear features in an image that are oriented in a particular direction. *See also* edge detection, digital image processing.

directory [COMPUTING] An area of a computer disk that holds a set of data files, other directories, or both. Operating systems use directories to organize data. Directories are arranged in a tree structure, in which each branch is a subdirectory of its parent branch. The location of a directory is specified with a path—for example, C:\gisprojects\shrinkinglemurhabitatgrids.

Dirichlet tessellation *See* Voronoi diagram.

dirty areas Regions surrounding features that have been altered after the initial topology validation process and require additional topology validation to be performed to find any errors.

disabled feature In geometric networks, an object or shape representing a geographic object through

D

which flow is impossible. *See also* barrier, enabled feature.

discrete data Data that represents phenomena with distinct boundaries. Property lines and streets are examples of discrete data. *See also* continuous data.

discrete digitizing A method of digitizing in which points are placed individually to define a feature's shape. *See also* digitizing, stream mode digitizing.

discrete raster A raster that typically represents phenomena that have clear boundaries with attributes that are descriptions, classes, or categories. Generally, integers are used for the cell values. In a raster of land cover, for example, the value 1 might represent forestland, the value 2 urban land, and so on. It is assumed that the phenomena that each value represents fill the entire area of the cell. Rasters representing land use, political boundaries or ownership are examples of discrete rasters. *See also* continuous raster.

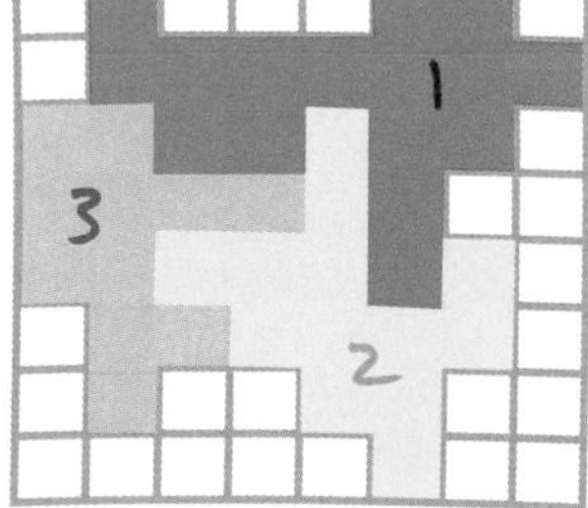

discrete raster

display scale The scale at which data is rendered on a computer screen or on a printed map. *See also* automation scale.

display unit The unit of measure used to render dimensions of shapes, distance tolerances, and offsets on a computer screen or on a printed map. Although they are stored with consistent units in the dataset, users can choose the units in which coordinates and measurements are displayed—for example, feet, miles, meters, or kilometers. *See also* unit of measure.

dissemination *See* diffusion.

dissolve The process of removing unnecessary boundaries between features, such as the edges of adjacent map sheets, after data has been captured.

distance The measure of separation between two entities or locations that may or may not be connected, such as two points. Distance is differentiated from length, which implies a physical connection between entities or locations.

distance decay A mathematical representation of the effect of distance on the accessibility of locations and the number of interactions between them, reflecting the notion that demand drops as distance increases. Distance decay can be expressed as a power function or as an exponential function.

distance unit [PHYSICS] The unit of measurement for distance, such as feet, miles, meters, and kilometers. *See also* unit of measure.

distortion [CARTOGRAPHY] On a map or image, the misrepresentation of shape, area, distance, or direction of or between geographic features when compared to their true measurements on the curved surface of the earth.

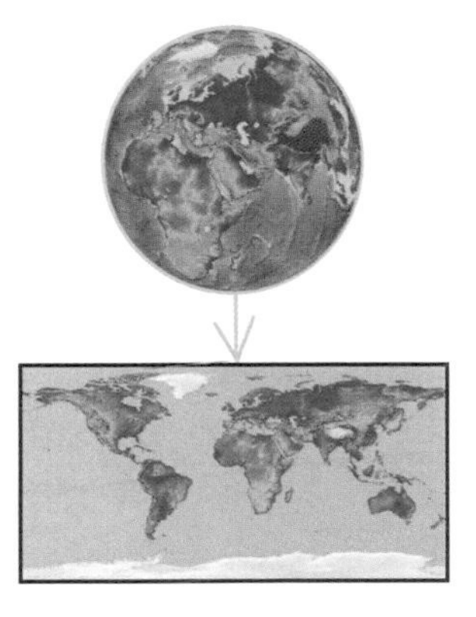

distortion

distortion circle *See* Tissot's indicatrix.

distributed database A database with records that are dispersed between two or more physical locations. Data distribution allows two or more people to be working on the same data in separate locations. *See also* database.

distribution 1. [STATISTICS] The frequency or amount at which a thing or things occur within a given area. 2. [STATISTICS] The set of probabilities that a variable will have a particular value.

dithering The approximation of shades of gray or colors in a computer image made by arranging pixels of black and white or other colors in alternate layers. The technique gives the appearance of a wider range of color or shades than is actually present in the image. It is widely used to improve the appearance of images displayed on devices with limited color palettes.

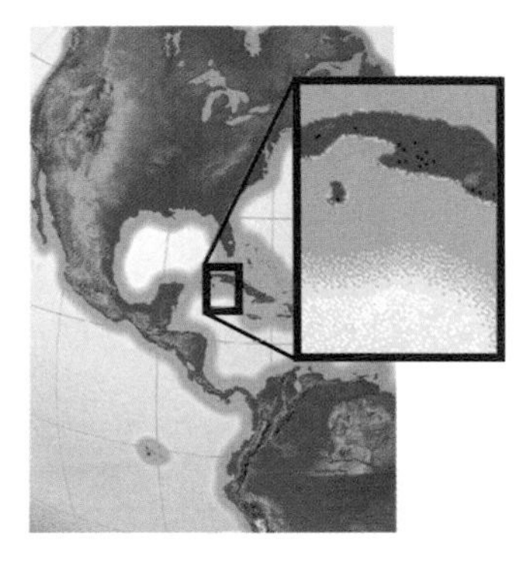

dithering

diurnal [ASTRONOMY] Daily, as in the revolution of the earth.

diurnal arc [ASTRONOMY] The apparent path from rise to set made by a heavenly body across the sky. *See also* diurnal.

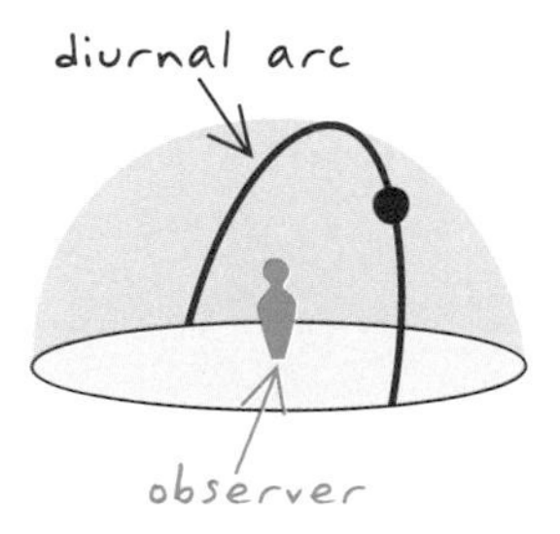

diurnal arc

D

D

DLG *Acronym for digital line graph.* Data files containing vector representations of cartographic information derived from USGS maps and related sources. DLGs include information from the USGS planimetric map base categories such as transportation, hydrography, contours, and public land survey boundaries.

DLL [INTERNET] *Acronym for dynamic-link library.* A type of file that stores shared code to be used by multiple programs (a "code library"). Programs access the shared code by linking to the DLL file when they run, a process referred to as dynamic linking. The DLL file must be registered for other programs to locate it. *See also* register.

DMS *See* degrees/minutes/seconds.

DNC [NAVIGATION] *Acronym for digital nautical chart.* A nautical database developed from existing hard-copy charts, digital data, bathymetric survey information, imagery, and various raster data. DNCs are used by the U.S. military and its allies for marine navigation. *See also* ENC.

DNS [INTERNET] *Acronym for domain name system.* The Internet distributed system that stores IP addresses and domain names to assist with the routing of network traffic. *See also* domain name.

docking [COMPUTING] Moving a floating toolbar or window to a fixed location in the graphical user interface.

documentation Supporting information for software data and tools. Documentation may be descriptive or instructional, and is published in a variety of formats, including user's guides and manuals, desktop help systems, embedded or context-sensitive help, tutorials, reports, and metadata.

domain **1.** The range of valid values for a particular metadata element. **2.** [COMPUTING] A group of computers and devices on a network that are administered as a unit with common rules and procedures. Within the Internet, a domain is defined by an IP address. All devices sharing a common part of the IP address are said to be in the same domain. *See also* range domain, spatial domain.

domain name [INTERNET] The unique name of a computer system on the Internet, such as "esri.com." *See also* domain.

domain name system *See* DNS.

DOP [GPS] *Acronym for dilution of precision.* An indicator of satellite geometry for a constellation of satellites used to determine a position. Positions with a lower DOP value generally constitute better measurement results than those with higher DOP. Factors determining the total GDOP (geometric DOP) for a set of satellites include PDOP (positional DOP), HDOP (horizontal DOP), VDOP (vertical DOP), and TDOP (time DOP).

Doppler-aided GPS [GPS] Signal processing that uses a measured Doppler shift to help the receiver track the GPS signal. *See also* Doppler shift.

Doppler shift [PHYSICS] The apparent change in frequency of sound or light waves caused by the relative motion between a source and an observer. As they approach one another, the frequency increases; as they draw apart, the frequency decreases. The Doppler shift is also known as the Doppler effect, and is named for Austrian physicist and mathematician Christian Andreas Doppler.

DOQ [REMOTE SENSING] *Acronym for digital orthophoto quadrangle.* A computer-generated, uniform-scale image created from an aerial photograph. Digital orthophoto quadrangles are true photographic maps in which the effects of tilt and relief are removed by a mathematical process called transformation or rectification. The uniform scale of a DOQ allows accurate measurement of distances. *See also* transformation, orthophotograph.

DOQ

DOQQ [NAVIGATION] *Acronym for digital orthophoto quarter quadrangle.* A digital orthophoto quadrangle (DOQ) divided into four quadrants. *See also* DOQ, orthophotograph.

dot density map [CARTOGRAPHY] A quantitative, thematic map on which dots of the same size are randomly placed in proportion to a numeric attribute associated with an area. Dot density maps convey the intensity of an attribute.

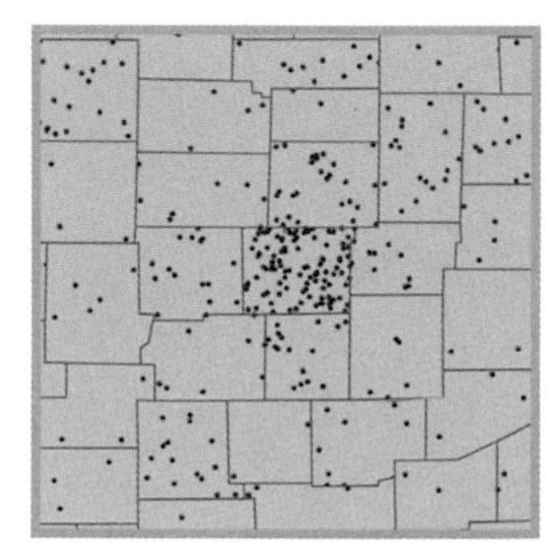

dot density map

dot distribution map [CARTOGRAPHY] A map that uses dots or other symbols to represent the presence, quantity, or value of a phenomenon or thing in a specific area. In a dot distribution map, the size of the dots is scaled in proportion to the intensity of the variable.

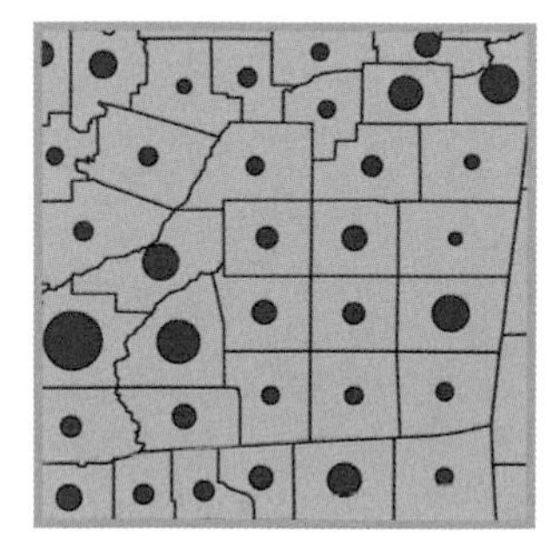

dot distribution map

D

dot screen A photographic film covered with uniformly sized, evenly spaced dots used to break up a solid color, producing an apparently lighter color. *See also* halftone image, continuous tone image.

dots per inch *See* dpi.

double-coordinate precision *See* double precision.

double precision [COMPUTING] The level of coordinate exactness based on the possible number of significant digits that can be stored for each coordinate. Datasets can be stored in either single or double precision. Double-precision geometries store up to 15 significant digits per coordinate (typically 13 to 14 significant digits), retaining the accuracy of much less than 1 meter at a global extent. *See also* single precision.

Douglas-Peucker algorithm An algorithm that simplifies complex lines by reducing the number of points used to represent them. The Douglas-Peucker algorithm was developed by the Canadian geographers David H. Douglas and Thomas K. Peucker.

downstream In network tracing, the direction along a line or edge that is the same as the direction of flow. *See also* upstream, directed network flow.

dpi *Acronym for dots per inch.* A measure of the resolution of scanners, printers, and graphic displays. The more dots per inch, the more detail can be displayed in an image.

drafting [CARTOGRAPHY] A method of drawing with pencil or pen and ink, used in cartographic reproduction.

drainage All map features associated with the movement and flow of water, such as rivers, streams, and lakes.

draping A perspective or panoramic rendering of a two-dimensional image superimposed onto a three-dimensional surface. For example, an aerial photograph might be draped over a digital elevation model (DEM) to create a realistic terrain visualization.

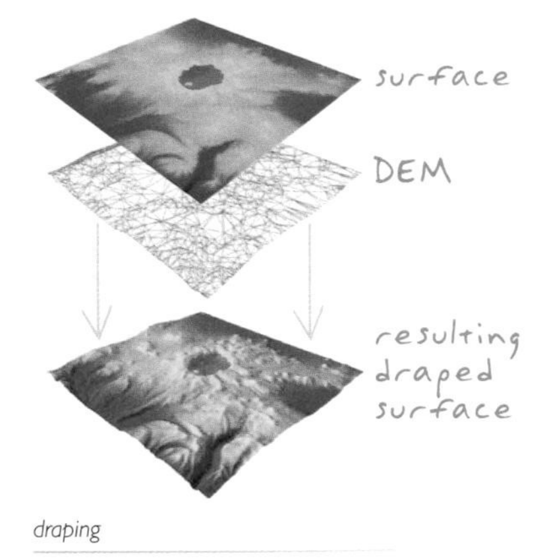

draping

DRG *Acronym for digital raster graphic.* A raster image of a scanned USGS standard series topographic map, usually including the original border information, referred to as the map collar, map surround, or marginalia. Source maps are georeferenced to the surface of the earth, fit to the universal transverse Mercator (UTM) projection, and scanned at a minimum resolution of 250 dpi. The accuracy and datum of a DRG matches the accuracy and datum of the source map. *See also* map surround.

drift [STATISTICS] The general pattern of z-values throughout a kriging model. The drift, or structure, forms the model's basic shape. *See also* kriging, z-value.

drive-time area A zone around a map feature measured in units of time needed for travel by car. For example, a store's ten-minute drive-time area defines the area in which drivers can reach the store in ten minutes or less.

drive-time area

drum scanner A type of scanner in which a hard-copy image or map is attached to a cylinder that spins while a sensor captures a digital image from the surface of the page. *See also* scanner, roller-feed scanner, flatbed scanner.

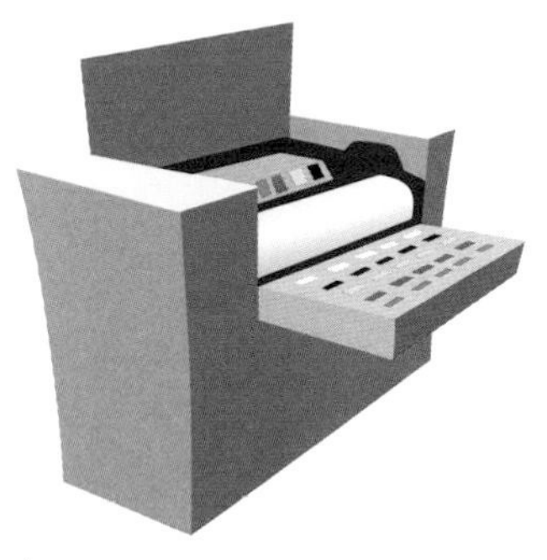

drum scanner

DTED *Acronym for digital terrain elevation data.* A format for elevation data, usually tiled in 1-degree cells, produced by the National Geospatial-Intelligence Agency and U.S. allies for military applications. *See also* DEM.

DTM *See* DEM.

Dual Independent Map Encoding *See* DIME.

dynamic binding *See* late binding.

dynamic HTML [INTERNET] An extension to HTML that allows Web designers to make elements on a Web page interactive, rather than changeable only when the page is loaded. *See also* HTML.

dynamic-link library *See* DLL.

dynamic segmentation The process of computing the map locations of linearly referenced data (for example, attributes stored in a table) at run time so they can be displayed on a map, queried, and analyzed using a GIS. The dynamic segmentation process enables multiple sets of attributes to be associated with any portion of a line feature without segmenting the underlying feature. In the transportation field, examples of such linearly referenced data might include accident sites, road quality, and traffic volume. *See also* linear referencing.

D

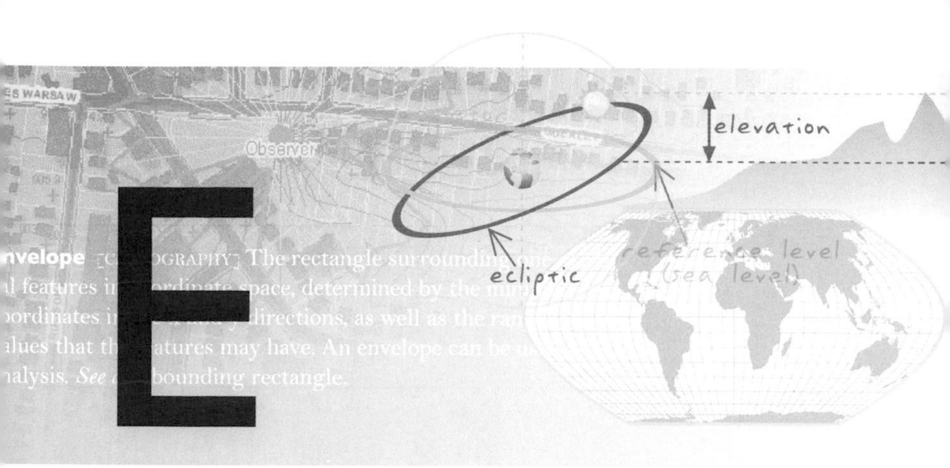

early binding [PROGRAMMING] A COM technique an application uses to access an object. In early binding, an object's properties and methods are discovered from an interface at compile time, instead of being checked at run time as in late binding. This difference often gives early binding performance benefits over late binding.

earth-centered datum *See* geocentric datum.

easting 1. [CARTOGRAPHY] The distance east of the origin that a point in a Cartesian coordinate system lies, measured in that system's units.

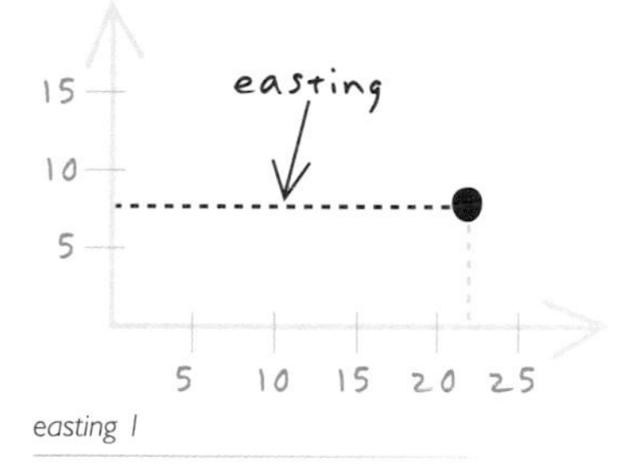

easting 1

2. [CARTOGRAPHY] The positive x-value in a rectangular coordinate system. *See also* northing.

eccentricity [MATHEMATICS] A measure of how much an ellipse deviates from a circle, expressed as the ratio of the distance between the center and one focus of an ellipsoid to the length of its semimajor axis. The square of the eccentricity (e^2) is commonly used with the semimajor axis *a* to define a spheroid.

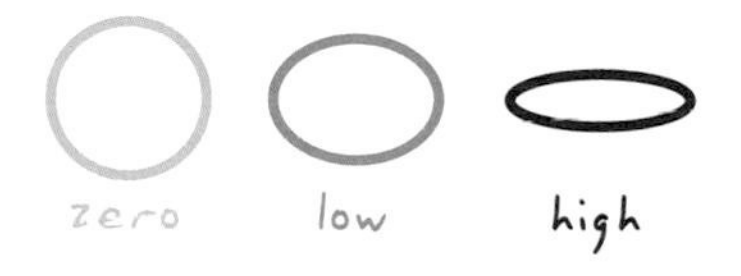

eccentricity

ecliptic 1. [ASTRONOMY] The great circle formed by the intersection of the plane of the earth's orbit around the sun (or apparent orbit of the sun around the earth) and the celestial sphere.

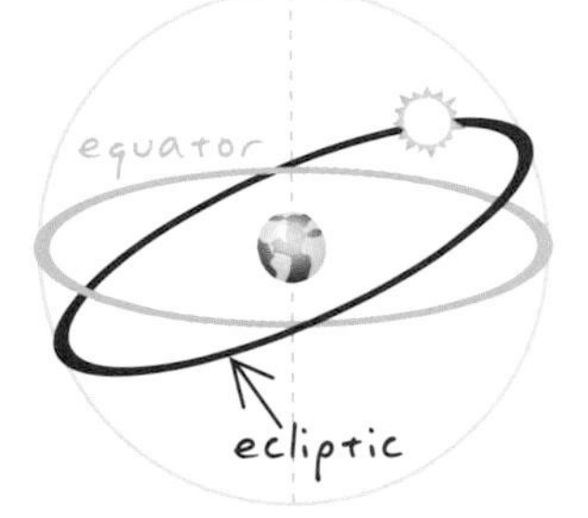

ecliptic

2. [ASTRONOMY] The mean plane of the earth's orbit around the sun. *See also* celestial sphere.

ecological fallacy [STATISTICS] The assumption that an individual from a specific group or area will exhibit a trait that is predominant in the group as a whole.

economic geography [GEOGRAPHY] The field of geography concerning the distribution and variation of economic factors by location, including how economic factors interact with geographic factors such as climate, land use, and geology. *See also* geography.

edge 1. A line between two points that forms a boundary. In a geometric shape, an edge forms the boundary between two faces. In an image, edges separate areas of different tones or colors. In topology, an edge defines lines or polygon boundaries.

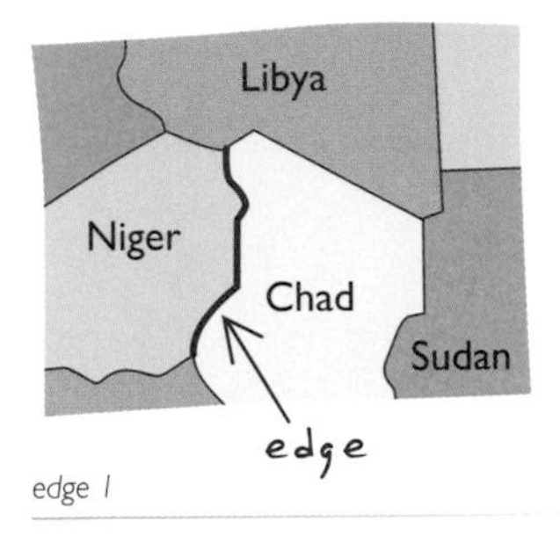

edge 1

2. In a network system, a line feature through which a substance, resource, or traffic flows. Examples include a street in a transportation network and a pipeline in a sewer system. In a geometric network, a network edge can be simple or complex. A simple edge is always connected to exactly two junction features, one at each end. A complex edge is always connected to at least two junction features at its endpoints, but it can also be connected to additional junction features along its length. In a network dataset, a network edge is only connected to two junctions at its endpoints.

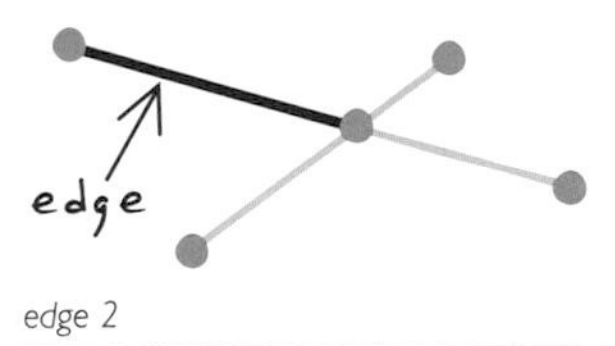

edge 2

3. In a TIN data model, a line segment between nodes (sample data points). Edges store topologic information about the faces that they border.

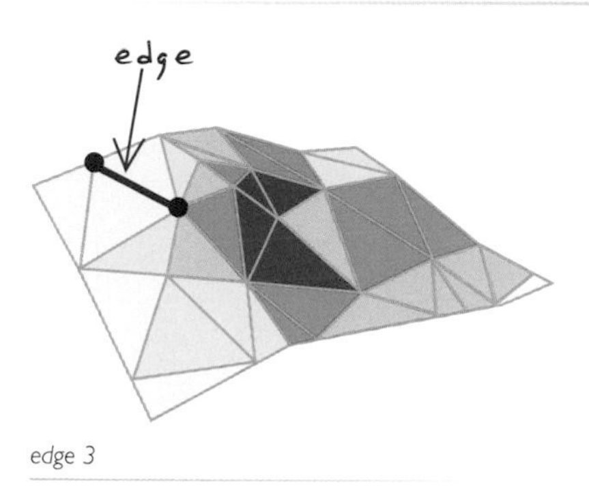

edge 3

edge detection [DIGITAL IMAGE PROCESSING] A technique for isolating edges in a digital image by examining it for abrupt changes in pixel value.

edge element *See* edge.

edge enhancement [DIGITAL IMAGE PROCESSING] A technique for emphasizing the appearance of edges and lines in an image. *See also* high-pass filter.

edgematching A spatial adjustment process that aligns features along the edge of an extent to the corresponding features in an adjacent extent. *See also* rubber sheeting.

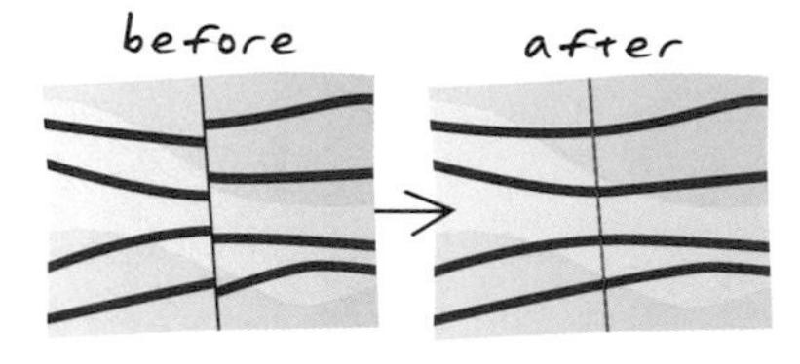

edgematching

EDMS *Acronym for electronic document management system.* A computer-based system for organizing, maintaining, and retrieving digital and hard-copy documents. An EDMS usually includes a check-in, check-out system for document tracking, versioning, and search-and-retrieval capabilities.

elastic transformation *See* rubber sheeting.

electromagnetic radiation [PHYSICS] Energy that moves through space at the speed of light as different wavelengths of time-varying electric and magnetic fields. Types of electromagnetic radiation include gamma, x, ultraviolet, visible, infrared, microwave, and radio.

electromagnetic spectrum [PHYSICS] The entire range of wavelengths (frequencies) over which electromagnetic radiation extends. *See also* electromagnetic radiation.

electronic atlas A mapping system that displays but does not allow for the spatial analysis of data.

electronic document management system *See* EDMS.

electronic navigational chart *See* ENC.

element [ESRI SOFTWARE] In geoprocessing in ArcGIS, a component of a model. Elements can be variables, such as input and derived data, or tools.

elevation [GEODESY] The vertical distance of a point or object above or below a reference surface or datum (generally mean sea level). Elevation generally refers to the vertical height of land. *See also* vertical geodetic datum, altitude.

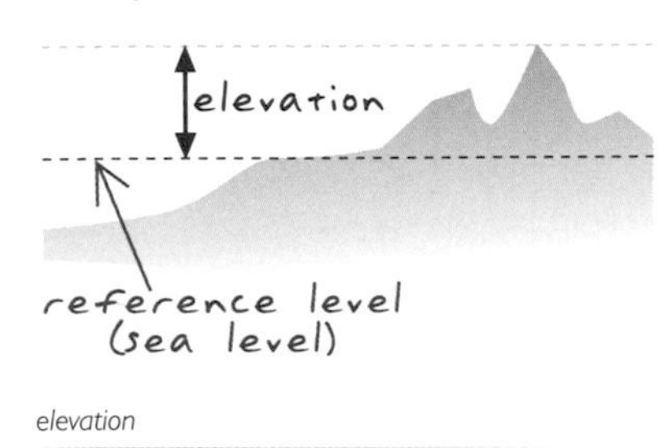

elevation

elevation guide [CARTOGRAPHY] A map element that displays a simplified representation of the terrain within a map's extent. Elevation guides are designed to provide a quick overview of topography, including the high and low points. *See also* elevation tints.

elevation tints [CARTOGRAPHY] Hypsometric tint bands based on elevation ranges used in an elevation guide. *See also* elevation guide, hypsometric tinting.

ellipse [MATHEMATICS] A geometric shape described mathematically as the collection of points whose distances from two given points (the foci) add up to the same sum. An ellipse is shaped like a circle viewed obliquely.

ellipsoid 1. [MATHEMATICS] A three-dimensional, closed geometric shape, all planar sections of which are ellipses or circles. An ellipsoid has three independent axes, and is usually specified by the lengths a,b,c of the three semi-axes. If an ellipsoid is made by rotating an ellipse about one of its axes, then two axes of the ellipsoid are the same, and it is called an ellipsoid of revolution, or spheroid. If the lengths of all three of its axes are the same, it is a sphere.

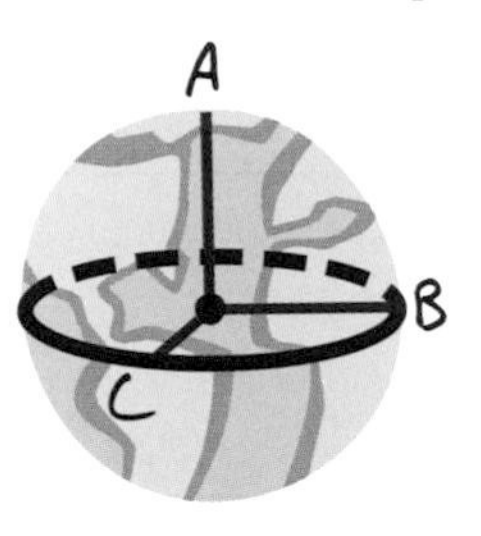

ellipsoid 1

2. [GEODESY] When used to represent the earth, an oblate ellipsoid of revolution, made by rotating an ellipse about its minor axis. *See also* geoid.

ellipticity *See* eccentricity.

empirical [STATISTICS] That property of a quantity that indicates that the quantity depends on data, observations, or experiment only; that is, it is not a model or part of a model. An empirical semivariogram is computed on data only, in contrast to a theoretical semivariogram model.

enabled feature [ESRI SOFTWARE] In geometric networks, a network feature that allows flow to pass through it. *See also* disabled feature.

ENC [NAVIGATION] *Acronym for electronic navigational chart.* A vector data product used for nautical navigation. ENC data is produced by nautical charting agencies throughout the world and uses the IHO (International Hydrographic Organization) S-57 standard for its database structure and attribution. *See also* DNC.

enclosure A file describing the contents of an item included in metadata. Enclosing files in metadata works the same way as enclosing files in an e-mail message. *See also* metadata.

encoding The recording or reformatting of data into a computer format. Data may be encoded to reduce storage, increase security, or to transfer it between systems using different file formats. In GIS, analog graphic data, such as paper maps and images, are encoded into computer formats by scanning or digitizing.

end offset [ESRI SOFTWARE] An adjustable value that dictates how far away from the end of a line an address location should be placed. Using an end offset prevents the point from being placed directly over the intersection of cross streets if the address happens to fall on the beginning or end of the street. *See also* side offset.

enhancement [REMOTE SENSING] Applying operations to raster data to improve appearance or usability by making specific features more detectable. Such operations can include contrast stretching, edge enhancement, filtering, smoothing, and sharpening.

enterprise GIS A geographic information system that is integrated through an entire organization so that a large number of users can manage, share, and use spatial data and related information to address a variety of needs, including data creation, modification, visualization, analysis, and dissemination. *See also* GIS.

envelope [CARTOGRAPHY] The rectangle surrounding one or more geographical features in coordinate space, determined by the minimum and maximum coordinates in the x and y directions, as well as the ranges of any z- or m-values that the features may have. An envelope can be used to filter data for analysis. *See also* bounding rectangle.

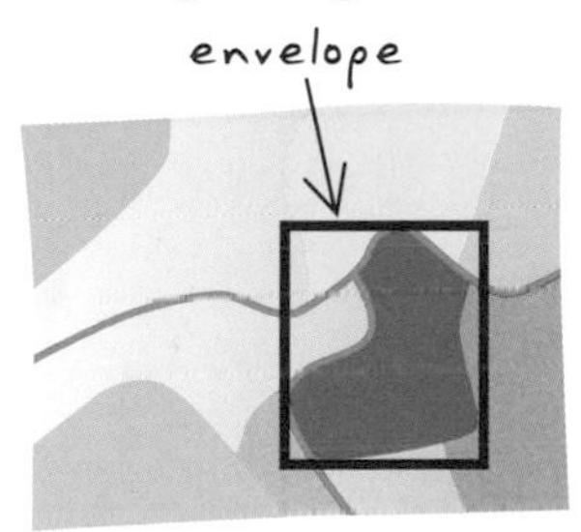

envelope

environmental model An abstract representation of a complex environmental process, emphasizing relationships and patterns in natural systems. Environmental models allow decision makers to better understand the effects of natural systems or the impact of human activities on natural systems.

ephemeris [GPS] A table of the predicted positions of a satellite within its orbit for each day of the year, or for other regular intervals.

equal-area classification [CARTOGRAPHY] A data classification method that divides polygon features into groups so that the total area of the polygons in each group is approximately the same. *See also* classification, equal-interval classification.

E

equal-area projection [CARTOGRAPHY] A projection in which the whole of the map as well as each part has the same proportional area as the corresponding part of the earth. An equal-area projection may distort shape, angle, scale, or any combination thereof. No flat map can be both equal-area and conformal. *See also* projection.

equal-area projection

equal competition area A trade area boundary set halfway between a store or service point and its neighboring stores or service points.

equal-interval classification [CARTOGRAPHY] A data classification method that divides a set of attribute values into groups that contain an equal range of values. *See also* classification, equal-area classification.

equator [GEODESY] The parallel of reference that is equidistant from the poles and defines the origin of latitude values. *See also* parallel, latitude.

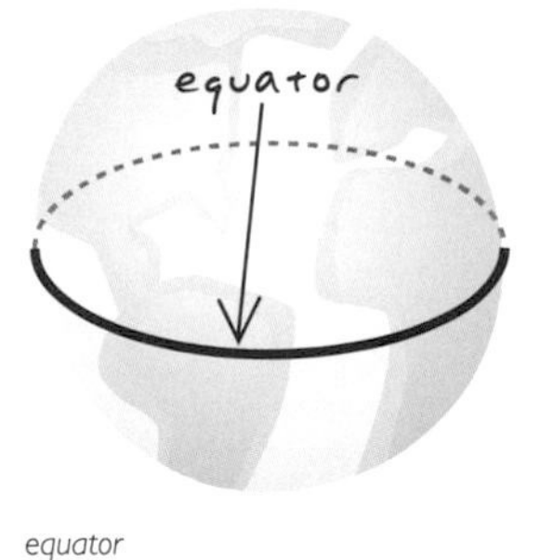

equator

equatorial aspect [CARTOGRAPHY] A planar (or azimuthal) projection with its central point located at the equator. *See also* projection.

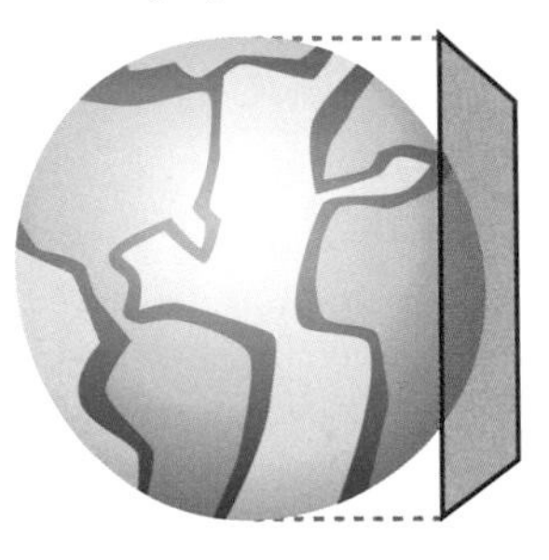

equatorial aspect

equidistant projection [CARTOGRAPHY] A projection that maintains scale along one or more lines, or from one or two points to all other points on the map. Lines along which scale (distance) is correct are the same proportional length as the lines they reference on the globe. In the sinusoidal projections, for example, the central meridian and all parallels are their true lengths. An azimuthal equidistant projection centered on Chicago shows the correct distance between Chicago and any other point on the projection, but not between any other two points. *See also* projection.

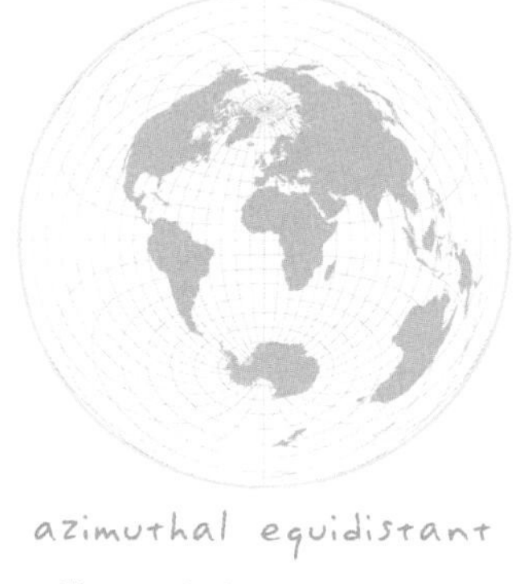

equidistant projection

error 1. A measured, observed, calculated, or interpreted value that differs from the true value or the value that would be obtained by a perfect observer using perfect equipment and perfect methods under perfect conditions. **2.** In a GIS database, a spatial or attribute value that differs from the true value. Error may also be understood as the totality of wrong or unreliable information in a database. Spatial errors are mainly errors in position (feature coordinates are wrong) and topology (features do not properly connect, intersect, or adjoin). Attribute errors are wrong quantities or descriptions associated with features, or missing or invalid values. Errors enter a GIS database through various processes, including data collection (for instance, flawed instruments); data conversion (for example, map digitizing mistakes); data entry and editing; data integration (for example, mixing data at different scales); spatial data processing (for example, inaccuracies caused by generalization); and data analysis (for example, features assigned to inappropriate categories on the basis of flawed criteria). **3.** [ESRI SOFTWARE] In geodatabase topology, violation of a topology rule detected during the validation process. **4.** [PROGRAMMING] A piece of code that prevents a program from compiling or running; also, program logic that makes a program end prematurely, go into an endless loop, or give incorrect results. *See also* error propagation, uncertainty.

error propagation In GIS data processing, the persistence of an error into new datasets calculated or created using datasets that originally contained errors. The study of error propagation is concerned with the effects of combined and accumulated errors throughout a series of data processing operations. *See also* error.

estimation [STATISTICS] In spatial modeling, the process of forming a statistic from observed data to assign optimal parameters in a model or distribution. *See also* prediction.

Euclidean distance [MATHEMATICS] The straight-line distance between two points on a plane. Euclidean distance, or distance "as the crow flies," can be calculated using the Pythagorean theorem.

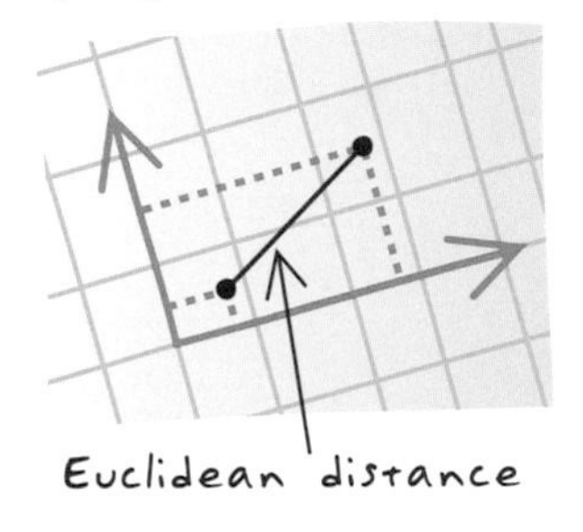

Euclidean distance

event 1. A geographic location stored in tabular rather than spatial form. Event types include address events, route events, x,y events, and temporal events. 2. [COMPUTING] An outcome or occurrence that happens when a user interacts with an application. For example, in a case in which clicking a button triggers the closing of a form, the event is the closing of the form.

event location *See* event.

executable file [COMPUTING] A binary file containing a program that can be run as a stand-alone application. In the Microsoft Windows program, executable files are designated with an .exe extension. *See also* data file.

explode An editing process that separates a multipart feature into its component features, which become independent features. *See also* multipart feature.

explode

exponent [MATHEMATICS] A number that indicates how many times a base value is multiplied by itself. Exponents are usually indicated with superscripts.

export [COMPUTING] To move data from one computer system to another, and often, in the process, from one file format to another.

exposure station [REMOTE SENSING] In aerial photography, each point in the flight path at which the camera exposes the film.

expression [MATHEMATICS] A sequence of operands and operators constructed according to the syntactic rules of a symbolic language that evaluates to a single number, string, or value.

extended postal code *See* ZIP+4 Code.

Extensible Markup Language *See* XML.

Extensible Style Language *See* XSL.

Extensible Style Language Transformations *See* XSLT.

extent The minimum bounding rectangle (xmin, ymin and xmax, ymax) defined by coordinate pairs of a data source. All coordinates for the data source fall within this boundary. *See also* map extent, minimum bounding rectangle.

extrapolation [STATISTICS] Using known or observed data to infer or calculate values for unobserved times, locations or other variables outside a sampled area. In the absence of data, extrapolation is a common method for making predictions, but it is not always accurate. For example, based on observed economic indicators, an economist can make predictions about the state of the economy at a future time. These predictions may not be accurate because they cannot take into account seemingly random events such as natural disasters.

extrusion The process of projecting features in a two-dimensional data source into a three-dimensional representation: points become vertical lines, lines become planes, and polygons become three-dimensional blocks. Uses of extrusion include showing the depth of well point features or the height of building-footprint polygons.

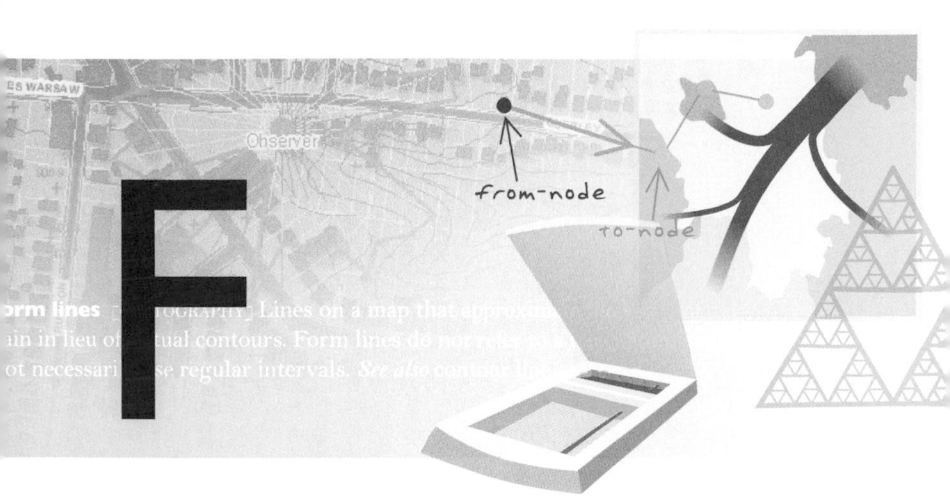

F *See* F statistic.

face 1. [MATHEMATICS] A planar surface of a geometric shape, bounded by edges.

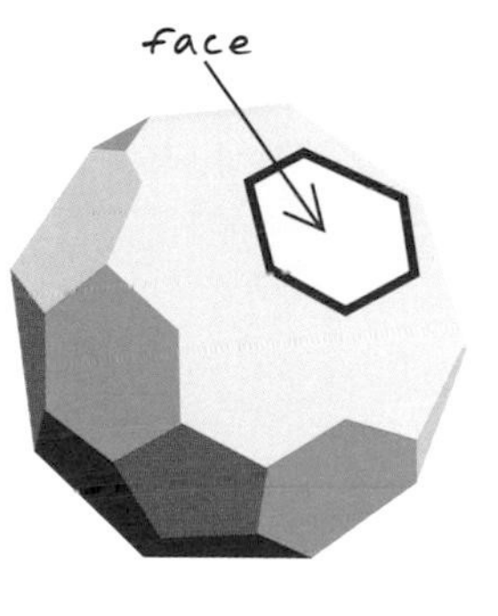

face 1

2. In a TIN, the flat surface of a triangle bounded by three edges and three nodes. Faces do not overlap; each face is adjacent to three other faces of the TIN. A face defines a plane with an aspect and slope.

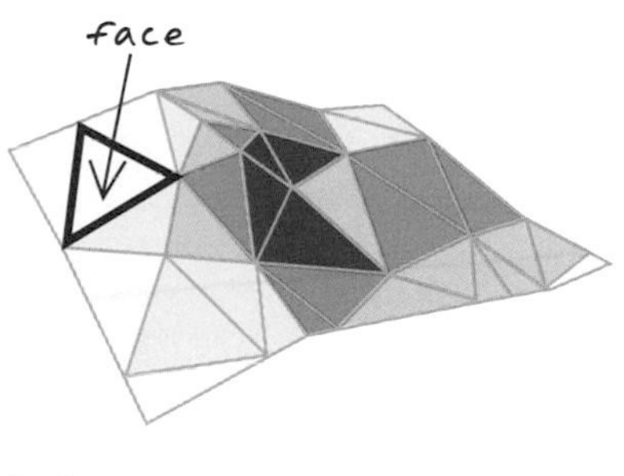

face 2

false easting [CARTOGRAPHY] The linear value added to all x-coordinates of a map projection so that none of the values in the geographic region being mapped are negative. *See also* easting, false northing.

false northing [CARTOGRAPHY] The linear value added to all y-coordinates of a map projection so that none of the values in the geographic region being mapped are negative. *See also* northing, false easting.

feature [CARTOGRAPHY] A representation of a real-world object on a map. *See also* point feature, line feature, polygon feature, multipoint feature, polyline feature, multipatch feature, multipart feature. *For *more information about features, see* How features and feature attributes are stored *on page 253.*

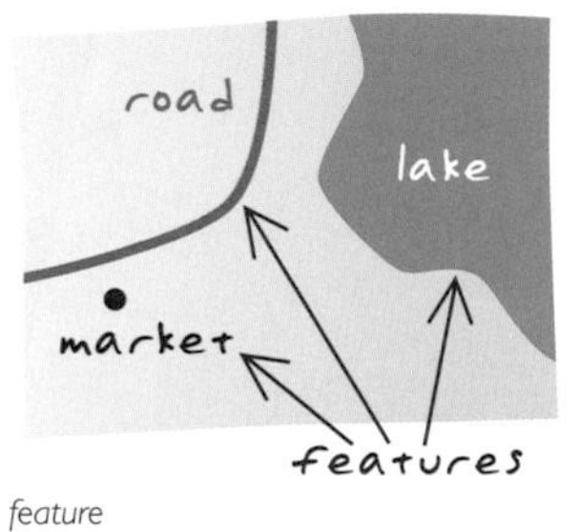

feature

feature attribute table *See* attribute table.

feature class [ESRI SOFTWARE] In ArcGIS, a collection of geographic features with the same geometry type (such as point, line, or polygon), the same attributes, and the same spatial reference. Feature classes can be stored in geodatabases, shapefiles, coverages, or other data formats. Feature classes allow homogeneous features to be grouped into a single unit for data storage purposes. For example, highways, primary roads, and secondary roads can be grouped into a line feature class named "roads." In a geodatabase, feature classes can also store annotation and dimensions. *See also* object class.

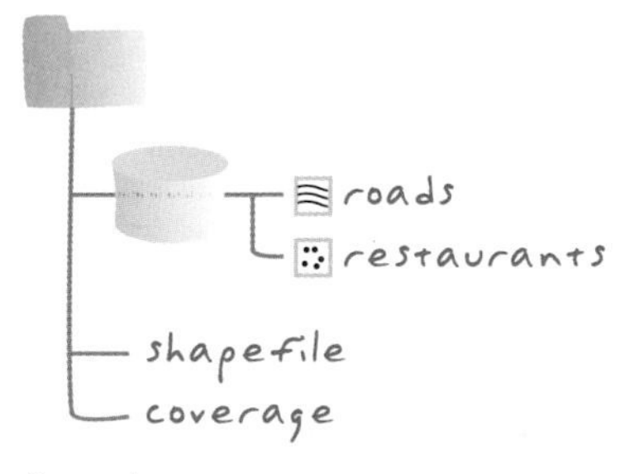

feature class

feature data Data that represents geographic features as geometric shapes.

feature dataset [ESRI SOFTWARE] In ArcGIS, a collection of feature classes stored together that share the same spatial reference; that is, they share a coordinate system, and their features fall within a common geographic area. Feature classes with different geometry types may be stored in a feature dataset. *See also* feature class.

feature displacement [CARTOGRAPHY] The movement of features that would otherwise overprint or conflict with other features. For example, if a river, a road, and a railway run through a narrow valley, it is necessary, at some scales, to displace at least one of the features that represent them on the map to keep their symbols distinct. *See also* generalization, line simplification.

feature extraction [DIGITAL IMAGE PROCESSING] A method of pattern recognition in which patterns within an image are measured and then classified as features based on those measurements.

Federal Geographic Data Committee *See* FGDC.

FGDC *Acronym for Federal Geographic Data Committee.* An organization established by the United States Federal Office of Management and Budget responsible for coordinating the development, use, sharing, and dissemination of surveying, mapping, and related spatial data. The committee is comprised of representatives from federal and state government agencies, academia, and the private sector. The FGDC defines spatial data metadata standards for the United States in its Content Standard for Digital Geospatial Metadata and manages the development of the National Spatial Data Infrastructure (NSDI). *See also* NSDI.

FGDC Clearinghouse *See* NSDI Clearinghouse Node.

FGDC standard *See* Content Standard for Digital Geospatial Metadata.

field **1.** A column in a table that stores the values for a single attribute.

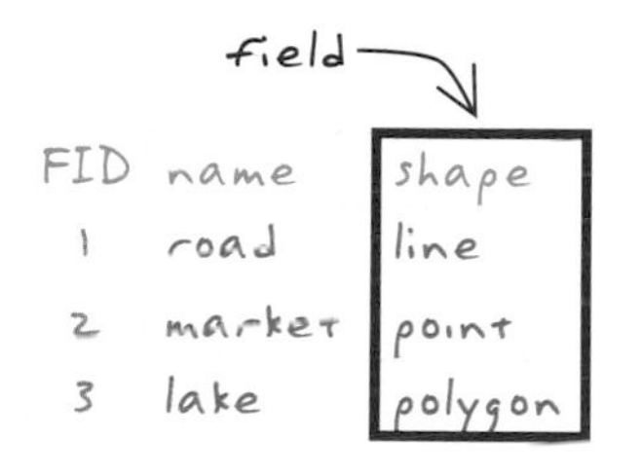

field 1

2. [COMPUTING] The place in a database record, or in a GUI, where data can be entered. **3.** A synonym for surface. *See also* attribute, column.

field precision [COMPUTING] The number of digits that can be stored in a field in a table. *See also* dataset precision.

field scale [ESRI SOFTWARE] The number of decimal places for float or double-type geodatabase table fields. *See also* floating point.

field view A philosophical view of geographic space in which space is completely filled by occurrences of phenomena, and in which phenomena are described by a range of values on a numeric scale. In this view, every spatial location is something, even if it is the zero value of a phenomenon. *See also* object view.

file [COMPUTING] A collection of uniquely named information stored on a drive, disk, or tape. A file generally resides within a directory.

file name [COMPUTING] The name that distinguishes a file from all other files in a particular directory. It can refer to the name of the file by itself (harold), the name plus the file extension (harold.shp), or the whole path of a file up to and including the file name extension (C:\mygisdata\shapefiles\harold.shp). *See also* file name extension.

file name extension [COMPUTING] The abbreviation following the final period in a file name that indicates the file's format, such as .shp, .zip, or .tif. File name extensions are usually one to four letters long. *See also* file name.

File Transfer Protocol *See* FTP.

fill The interior of a polygon; the area inside the perimeter.

fillet A segment of a circle used to connect two intersecting lines. Fillets are used to create smoothly

F

curving connections between lines, such as pavement edges at street intersections or rounded corners on parcel features.

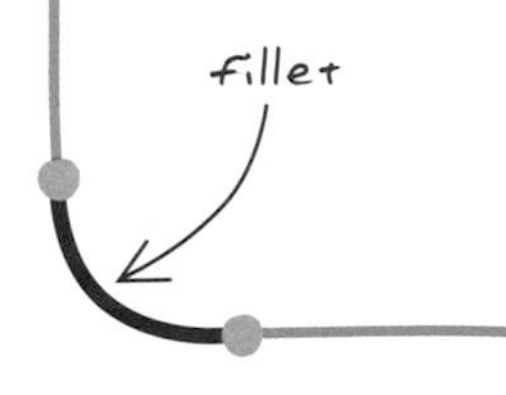

fillet

fill symbol [CARTOGRAPHY] A color or pattern used to fill polygons on a map. *See also* marker symbol.

fill symbol

filter 1. On a raster, an analysis boundary or processing window within which cell values affect calculations and outside which they do not. Filters are used mainly in cell-based analysis where the value of a center cell is changed to the mean, the sum, or some other function of all cell values inside the filter. A filter moves systematically across a raster until each cell has been processed. Filters can be of various shapes and sizes, but are most commonly three-cell by three-cell squares.

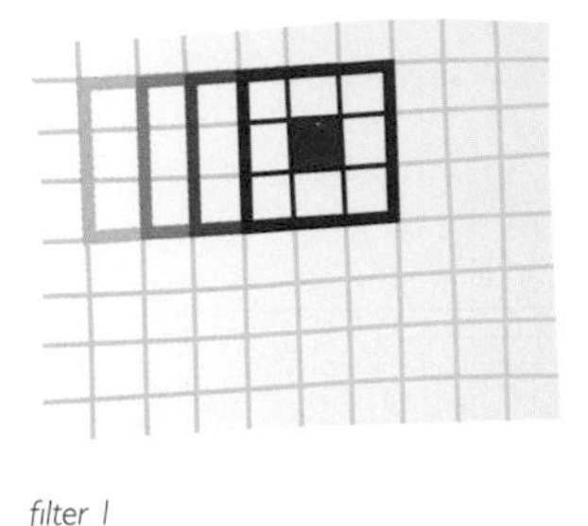

filter 1

2. A desktop GIS operation used to hide (but not delete) features in a map document or attribute table. 3. A constraint used to define a subset of data.

first normal form The first level of guidelines for designing table and data structures in a relational database. The first normal form guideline recommends creating a unique key for every row in a database table, eliminating duplicate columns from a table, and creating separate tables to contain related data. A database that follows these guidelines is said to be in first normal form. *See also* normal form.

fitness for use The degree to which a dataset is suitable for a particular application or purpose, encompassing factors such as data quality, scale, interoperability, cost, data format, and so on. *See also* metadata.

fix [GEODESY] A single position obtained by surveying, GPS, or astronomical measurements, usually given with altitude, time, date, and latitude-longitude or grid position.

fixed use *See* single use.

flatbed scanner A type of scanner with a flat, clear surface on which a map or image remains stationary while a sensor beam moves across it and captures a digital image. *See also* scanner, drum scanner, roller-feed scanner.

flatbed scanner

flattening [GEODESY] A measure of how much an oblate spheroid differs from a sphere. The flattening equals the ratio of the semimajor axis minus the semiminor axis to the semimajor axis. *See also* oblate ellipsoid.

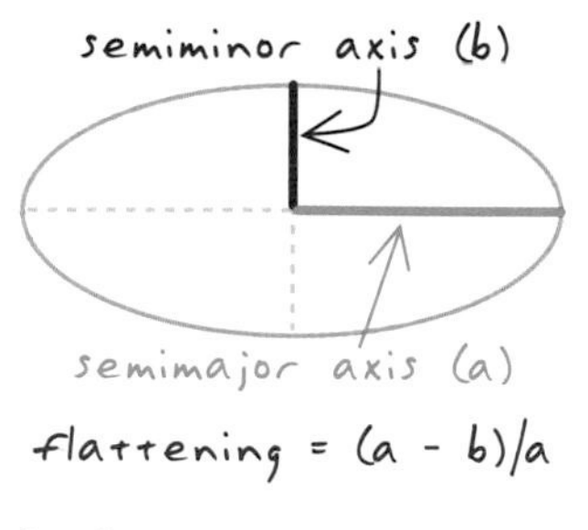

flattening

floating point [COMPUTING] A type of numeric field for storing real numbers with a decimal point. The decimal point can be in any position in the field and, thus, may "float" from one location to another for different values stored in the field. For example, a floating-point field can store the numbers 23.632, 0.000087, and -96432.15.

flow direction The route or course followed by commodities proceeding through edge elements in a network.

flow map [CARTOGRAPHY] A map that uses line symbols of variable thickness to show the proportion of traffic or flow within a network.

flow map

focal analysis The computation of an output raster where the output value at each cell location is a function of the value at that cell location and the values of the cells within a specified neighborhood around the cell.

focal functions *See* focal analysis.

folder [COMPUTING] A location on a disk containing a set of files, other folders, or both. *See also* directory.

font [CARTOGRAPHY] A single typeface or a set of related patterns representing characters or symbols at one size. *See also* point size.

foreground 1. In a scene or display, the area that appears to be closest to an observer. 2. [ESRI SOFTWARE] The area in a raster layer where cells are eligible for selection and vectorization.

F

foreign key An attribute or set of attributes in one table that match the primary key attributes in another table. Foreign keys and primary keys are used to join tables in a database. *See also* key.

form lines [CARTOGRAPHY] Lines on a map that approximate the shape of terrain in lieu of actual contours. Form lines do not refer to a true datum and do not necessarily use regular intervals. *See also* contour line.

form lines

fractal [MATHEMATICS] A geometric pattern that repeats itself, at least roughly, at ever smaller scales to produce self-similar, irregular shapes and surfaces that cannot be represented using classical geometry. If a fractal curve of infinite length serves as the boundary of a plane region, the region itself will be finite. Fractals can be used to model complex natural shapes such as clouds and coastlines.

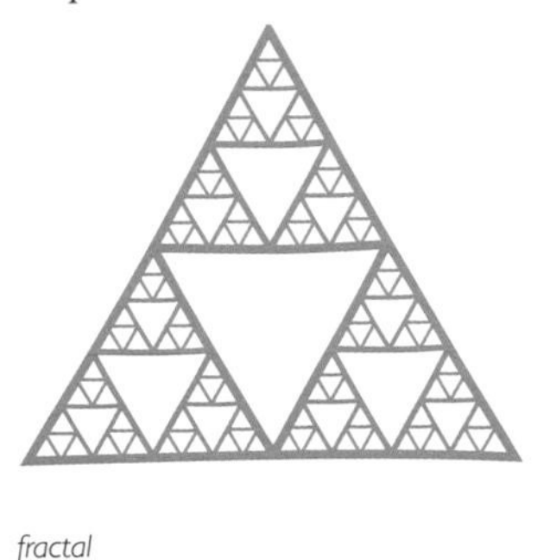

fractal

frequency [PHYSICS] The number of oscillations per unit of time in a wave of energy, or the number of wavelengths that pass a point in a given amount of time.

from-node Of an arc's two endpoints, the first one digitized. From- and to-nodes give an arc left and right sides and, therefore, direction. *See also* to-node.

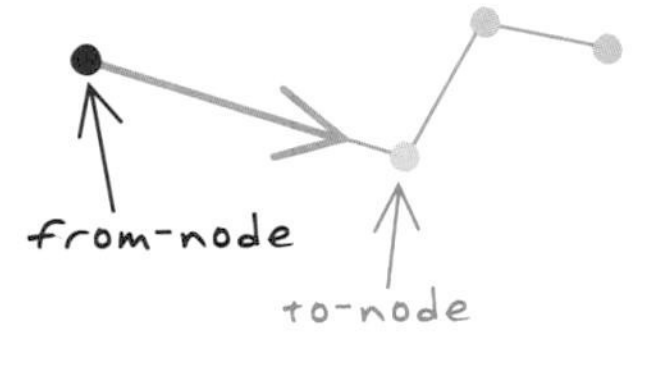

from-node

F statistic [STATISTICS] A ratio of variances, calculated from a sample of data and used to provide information about a whole dataset. For example, statistic F may be used to provide estimates of variance, or differences, in a population, based on observations from two or more random samples. *See also* F test.

F test [STATISTICS] A statistical test for determining the probability that the variances of two different samples are the same. The F test uses a statistic known as statistic F to test statistical hypotheses about the variances of distributions from which samples have been drawn. *See also* F statistic.

FTP *Acronym for File Transfer Protocol.* A protocol that allows the transmission of files between computers over a network.

function An operation. In GIS, functions include data input, editing, and management; data query, analysis, and visualization; and output operations. *See also* tool.

fuzzy boundary A boundary that has a vague or indeterminate location, or that is a gradual transition between two zones.

fuzzy classification Any method for classifying data that allows attributes to apply to objects by membership values, so that an object may be considered a partial member of a class. Class membership is usually defined on a continuous scale from zero to one, where zero is nonmembership and one is full membership. Fuzzy classification may also be applied to geographic objects themselves, so that an object's boundary is treated as a gradated area rather than an exact line. In GIS, fuzzy classification has been used in the analysis of soil, vegetation, and other phenomena that tend to change gradually in their physical composition and for which attributes are often partly qualitative in nature. *See also* fuzzy set, fuzzy boundary, vagueness.

fuzzy set [MATHEMATICS] In mathematics, a collection of elements that belong together based on specified criteria, so that elements with partial or uncertain degrees of membership may be included in the collection.

fuzzy tolerance [ESRI SOFTWARE] The distance within which coordinates of nearby features are adjusted to coincide with each other when topology is being constructed or polygon overlay is performed. Nodes and vertices within the fuzzy tolerance are merged into a single coordinate location. Fuzzy tolerance is a very small distance, usually from 1/1,000,000 to 1/10,000 times the width of the coverage extent, and is generally used to correct inexact intersections. *See also* cluster tolerance.

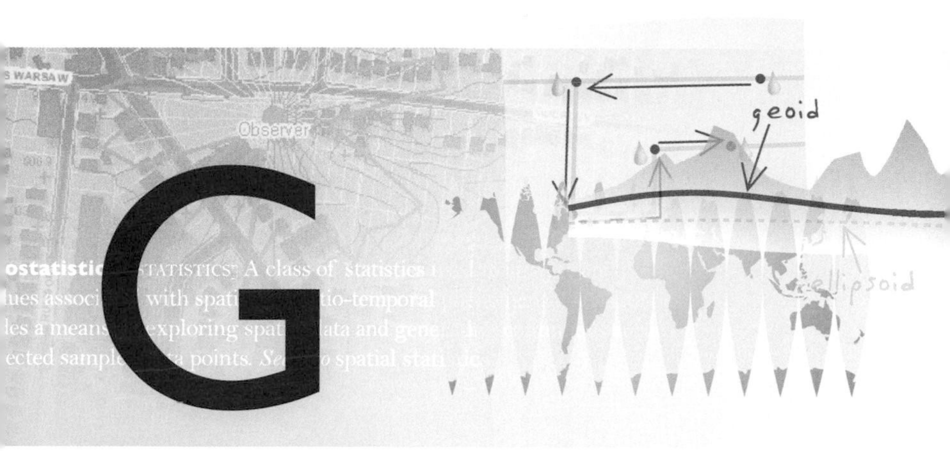

Gaussian distribution *See* normal distribution.

Gauss-Krüger projection [CARTOGRAPHY] A projected coordinate system that uses the transverse Mercator projection to divide the world into standard zones 6 degrees wide. Used mainly in Europe and Asia, the Gauss-Krüger coordinate system is similar to the universal transverse Mercator coordinate system. The Gauss-Krüger projection is named for the German mathematician and scientist Karl Friedrich Gauss and the German geodesist and mathematician Johann Heinrich Louis Krüger.

gazetteer [CARTOGRAPHY] A list of geographic place names and their coordinates. Entries may include other information as well, such as area, population, or cultural statistics. Atlases often include gazetteers, which are used as indexes to their maps. Well-known digital gazetteers include the U.S. Geological Survey Geographic Names Information System (GNIS) and the Alexandria Digital Library Gazetteer.

GBF/DIME *Acronym for Geographic Base Files/Dual Independent Map Encoding.* Vector geographic base files made for the 1970 and 1980 U.S. censuses, containing address ranges, ZIP Codes, and the coordinates of street segments and intersections for most metropolitan areas in the United States. TIGER files replaced DIME files for the 1990 and subsequent censuses. *See also* TIGER, DIME.

GDB *See* geodatabase.

generalization **1.** [CARTOGRAPHY] The abstraction, reduction, and simplification of features for change of scale or resolution. **2.** The process of reducing the number of points in a line without losing the line's essential shape. **3.** The process of enlarging and resampling cells in a raster format. *See also* cartographic generalization, database generalization.

genetic algorithm [COMPUTING] A search algorithm inspired by genetics and Darwin's theory of natural selection. The algorithm goes through an iterative process of applying genetic operators, such as reproduction, mutation, and crossover, to a collection of data over several stages. At each stage the fitness of the results is evaluated and the best of the results population is retained, until the results present an optimal solution.

geocentric **1.** [GEODESY] Measured from the earth or the earth's center. **2.** [ASTRONOMY] Having the earth as a center.

geocentric coordinate system [CARTOGRAPHY] A three-dimensional, earth-centered reference system in which locations are identified by their x-, y-, and z-values. The x-axis is in the equatorial plane and intersects the prime meridian (usually Greenwich). The y-axis is also in the equatorial plane; it lies at right angles to the x-axis and intersects the 90-degree meridian. The z-axis coincides with the polar axis and is positive toward the north pole. The origin is located at the center of the sphere or spheroid.

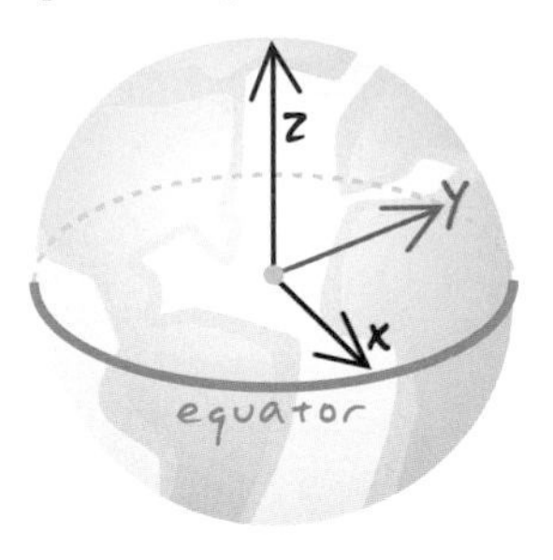

geocentric coordinate system

geocentric datum [GEODESY] A horizontal geodetic datum based on an ellipsoid that has its origin at the earth's center of mass. Examples are the World Geodetic System of 1984, the North American Datum of 1983, and the Geodetic Datum of Australia of 1994. The first uses the WGS84 ellipsoid; the latter two use the GRS80 ellipsoid. Geocentric datums are more compatible with satellite positioning systems, such as GPS, than are local datums. *See also* datum, horizontal geodetic datum, local datum, GPS.

geocentric latitude [CARTOGRAPHY] The angle between the equatorial plane and a line from a point on the surface to the center of the sphere or spheroid. On a sphere, all lines of latitude are geocentric. Latitude generally refers to geodetic latitude. *See also* geodetic latitude.

geocentric longitude [CARTOGRAPHY] The angle between the prime meridian and a line drawn from a point on the surface to the center of a sphere or spheroid. For an ellipsoid of revolution (such as the earth), geocentric longitude is the same as geodetic longitude. *See also* geodetic longitude.

geocode **1.** To assign a street address to a location. **2.** A code representing the location of an object, such as an address, a census tract, a postal code, or x,y coordinates. *See also* geocoding.

geocoding A GIS operation for converting street addresses into spatial data that can be displayed as features on a map, usually by referencing address information from a street segment data layer.

geocoding process The steps involved in translating an address entry, searching for the address in the reference data embedded in an address locator, and delivering the best candidate or candidates. These steps include parsing the address, standardizing abbreviated values, assigning each address element to a category known as a match key, indexing the needed categories, searching the reference data, assigning a score to each potential candidate, filtering the list of candidates based on the minimum match score, and delivering the best match. The process requires reference files, input address records, address locators, and software. *See also* address locator.

geocoding reference data Data that a geocoding service uses to determine the geometric representations for locations.

geocoding service *See* address locator.

geocomputation [COMPUTING] The application of computer technology to spatial problems, including problems of collecting, storing, visualizing, and analyzing spatial data, and of modeling spatial system dynamics. *See also* quantitative geography.

geodatabase [ESRI SOFTWARE] A collection of geographic datasets for use by ArcGIS. There are various types of geographic datasets, including feature classes, attribute tables, raster datasets, network datasets, topologies, and many others. *See also* database.

geodatabase data model [ESRI SOFTWARE] The schema for the various geographic datasets and tables in an instance of a geodatabase. The schema defines the GIS objects, rules, and relationships used to add GIS behavior and integrity to the datasets in a collection. *See also* database.

geodataset [ESRI SOFTWARE] Any organized collection of data in a geodatabase with a common theme.

geodesic [MATHEMATICS] The shortest distance between two points on the surface of a spheroid. Any two points along a meridian form a geodesic.

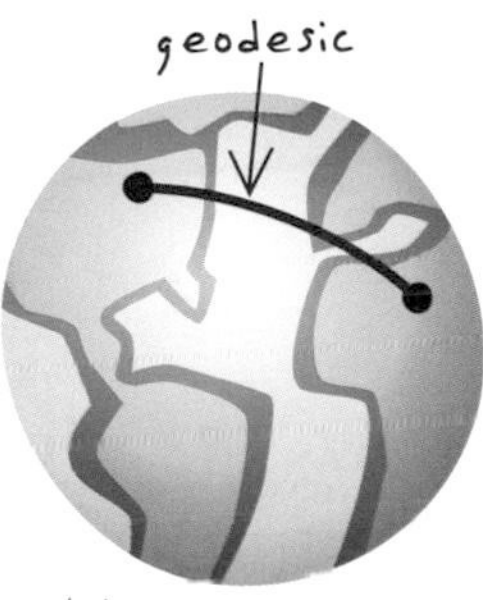

geodesic

geodesy The science of measuring and representing the shape and size of the earth, and the study of its gravitational and magnetic fields.

geodetic datum [GEODESY] A datum that is the basis for calculating positions on the earth's surface or heights above or below the earth's surface. *See also* datum, horizontal geodetic datum, vertical geodetic datum.

geodetic latitude [CARTOGRAPHY] The angle that a line drawn perpendicular

G

to the surface through a point on a spheroid makes with the equatorial plane. *See also* geocentric latitude.

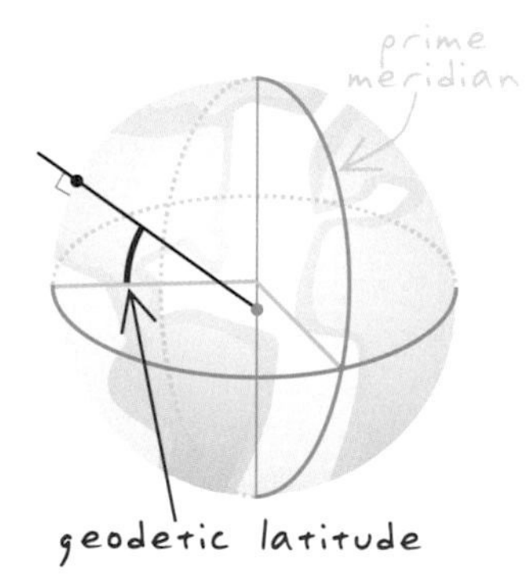

geodetic latitude

geodetic longitude [CARTOGRAPHY] The angle between the plane of the meridian that passes through a point on the surface of the spheroid and the plane of a prime meridian, usually the Greenwich meridian. *See also* geocentric longitude, longitude, prime meridian.

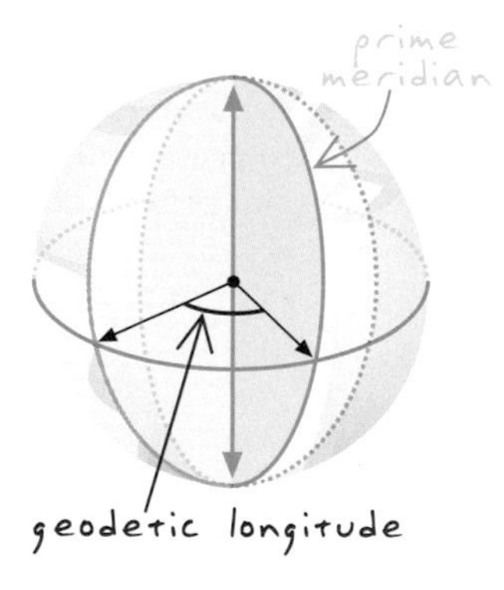

geodetic longitude

Geodetic Reference System of 1980 *See* GRS80.

geodetic survey [GEODESY] A survey that takes the shape and size of the earth into account, used to precisely locate horizontal and vertical positions suitable for controlling other surveys. *See also* surveying.

geodetic transformation *See* geographic transformation.

geographic [GEOGRAPHY] Of or relating to the earth. *See also* spatial.

geographic constraint *See* extent.

geographic coordinates [CARTOGRAPHY] A measurement of a location on the earth's surface expressed in degrees of latitude and longitude. *See also* geographic coordinate system.

geographic coordinate system [CARTOGRAPHY] A reference system that uses latitude and longitude to define the locations of points on the surface of a sphere or spheroid. A geographic coordinate system definition includes a datum, prime meridian, and angular unit. *See also* datum, prime meridian, angular unit.

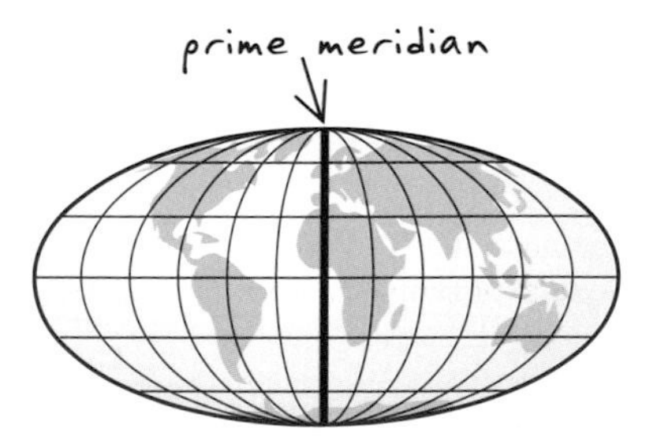

geographic coordinate system

G

geographic data Information describing the location and attributes of things, including their shapes and representation. Geographic data is the composite of spatial data and attribute data. *See also* spatial data, attribute data.

geographic information science *See* GIScience.

geographic information system *See* GIS.

geographic north *See* true north.

geographic transformation [CARTOGRAPHY] A systematic conversion of the latitude-longitude values for a set of points from one geographic coordinate system to equivalent values in another geographic coordinate system. Depending on the geographic coordinate systems involved, the transformation can be accomplished in various ways. Typically, equations are used to model the position and orientation of the "from" and "to" geographic coordinate systems in three-dimensional coordinate space; the transformation parameters may include translation, rotation, and scaling. Other methods, including one used in transformations between NAD 1927 and NAD 1983, use files in which the differences between the two geographic coordinate systems are given for a set of coordinates; the values of other points are interpolated from these. *See also* geographic coordinate system.

geography **1.** [GEOGRAPHY] The study of the earth's surface, encompassing the description and distribution of the various physical, biological, economic, and cultural features found on the earth and the interaction between those features. **2.** [GEOGRAPHY] The arrangement of the geographic features of an area. *See also* biogeography, cultural geography, economic geography, human geography, physical geography, quantitative geography, urban geography.

geography level A division of statistical geographic data, such as country, province, postal code, tract, or block group.

Geography Markup Language *See* GML.

geoid [GEODESY] A hypothetical surface representing the form the earth's oceans would take if there were no land and the water were free to respond to the earth's gravitational and centrifugal forces. The resulting geoid is irregular and varies from a perfect sphere by as much as 75 meters above and 100 meters below its surface. *See also* ellipsoid, spheroid.

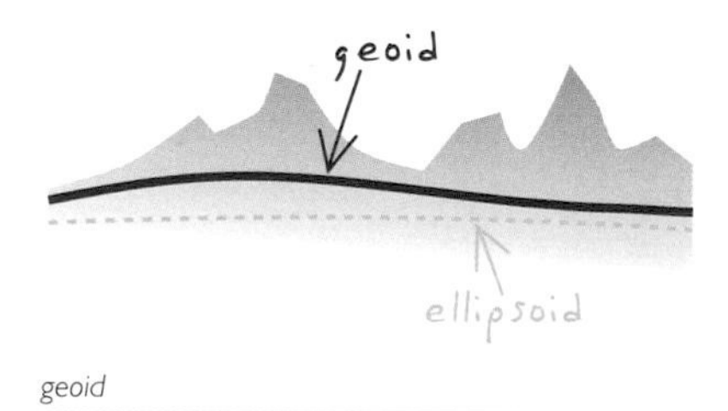

geoid

G

geoid-ellipsoid separation [GEODESY] The distance from the surface of an ellipsoid to the surface of the geoid, measured along a line perpendicular to the ellipsoid. The separation is positive if the geoid lies above the ellipsoid, negative if it lies below. *See also* geoid, geoid height.

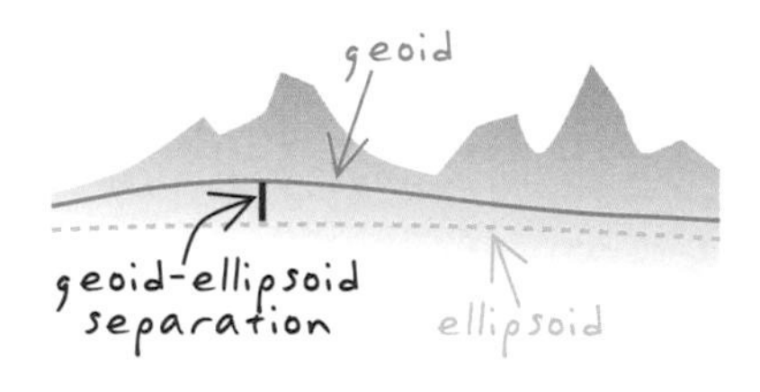

geoid-ellipsoid separation

geoid height [GEODESY] The height of the geoid above the ellipsoid. *See also* geoid-ellipsoid separation.

geolocation The process of creating geographic features from tabular data by matching the tabular data to a spatial location. An example of geolocation is creating point features from a table of x,y coordinates.

geometric coincidence [ESRI SOFTWARE] The distance within which features in a geometric network are deemed to be coincident and, therefore, connected. *See also* coincident.

geometric correction [REMOTE SENSING] The correction of errors in remotely sensed data, such as those caused by satellites or aircraft not staying at a constant altitude or by sensors deviating from the primary focus plane. Images are often compared to ground control points on accurate basemaps and resampled, so that exact locations and appropriate pixel values can be calculated.

geometric dilution of precision *See* DOP.

geometric element [MATHEMATICS] One of the most basic parts or components of a geometric figure: that is, a surface, shape, point, line, angle, or solid.

geometric network [ESRI SOFTWARE] Edge and junction features that represent a linear network, such as a utility or hydrologic system, in which the connectivity of features is based on their geometric coincidence. A geometric network does not contain information about the connectivity of features; this information is stored within a logical network. Geometric networks are typically used to model directed flow systems. *See also* logical network, directed network flow.

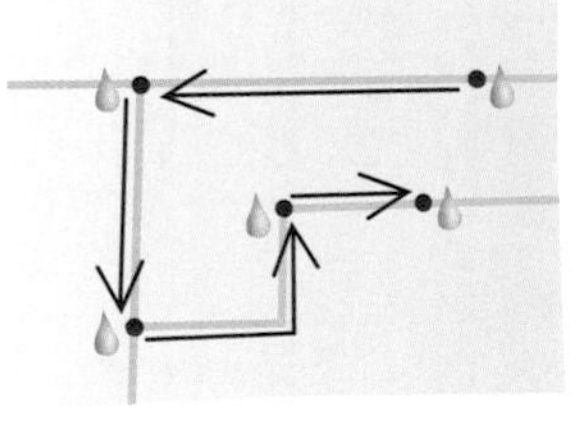

geometric network

geometric transformation [CARTOGRAPHY] The process of rectifying a raster dataset to map coordinates or converting a raster dataset from one coordinate system to another. *See also* transformation.

geometry 1. [MATHEMATICS] The measures and properties of points, lines, and surfaces. In a GIS, geometry is used to represent the spatial component of geographic features. **2.** [MATHEMATICS] The branch of mathematics concerning points, lines, and polygons, and their properties and relationships. **For more information about geometry, see* Geometry: Creating map features with points, lines, and polygons *on page 249.*

geomorphology [GEOGRAPHY] The study of the nature and origin of landforms, including relationships to underlying structures and processes of formation.

geoprocessing A GIS operation used to manipulate GIS data. A typical geoprocessing operation takes an input dataset, performs an operation on that dataset, and returns the result of the operation as an output dataset. Common geoprocessing operations include geographic feature overlay, feature selection and analysis, topology processing, raster processing, and data conversion. Geoprocessing allows for definition, management, and analysis of information used to form decisions.

georectification The digital alignment of a satellite or aerial image with a map of the same area. In georectification, a number of corresponding control points, such as street intersections, are marked on both the image and the map. These locations become reference points in the subsequent processing of the image. *See also* georeferencing, orthorectification, control point.

georeferencing [CARTOGRAPHY] Aligning geographic data to a known coordinate system so it can be viewed, queried, and analyzed with other geographic data. Georeferencing may involve shifting, rotating, scaling, skewing, and in some cases warping, rubber sheeting, or orthorectifying the data. *See also* transformation.

georelational data model A geographic data model that represents geographic features as an interrelated set of spatial and attribute data. The georelational model is the fundamental data model used in coverages. *See also* spatial data, attribute data.

geospatial data clearinghouse *See* NSDI Clearinghouse Network.

geospatial technology A set of technological approaches, such as GIS, photogrammetry, and remote sensing, for acquiring and manipulating geographic data.

geospecific model [CARTOGRAPHY] A model used to represent a real-world feature. For example, a geospecific model for the White House would look exactly like the White House and be used to represent the White House on a map of Washington, D.C. *See also* geotypical model.

geostationary [ASTRONOMY] Positioned in an orbit above the earth's equator with an angular velocity the same as that of the earth and an inclination and eccentricity approaching zero. A geostationary satellite will orbit as fast as the earth rotates on its axis, so that it remains effectively stationary above a point on

G

the equator. A geostationary satellite is geosynchronous, but geosynchronous satellites are not necessarily geostationary. *See also* geosynchronous.

geostatistics [STATISTICS] A class of statistics used to analyze and predict the values associated with spatial or spatiotemporal phenomena. Geostatistics provides a means of exploring spatial data and generating continuous surfaces from selected sampled data points. *See also* spatial statistics.

geosynchronous [ASTRONOMY] Positioned in an orbit moving west to east with an orbital period equal to the earth's rotational period. If a satellite is in a geosynchronous orbit that is circular and lies in the equatorial plane, it is geostationary because it remains over one point on the equator. If not, the satellite appears to make a figure eight once a day between the latitudes that correspond to its angle of inclination over the equator. *See also* geostationary.

geotypical model [CARTOGRAPHY] A symbolic representation for a class of map features, such as government buildings. For example, on a map of the United States, a white building with a dome on top could be used as a geotypical model for all state capitols. *See also* geospecific model.

GIF *Acronym for Graphic Interchange Format.* A low-resolution file format for image files, commonly used on the Internet. It is well-suited for images with sharp edges and reduced numbers of colors. *See also* JPEG.

GIS *Acronym for geographic information system.* An integrated collection of computer software and data used to view and manage information about geographic places, analyze spatial relationships, and model spatial processes. A GIS provides a framework for gathering and organizing spatial data and related information so that it can be displayed and analyzed. *See also* spatial data, analysis, spatial analysis, model.

GIScience *Abbreviation for geographic information science.* The field of research that studies the theory and concepts that underpin GIS. It seeks to establish a theoretical basis for the technology and use of GIS, study how concepts from cognitive science and information science might apply to GIS, and investigate how GIS interacts with society.

global analysis The computation of an output raster where the output value at each cell location may be a function of all the cells in the input raster. *See also* local analysis.

global functions *See* global analysis.

Global Navigation Satellite System *See* GLONASS.

Global Positioning System *See* GPS.

global spatial data infrastructure *See* GSDI.

globe [CARTOGRAPHY] A sphere on which a map of the earth or a celestial body is represented. Since the earth's natural shape is similar to a sphere, globes distort the earth's features far less than flat maps.

GLONASS [GPS] *Acronym for Global Navigation Satellite System.* The Russian counterpart of the United States' Global Positioning System. *See also* GPS.

glyph [CARTOGRAPHY] The geometric shape of a character in a font.

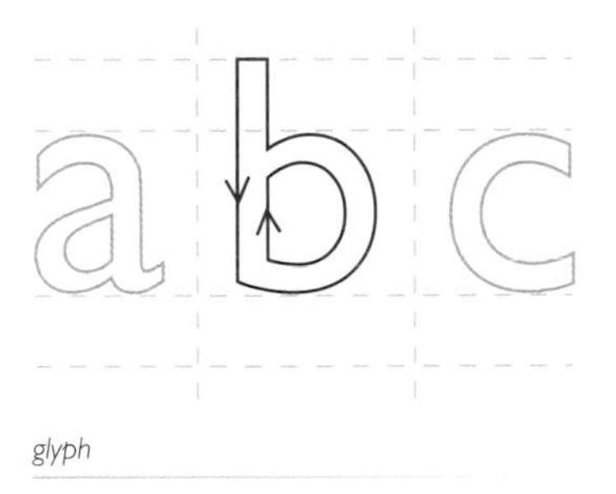

glyph

GML *Acronym for Geography Markup Language.* An OpenGIS Implementation Specification designed to store and transport geographic information. GML is a profile (encoding) of XML. *See also* XML.

GMT *See* Greenwich mean time.

gnomonic projection [CARTOGRAPHY] A planar projection, tangent to the earth at one point, projected from the center of the globe. All great circles appear as straight lines on this projection, so that the shortest distance between two points is a straight line. The gnomonic projection is useful in navigation. The gnomonic projection was used by Thales of Miletus, an ancient Greek astronomer and philosopher, to chart the heavens. It is possibly the oldest map projection. *See also* projection.

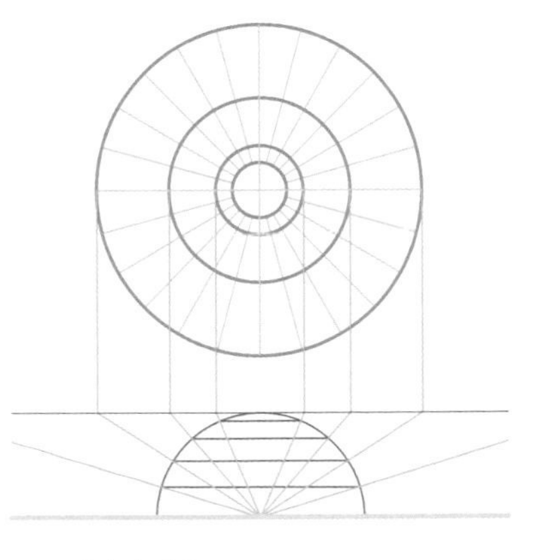

gnomonic projection

gon *See* gradian.

goodness of fit In modeling, the degree to which a model predicts observed data; a measure of predictive power. *See also* model.

gore [CARTOGRAPHY] A map, shaped like the area between a pair of parentheses, of an area that lies between two lines of longitude. A gore can be fitted to the surface of a globe with little distortion.

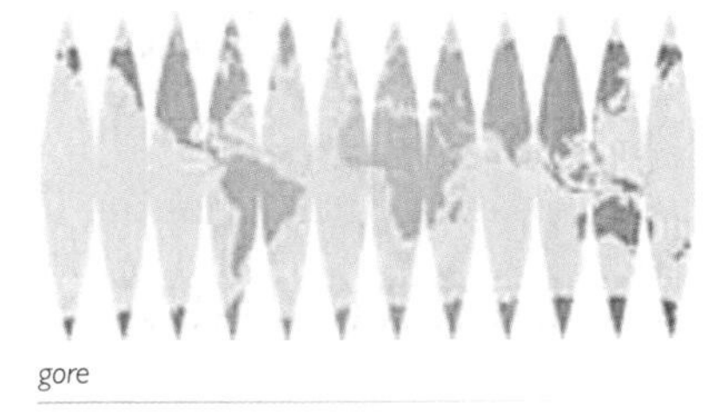

gore

GPS [GPS] *Acronym for Global Positioning System.* A system of radio-emitting and -receiving satellites

G

used for determining positions on the earth. The orbiting satellites transmit signals that allow a GPS receiver anywhere on earth to calculate its own location through trilateration. Developed and operated by the U.S. Department of Defense, the system is used in navigation, mapping, surveying, and other applications in which precise positioning is necessary. *See also* GLONASS.

grad *See* gradian.

gradian [MATHEMATICS] A unit of angular measurement in which the angle of a full circle is 400 gradians and a right angle is 100 gradians. The common abbreviation for gradian is grad.

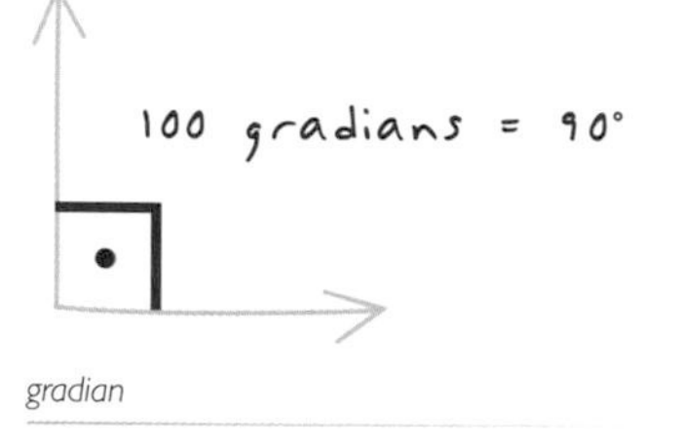

gradian

gradient 1. [GEODESY] The ratio between vertical distance (rise) and horizontal distance (run), often expressed as a percentage. A 10-percent gradient rises 10 feet for every 100 feet of horizontal distance. 2. [GEODESY] The inclination of a surface in a given direction. 3. [PHYSICS] The rate at which a quantity such as temperature or pressure changes in value. *See also* gradient of gravity, slope.

gradient of gravity [GEODESY] The direction of the maximum increase in gravity in a horizontal plane. *See also* gradient.

graduated color map [CARTOGRAPHY] A map on which a range of colors indicates a progression of numeric values. For example, increases in population density might be represented by the increased saturation of a single color, or temperature differences by a sequence of colors from blue to red.

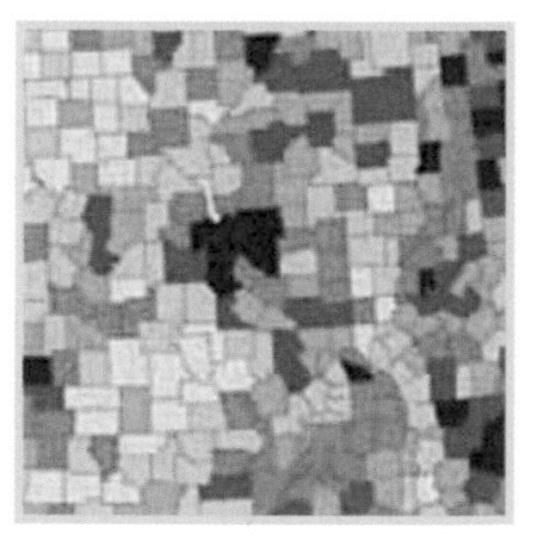

graduated color map

graduated symbol map [CARTOGRAPHY] A map with symbols that change in size according to the value of the attribute they represent. For example, denser populations might be represented by larger dots, or larger rivers by thicker lines.

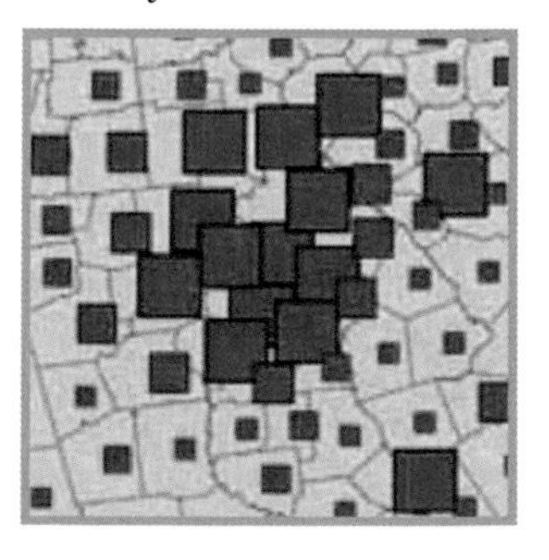

graduated symbol map

granularity 1. The coarseness or resolution of data. Granularity describes the clarity and detail of data during its capture and visualization. 2. [REMOTE SENSING] The objective measure of the random groupings of silver halide grains into denser and less dense areas in a photographic image.

graph *See* chart.

graphic An image produced by and stored in a computer as data for display. *See also* symbol.

graphical user interface *See* GUI.

graticule [CARTOGRAPHY] A network of longitude and latitude lines on a map or chart that relates points on a map to their true locations on the earth. *See also* latitude, longitude, grid.

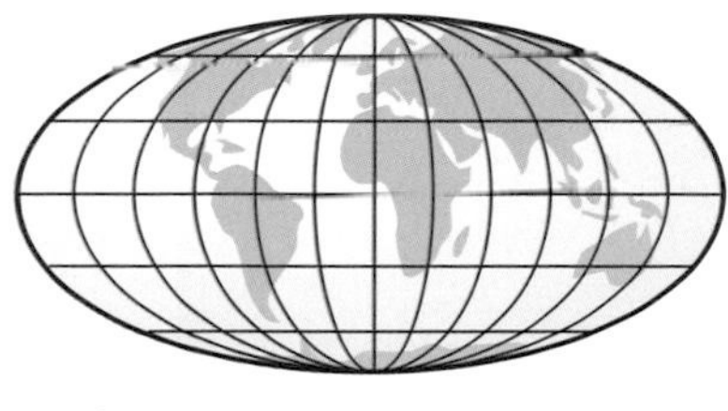

graticule

graticule alignment of labels [CARTOGRAPHY] A label positioning method in which labels are oriented along the graticule of the data frame. This is useful for maps of large areas, for cartographic or stylistic reasons. *See also* label orientation.

gravimeter [GEODESY] A device used to measure small variations in the earth's gravitational field between two or more points.

tension spring
weighted mass
gravitational pull (variable)
1
2
3
4
5
6
difference measure

gravimeter

gravimetric geodesy [GEODESY] The science of deducing the size and shape of the earth by measuring its gravitational field.

gravity model [GEOGRAPHY] A model that assumes that the influence of phenomena or populations on each other varies inversely with the distance between them.

gray scale 1. [COMPUTING] All the shades of gray from white to black.

gray scale 1

2. Levels of brightness used to display information on a monochrome display device.

great circle [GEODESY] Any circle or near circle produced by the intersection of the surface of a sphere and a flat plane that passes through the center of the sphere. The equator and all lines of longitude are great circles. Great circles are used in navigation, since the shortest path between any two points on the earth's surface lies on a great circle.

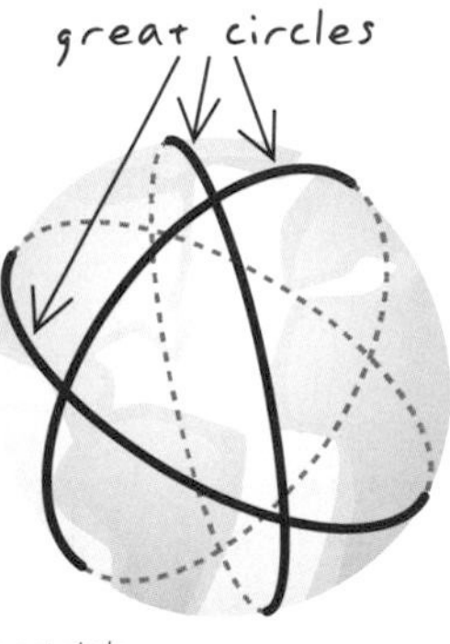

great circle

Greenwich mean time [ASTRONOMY] The time at the prime meridian, which runs through the Royal Observatory in Greenwich, England. From 1884 to 1928, Greenwich mean time was the official name (and is still the popular name) for universal time. It sometimes also refers to coordinated universal time. *See also* universal time, coordinated universal time, Greenwich meridian, prime meridian.

Greenwich meridian [ASTRONOMY] The meridian adopted by international agreement in 1884 as the prime meridian, the 0-degree meridian from which all other longitudes are calculated. The Greenwich prime meridian runs through the Royal Observatory in Greenwich, England. *See also* prime meridian, meridian.

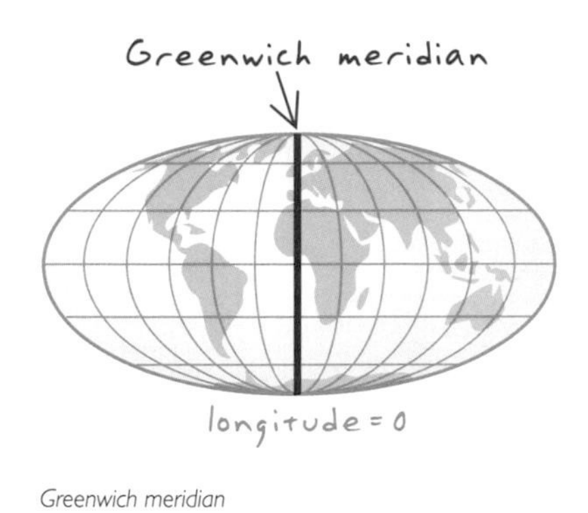

Greenwich meridian

grid [CARTOGRAPHY] Any network of parallel and perpendicular lines superimposed on a map and used for reference. These grids are usually referred to by the map projection or coordinate system they represent, such as universal transverse Mercator grid.

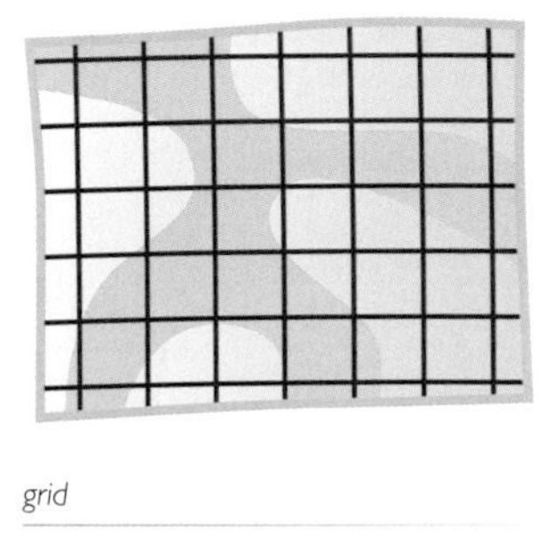

grid

grid cell *See* cell.

grid north [CARTOGRAPHY] The direction north along the north-south grid lines of a map projection. *See also* true north, magnetic north.

ground control [GEODESY] A system of points with known positions, elevations, or both, used as fixed references in georeferencing map features, aerial photographs, or remotely sensed images.

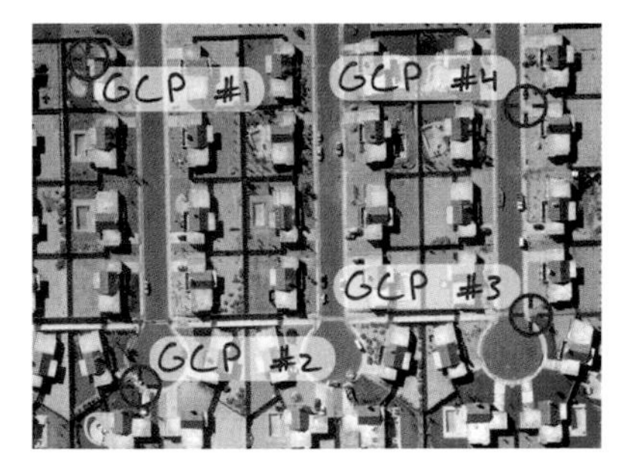

ground control

ground control point *See* control point.

ground receiving station [REMOTE SENSING] Communications equipment for receiving and transmitting signals to and from satellites such as Landsat.

ground truth The accuracy of remotely sensed or mathematically calculated data based on data actually measured in the field.

GRS80 [GEODESY] *Acronym for Geodetic Reference System of 1980.* The standard measurements of the earth's shape and size adopted by the International Union of Geodesy and Geophysics in 1979.

GSDI *Acronym for global spatial data infrastructure.* A global framework of technologies, policies, standards, and human resources necessary to acquire, process, store, distribute, and improve the use of geospatial data across multiple countries and organizations. *See also* SDI, NSDI.

GUI [COMPUTING] *Acronym for graphical user interface.* A software display of program options that allows a user to choose commands by pointing to icons, dialog boxes, and lists of menu items on the screen, typically using a mouse. This contrasts with a command line interface in which control is accomplished via the exchange of strings of text. *See also* command line interface.

G

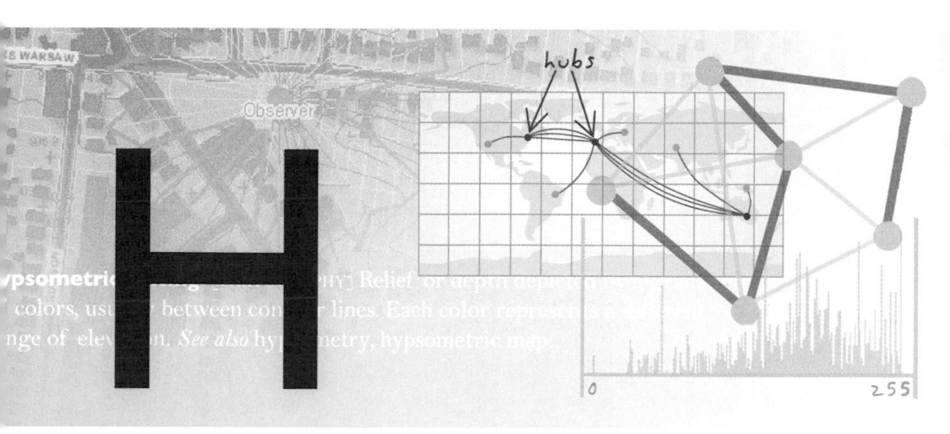

hachure [CARTOGRAPHY] A short line on a map that indicates the direction and steepness of a slope. Hachures that represent steep slopes are short and close together; hachures that represent gentle slopes are longer, lighter, and farther apart. Contours, shading, and hypsometric tints have largely replaced hachuring on modern maps.

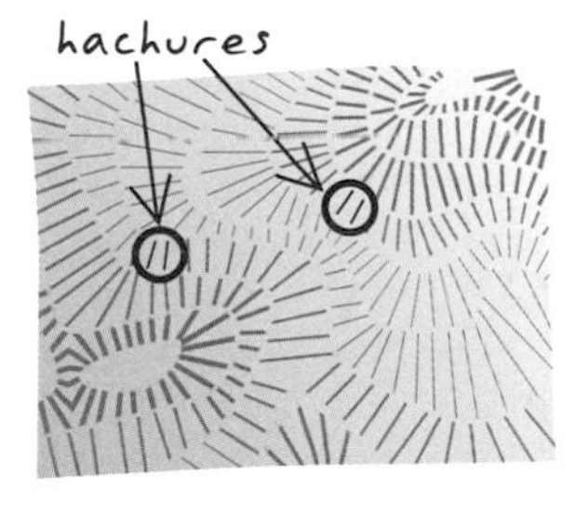

hachure

hachured contour [CARTOGRAPHY] On a topographic map, concentric contour lines drawn with hachures to indicate a closed depression or basin. Concentric contour lines drawn without hachure marks indicate a hill.

hachured contour

halftone image A continuous tone image photographed through a fine screen that converts it into uniformly spaced dots of varying size while maintaining the gradations of highlight and shadow. The size of the dots varies in proportion to the intensity of the light passing through them. *See also* dot screen, continuous tone image.

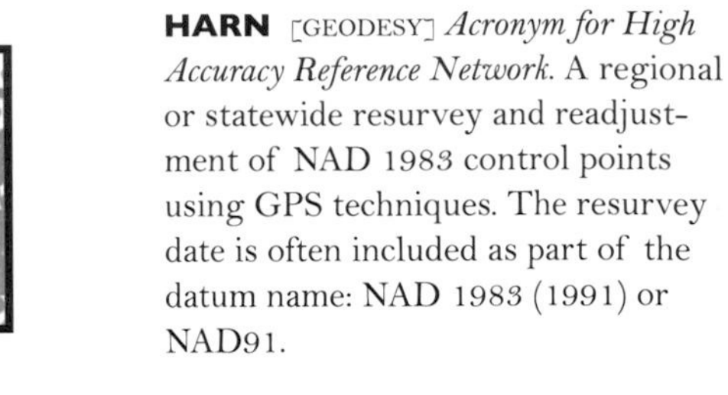

halftone image

Hamiltonian circuit [MATHEMATICS] A path through a network that visits each junction in the network only once and then returns to its point of origin. Hamiltonian circuits are named after the Irish mathematician, physicist, and astronomer William Rowan Hamilton.

Hamiltonian path A path through a network that visits each junction in the network only once without returning to its point of origin. Hamiltonian paths are named after the Irish mathematician, physicist, and astronomer William Rowan Hamilton.

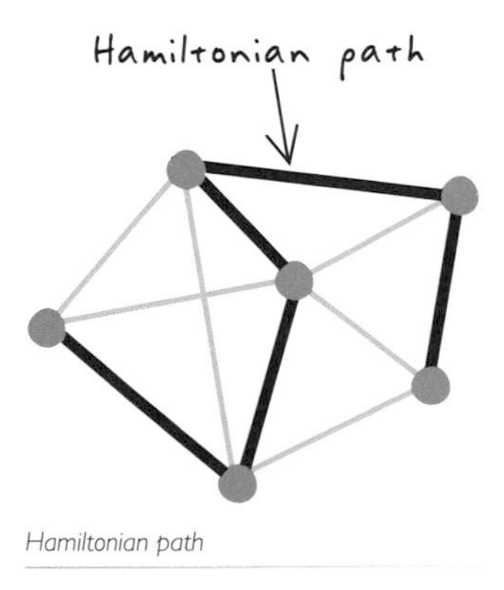

Hamiltonian path

HARN [GEODESY] *Acronym for High Accuracy Reference Network.* A regional or statewide resurvey and readjustment of NAD 1983 control points using GPS techniques. The resurvey date is often included as part of the datum name: NAD 1983 (1991) or NAD91.

hatches In linear referencing, a series of vertical line or marker symbols displayed on top of features at an interval specified in route measure units.

hatching In linear referencing, a type of labeling that posts and labels hatches or symbols at a regular interval along measured line features. *See also* hatches.

HDOP [GEODESY] *Acronym for horizontal dilution of precision.* A measure of the geometric quality of a GPS satellite configuration in the sky. HDOP is a factor in determining the relative accuracy of a horizontal position. The smaller the DOP number, the better the geometry. *See also* DOP.

heading [NAVIGATION] The direction of a moving object, expressed as an angle from a known direction, usually north. *See also* bearing.

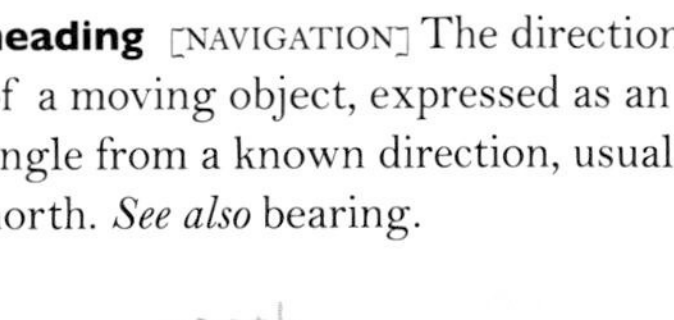

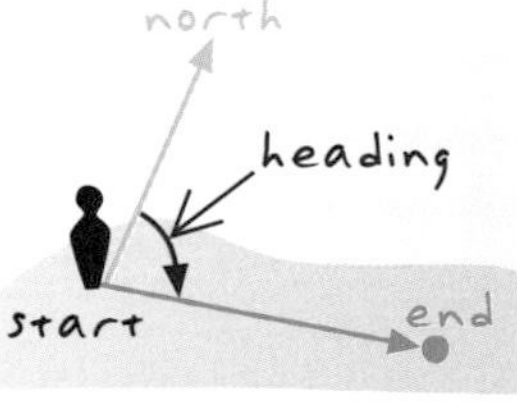

heading

heads-up digitizing Manual digitization by tracing a mouse over features displayed on a computer monitor, used as a method of vectorizing raster data. *See also* digitizing, autovectorization.

heap [COMPUTING] In computer programming, a variation on the binary tree data structure in which each node's value is greater than the value of its leaf nodes. Sorting data in a heap allows an element to be located more quickly than it could be found in an ordinary list. *See also* node.

height [MATHEMATICS] The vertical distance between two points, or above a specified datum.

Helmert transformation A geometric transformation that scales, rotates, or translates images or coordinates between any two Euclidean spaces. It is commonly used in GIS to transform maps between coordinate systems. In a Helmert transformation, parallel lines remain parallel. The midpoint of a line segment remains a midpoint, and all points on a straight line remain on a straight line. The Helmert transformation is named for the German mathematician and geodesist Friedrich Robert Helmert (1843–1917). *See also* transformation, affine transformation.

hemisphere 1. [ASTRONOMY] Half of a celestial body, such as the earth. 2. [MATHEMATICS] Half of a sphere. *See also* sphere.

heuristic 1. [COMPUTING] An algorithm that incorporates a shortcut or simplification for solving a programming problem, such as searching. While a heuristic may run faster than a more rigorous algorithm, there is no guarantee that it will find the best solution. 2. [MATHEMATICS] In graph theory, a function used to determine the lowest cost or shortest path between two given nodes in a tree. *See also* algorithm, hierarchical database.

hexadecimal [MATHEMATICS] A number system using base 16 notation, usually comprised of the digits 0–9 and the letters A–F or a–f.

hierarchical database [COMPUTING] A database that stores related information in a tree-like structure, where records can be traced to parent records, which in turn can be traced to a root record. *See also* database.

High Accuracy Reference Network *See* HARN.

high-level language [PROGRAMMING] A programming language that uses keywords and statements that are similar to expressions in human language or mathematics and is, therefore, easier for people to comprehend and use. A high-level language is named for the high level of abstraction it affords developers over low-level processor functions such as memory access and register storage, meaning such operations do not demand the developer's attention. *See also* low-level language.

high-pass filter [DIGITAL IMAGE PROCESSING] A spatial filter that blocks

H

low-frequency (long-wave) radiation, resulting in a sharpened image. *See also* band-pass filter, low-pass filter, edge enhancement.

High Precision Geodetic Network *See* HARN.

hillshading **1.** [CARTOGRAPHY] Shadows drawn on a map to simulate the effect of the sun's rays over the varied terrain of the land. **2.** [CARTOGRAPHY] The hypothetical illumination of a surface according to a specified azimuth and altitude for the sun. Hillshading creates a three-dimensional effect that provides a sense of visual relief for cartography, and a relative measure of incident light for analysis.

hillshading 2

histogram [STATISTICS] A graph showing the distribution of values in a set of data. Individual values are displayed along a horizontal axis, and the frequency of their occurrence is displayed along a vertical axis.

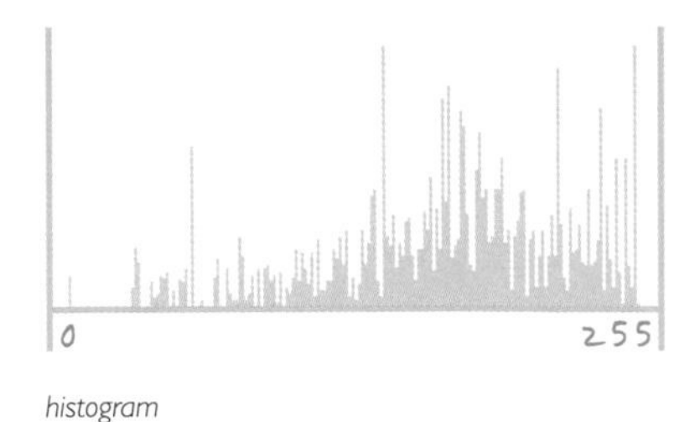

histogram

histogram equalization [DIGITAL IMAGE PROCESSING] The redistribution of pixel values in an image so that each range contains approximately the same number of pixels. A histogram showing this distribution of values would be nearly flat.

hole A small gap in a raster line feature, usually considered to be an error caused by the poor quality of a source document or by the scanning process.

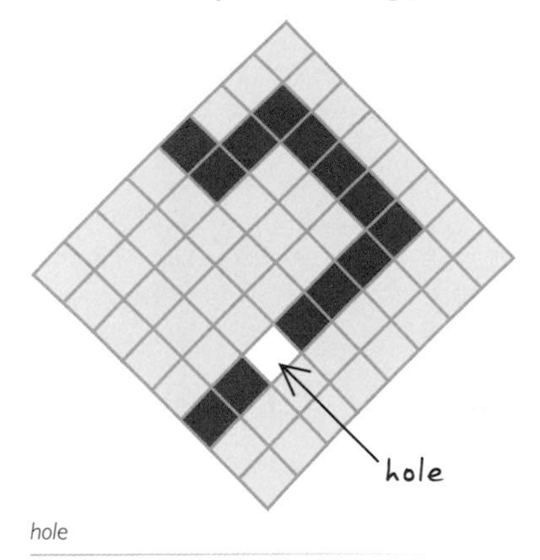

hole

horizon **1.** [NAVIGATION] The apparent or visible junction of land and sky. **2.** [ASTRONOMY] The horizontal plane tangent to the earth's surface and perpendicular to the line through an observer's position and the zenith of that position. The apparent or visible horizon approximates the true horizon only when the point of vision is very close to sea level. **3.** [ASTRONOMY] The great circle in which an observer's horizon meets the celestial sphere. **4.** [CARTOGRAPHY] The edge of a map projection. *See also* horizon circle, zenith, celestial sphere.

horizon circle [CARTOGRAPHY] The circle containing all points equidistant from the center of an azimuthal projection. *See also* azimuthal projection.

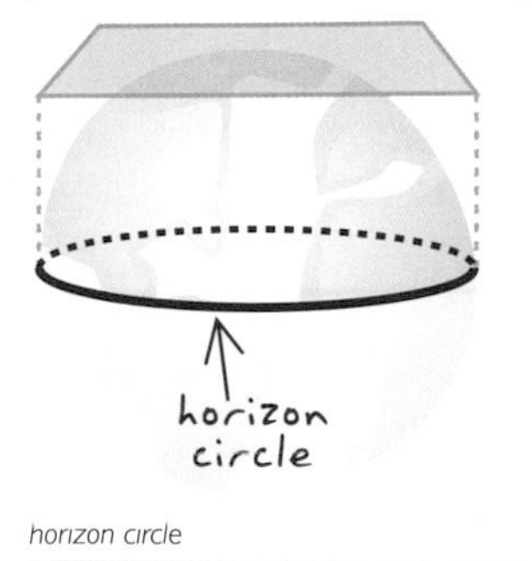

horizon circle

horizontal angle [NAVIGATION] The angle formed by the intersection of two lines in a horizontal plane.

horizontal control [GEODESY] A network of known horizontal geographic positions, referenced to geographic parallels and meridians or to other lines of orientation such as plane coordinate axes. *See also* horizontal geodetic datum.

horizontal control datum *See* horizontal geodetic datum.

horizontal datum *See* horizontal geodetic datum.

horizontal dilution of precision *See* HDOP.

horizontal geodetic datum [GEODESY] A geodetic datum for any extensive measurement system of positions, usually expressed as latitude-longitude coordinates, on the earth's surface. A horizontal geodetic datum may be local or geocentric. If it is local, it specifies the shape and size of an ellipsoid representing the earth, the location of an origin point on the ellipsoid surface, and the orientation of x- and y-axes relative to the ellipsoid. If it is geocentric, it specifies the shape and size of an ellipsoid, the location of an origin point at the intersection of x-, y-, and z-axes at the center of the ellipsoid, and the orientation of the x-, y-, and z-axes relative to the ellipsoid. Examples of local horizontal geodetic datums include the North American Datum of 1927, the European Datum of 1950, and the Indian datum of 1960; examples of geocentric horizontal geodetic datums include the North American Datum of 1983 and the World Geodetic System of 1984. *See also* datum, geodetic datum, geocentric datum, local datum.

host **1.** [COMPUTING] In a computer network, the computer that contains data being accessed by other computers. **2.** [COMPUTING] A computer connected to a TCP/IP network such as the Internet. Each host has a

H

unique IP address. *See also* TCP/IP, IP address.

HPGN *See* HARN.

HTML [INTERNET] *Acronym for HyperText Markup Language.* A markup language used to create Web pages for publication on the Internet. HTML is a system of tags that define the function of text, graphics, sound, and video within a document, and is now an Internet standard maintained by the World Wide Web Consortium.

HTML document [INTERNET] A computer file formatted with HTML tags so that it may be viewed in a Web browser and published on the World Wide Web. An HTML document may incorporate text, images, sound, video, and other media components. Characteristically, it also has embedded references, called hypertext links, to other HTML documents. These links enable a person viewing a document in a Web browser to open other documents—which may be stored on other computers anywhere in the world—by clicking on the link using a mouse. *See also* HTML, Web browser, World Wide Web.

HTTP [INTERNET] *Acronym for HyperText Transfer Protocol.* The protocol maintained by the World Wide Web Consortium for communicating between servers and clients to exchange HTML documents across the Internet.

HTTPS [INTERNET] *Acronym for HyperText Transfer Protocol (Secure).* A variant of HTTP enhanced by a security mechanism. It allows transactions such as e-commerce and data sharing to take place on the World Wide Web in a protected way. *See also* HTTP.

hub A central node in a network for routing goods to their destinations.

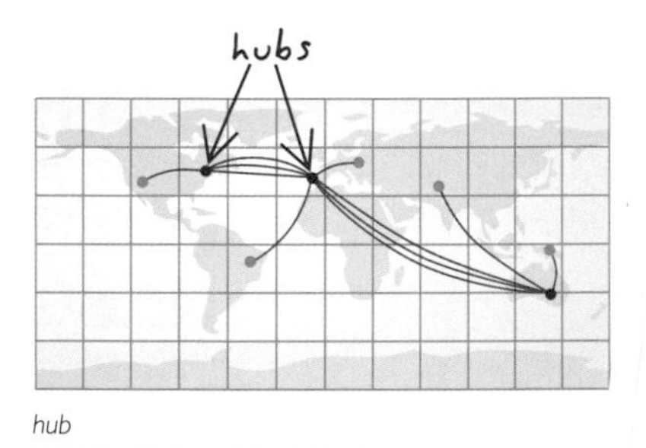

hub

hue [PHYSICS] The dominant wavelength of a color, by which it can be distinguished as red, green, yellow, blue, and so forth. *See also* saturation, value, intensity, chroma.

human geography [GEOGRAPHY] The field of geography concerning a range of social, cultural, and political aspects of human life as related to their distribution throug[h] physical space.

hydrographic datum [SURVEYIN[G]] A plane of reference for depths, d[e]pth contours, and elevations of foresh[o]re and offshore features. *See also* dat[u]m, hydrography, hydrographic surv[e]y.

hydrographic survey [GEODESY] A survey of a water body, particularly of its currents, depth, submarine relief, and adjacent land. *See also* surveying, hydrography, hydrographic datum.

hydrography [GEODESY] The measurement and description of water features and their related land areas for the purposes of safe marine navigation.

hydrologic cycle [GEOGRAPHY] The circulation of water from the earth through the atmosphere and back again. Its major stages are evaporation, condensation, precipitation, run-off, transpiration, infiltration, and percolation.

hydrology [GEOGRAPHY] The study of water, its behavior, and its movements across and below the surface of the earth, and through the atmosphere.

hyperlink [INTERNET] A reference (link) from one point in an electronic document to another document or another location in the same document (the target). Activating the link, usually by clicking it with the mouse, causes the browser to display the target of the link.

HyperText Markup Language *See* HTML.

HyperText Transfer Protocol *See* HTTP.

hypsography 1. [CARTOGRAPHY] The study and representation of elevation and the earth's topography. 2. [CARTOGRAPHY] The representation of relief features on a map.

hypsometric curve [CARTOGRAPHY] A curve showing the relationship of area to elevation for specified terrain. A hypsometric curve is plotted on a graph on which the x-axis represents surface area and the y-axis represents elevation above or below a datum (normally sea level). The curve shows how much area lies above and below marked elevation intervals. *See also* hypsometry.

hypsometric map [CARTOGRAPHY] A map showing relief, whether by contours, hachures, shading, or tinting. *See also* hypsometry, hypsometric tinting.

hypsometric tinting [CARTOGRAPHY] Relief or depth depicted by a gradation of colors, usually between contour lines. Each color represents a different range of elevation. *See also* hypsometry, hypsometric map.

hypsometric tinting

H

hypsometry **1.** [GEODESY] The science that determines the spatial distribution of elevations above an established datum, usually sea level. **2.** [GEODESY] The determination of terrain relief, by any method.

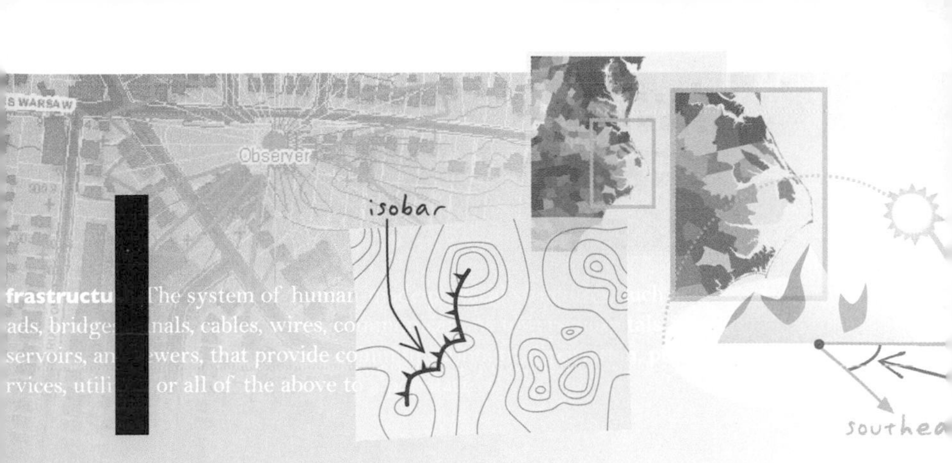

icon [COMPUTING] A graphic symbol on a computer screen that represents an operation, action, program, or file. A user can select an icon, usually by clicking it with a mouse, to perform an action or open a file or program.

ID *See* identifier.

identifier [COMPUTING] A unique character string or numeric value associated with a particular object.

identity In geoprocessing, a topological overlay that computes the geometric intersection of two datasets. The output dataset preserves all the features of the first dataset plus those portions of the second (polygon) dataset that overlap the first. For example, a road passing through two counties would be split into two arc features, each with the attributes of the road and the county it passes through. *See also* intersect, union.

identity link An anchor that prevents the movement of features during rubber sheeting. *See also* rubber sheeting.

IDW *See* inverse distance weighted interpolation.

IFSAR [REMOTE SENSING] *Acronym for interferometric synthetic aperture radar.* A dual-antenna radar sensor mounted on an airborne or space-borne platform that collects a remotely sensed radar image, called an interferogram. There is a measured energy shift between the signals received by each antenna, and this interference can be colorized to measure elevation or changes in the topography on the earth's surface. *See also* radar interferometry, interferogram.

illumination [CARTOGRAPHY] The light incident on a surface or object, either natural or artificial, as determined by the surface's slope and aspect and by the sun's azimuth and altitude. *See also* azimuth, altitude.

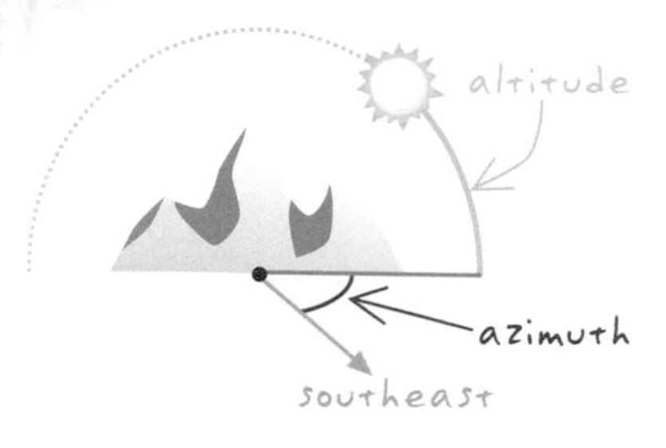

illumination

image A representation or description of a scene, typically produced by an optical or electronic device, such as a camera or a scanning radiometer. Common examples include remotely sensed data (for example, satellite data), scanned data, and photographs. *See also* analog image, digital image, raster, aerial photograph.

image coordinate An x,y coordinate pair specifying the location of a pixel, or cell, in terms of its row and column position. The x-coordinate gives the column number (commonly starting from 0 at the left edge of the data), and the y-coordinate gives the row number (commonly starting from 0 at the top of the data). *See also* raster, pixel.

image data Data produced by scanning a surface with an optical or electronic device. Common examples include scanned documents, remotely sensed data (for example, satellite images), and aerial photographs. An image is stored as a raster dataset of binary or integer values that represent the intensity of reflected light, heat, or other range of values on the electromagnetic spectrum.

image division [DIGITAL IMAGE PROCESSING] A technique for increasing the contrast between features in an image by dividing the pixel values in the image by the values of corresponding pixels in a second image. Image division is normally used to identify concentrations of vegetation.

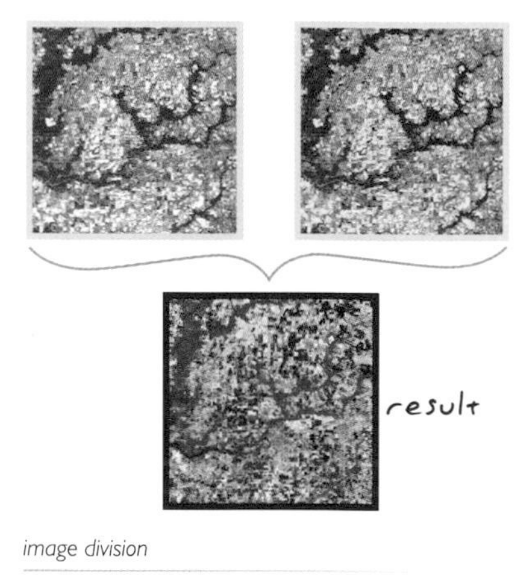

image division

imager [REMOTE SENSING] Any satellite or aerial instrument that measures and maps the earth and its atmosphere.

image scale The ratio between a distance in an image and the actual distance on the ground, calculated as focal length divided by the flying height above mean ground elevation. Image scale can vary in a single image from point to point due to surface relief and the tilt of the camera lens.

image space The x,y coordinate space defined by the number of columns and rows in a raster dataset. The origin of image space is commonly the center of the top left pixel of the data and is labeled (0,0). The x-axis corresponds to the number of columns in the raster, and the y-axis to the number of rows. For raster data to be used in GIS software, image space must be transformed to a real-world coordinate system through georeferencing. *See also* raster, pixel, coordinate system, georeferencing.

impedance A measure of the amount of resistance, or cost, required to traverse a path in a network, or to move from one element in the network to another. Resistance may be a measure of travel distance, time, speed of travel multiplied by distance, and so on. Higher impedance values indicate more resistance to movement, and a value of zero indicates no resistance. An optimum path in a network is the path of lowest impedance, also called the least-cost path. *See also* stop impedance, least-cost path.

impersonation [INTERNET] A process by which a Web application assumes the identity of a particular user and thus gains all the privileges to which that user is entitled.

import To bring data from one computer system or application into another. Importing often involves some form of data conversion.

incident energy [PHYSICS] Electromagnetic radiation that strikes a surface.

INCITS *Acronym for International Committee for Information Technology Standards.* An ANSI-accredited forum that creates and maintains information and communications technology standards through the participation and consensus of its industry members. *See also* ANSI.

indeterminate flow direction In networks, a flow direction that is unknown or undiscoverable. Indeterminate flow direction occurs when flow direction cannot be determined from the connectivity of the network, the locations of sources and sinks, and the enabled or disabled states of features. *See also* determinate flow direction.

index [COMPUTING] A data structure, usually an array, used to speed the search for records in a database or for spatial features in geographic datasets. In general, unique identifiers stored in a key field point to records or files holding more detailed information.

index contour line [CARTOGRAPHY] On a topographic map, a contour line that is thicker than the rest and usually labeled with the elevation that it represents. Depending on the contour interval, every fourth or fifth contour line may be an index contour. *See also* contour line.

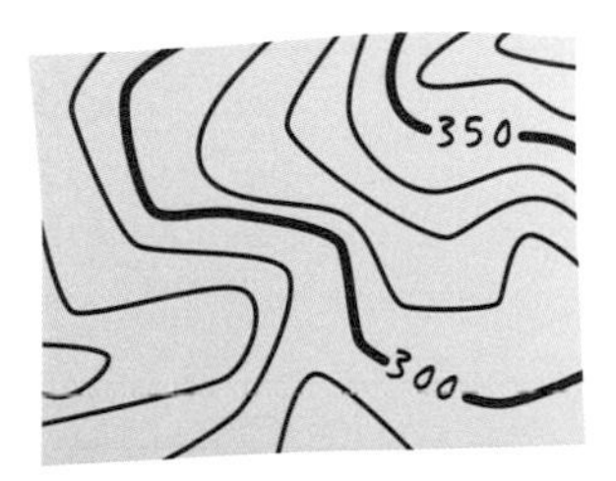

index contour line

index map [CARTOGRAPHY] A schematic map used as a reference for a collection of map sheets, outlining the total area covered along with the coverage extent of, and usually a name or reference for, each map sheet.

INFO database [ESRI SOFTWARE] A tabular database management system used by ArcInfo Workstation to store and manipulate attributes of a GIS dataset in ArcInfo Workstation format. INFO databases are stored inside a workspace folder with subdirectories containing files that represent the geometry and topology that make up a coverage.

information space A geometric representation of relationships between elements in a data domain, in which relative position indicates the degree of similarity between elements. Information spaces are often based on geographic metaphors and are used to provide more intuitive views of a complex, multidimensional data domain.

infrared scanner A device that detects infrared radiation and converts it into an electrical signal that can be recorded on film or magnetic tape.

infrastructure The system of human-made physical structures, such as roads, bridges, canals, cables, wires, communications towers, hospitals, pipes, reservoirs, and sewers, that provide communication, transportation, public services, utilities, or all of the above to a populace.

inheritance [COMPUTING] In object-oriented programming, the acquisition of methods and properties by child classes or interfaces from their previously existing parent classes or interfaces. Inheritance is one of the defining characteristics of an object-oriented system.

input data Data that is entered into a computer, device, program, or process. *See also* intermediate data, output data.

inset map [CARTOGRAPHY] A small map set within a larger map. An inset map might show a detailed part of the map at a larger scale, or the extent of the existing map drawn at a smaller scale within the context of a larger area. *See also* overview map.

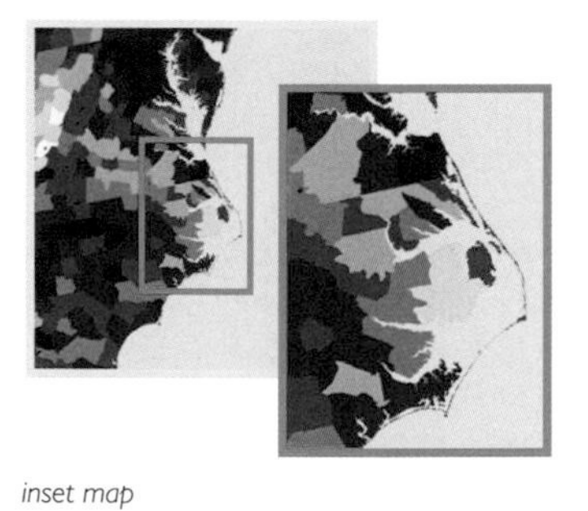

inset map

instance [COMPUTING] In object-oriented programming, a single object created based on the template or definition of the class to which it belongs. *See also* instantiation.

instantiation [COMPUTING] In programming, the process of creating a single object based on the template or definition of the class to which it belongs. *See also* instance.

integer data *See* discrete data.

integration A high degree of interconnection between two or more programs or datasets, in which they share a common schema, ontology, semantic approach, or method that allows information to be passed between them without being fully processed.

intensity In the IHS (intensity, hue, saturation) color model, the relative brightness of a color. *See also* saturation, hue, value, chroma.

intensity

interactive vectorization A manual process for converting raster data into vector features that involves tracing raster cells. *See also* vectorization.

interchange format A file format that allows the easy exchange of data between different software programs.

interferogram [REMOTE SENSING] A radar image that records interference patterns captured by two antennae a short distance apart. *See also* IFSAR, radar interferometry.

intermediate data Any data in a process that did not exist before the process existed and that will not be maintained after the process executes. *See also* input data.

international date line [CARTOGRAPHY] An imaginary line, generally following the meridian of longitude lying 180 degrees east and west of the Greenwich meridian, where the date changes. The time zone east of the international date line is twelve hours ahead of Greenwich mean time; the time zone west of the international date line is twelve hours behind Greenwich mean time. A traveler going west across the date line adds a day; a traveler going east across it subtracts a day. *See also* Greenwich mean time.

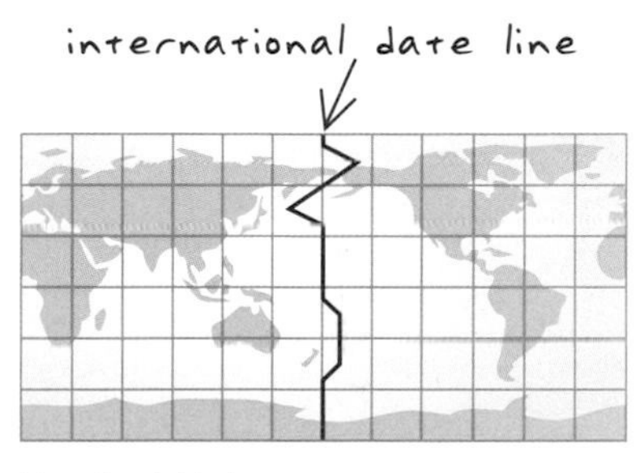

international date line

internationalization The process of creating software that can be adapted to the requirements of different languages and cultures without substantive changes to the source code. *See also* localization.

international meridian *See* Greenwich meridian.

International Organization for Standardization *See* ISO.

Internet [INTERNET] The global network of computers that communicate through common protocols, such as TCP/IP.

interoperability The capability of components or systems to exchange data with other components or systems, or to perform in multiple environments. In GIS, interoperability is required for a GIS user using software from one vendor to study data compiled with GIS software from a different provider. *See also* OGC.

interpolation **1.** [MATHEMATICS] The estimation of surface values at unsampled points based on known surface values of surrounding points. Interpolation can be used to estimate elevation, rainfall, temperature, chemical dispersion, or other spatially-based phenomena. Interpolation is commonly a raster operation, but it can also be done in a vector environment using a TIN surface model. There are several well-known interpolation techniques, including spline and kriging.

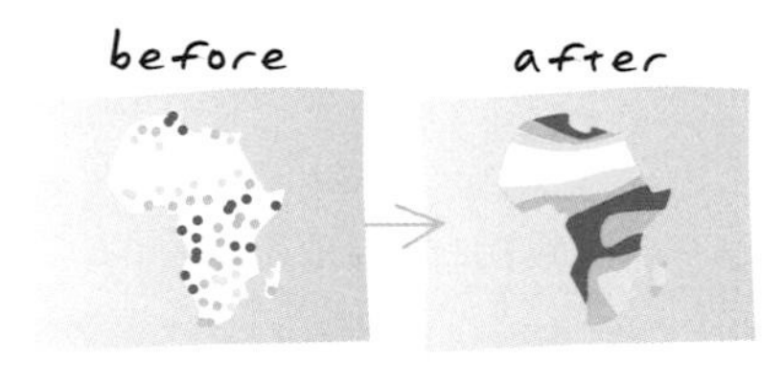

interpolation 1

2. [ESRI SOFTWARE] In the context of linear referencing, the calculation of measure values for a route between two known measure values. *See also* inverse distance weighted interpolation, kriging, natural neighbors, trend surface analysis.

interrupted projection [CARTOGRAPHY] A world projection that reduces distortion by dividing the projected area into gores, each with its own central meridian. *See also* projection, gore.

intersect A geometric integration of spatial datasets that preserves features or portions of features that fall within areas common to all input datasets. *See also* identity, union.

intersection The point where two lines cross. In geocoding, most often a street crossing.

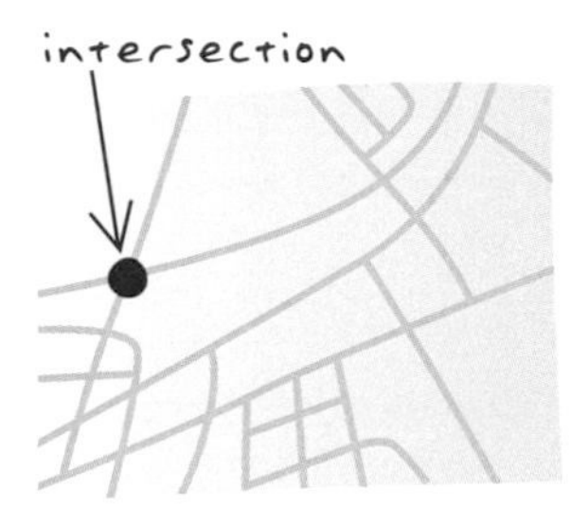

intersection

intranet [INTERNET] A computer network, often using the same software and serving the same functions as those found on the Internet, that is restricted to users within an organization.

intrinsic stationarity [STATISTICS] In spatial statistics, the assumption that a set of data comes from a random process with a constant mean and a semivariogram that depends only on the distance and direction separating any two locations. *See also* stationarity, semivariogram.

inverse distance [STATISTICS] One divided by distance, often raised to some power (1/D or 1/D2, for example), where D is a distance value. By inverting the distance among spatial features, and using that inverted value as a weight, near things have a larger weight or influence than things that are farther away. *See also* inverse distance weighted interpolation.

inverse distance weighted interpolation [MATHEMATICS] An interpolation technique that estimates cell values in a raster from a set of sample points that have been weighted so that the farther a sampled point is from the cell being evaluated, the less weight it has in the calculation of the cell's value. *See also* interpolation, weight.

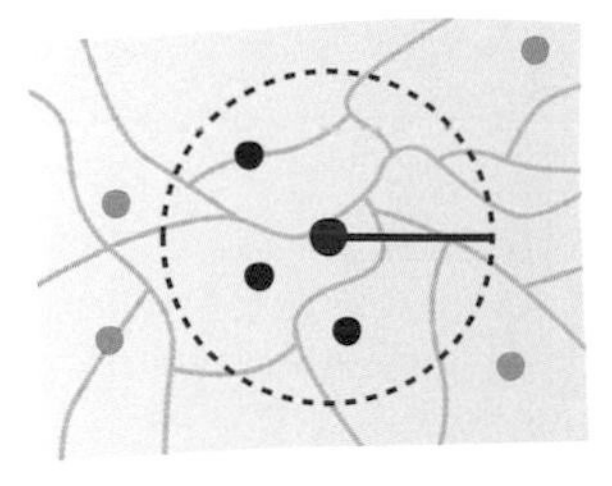

inverse distance weighted interpolation

IP address [INTERNET] *Acronym for Internet protocol address.* A unique number, such as 10.48.6.8, that identifies each computer on the Internet. IP addresses are similar to phone numbers, and allow data to travel between one computer and another via the Internet.

irregular triangular mesh *See* TIN.

irregular triangular surface model *See* TIN.

isanomal [CARTOGRAPHY] A line on a map connecting points of equal difference from a normal value, usually a meteorological value such as average temperature. *See also* isoline.

isarithm 1. [CARTOGRAPHY] An isoline drawn according to values that can occur at points; an isometric line. 2. [CARTOGRAPHY] A line connecting points of equal value on a map; an isoline. *See also* isoline.

ISO Abbreviation for *International Organization for Standardization.* A federation of national standards institutes from 145 countries that works with international organizations, governments, industries, businesses, and consumer representatives to define and maintain criteria for international standards.

isobar [CARTOGRAPHY] A line on a weather map connecting places of equal barometric pressure. *See also* isoline.

isobar

isochrone 1. [CARTOGRAPHY] A line on a map connecting points of equal elapsed time; especially, travel time to or from a given location. 2. [CARTOGRAPHY] A line on a map connecting points at which an event occurs, or a state of affairs exists, at the same time. *See also* isoline.

isohyet [CARTOGRAPHY] A line on a map connecting points of equal rainfall. *See also* isoline.

isoline [CARTOGRAPHY] A line connecting points of equal value on a map. Isolines fall into two classes: those in which the values actually exist at points, such as temperature or elevation values, and those in which the values are ratios that exist over areas, such as population per square kilometer or crop yield per acre. The first type of isoline is specifically called an isometric line or isarithm; the second type is called an isopleth. *See also* isometric line, isopleth.

isometric line [CARTOGRAPHY] An isoline drawn according to known values, either sampled or derived, that can occur at points. Examples of sampled quantities that can occur at points are elevation above sea level, an actual temperature, or an actual depth of precipitation. Examples of derived values that can occur at points are the average of temperature over time for one point or the ratio of smoggy days to clear days for one point. *See also* isoline, isopleth.

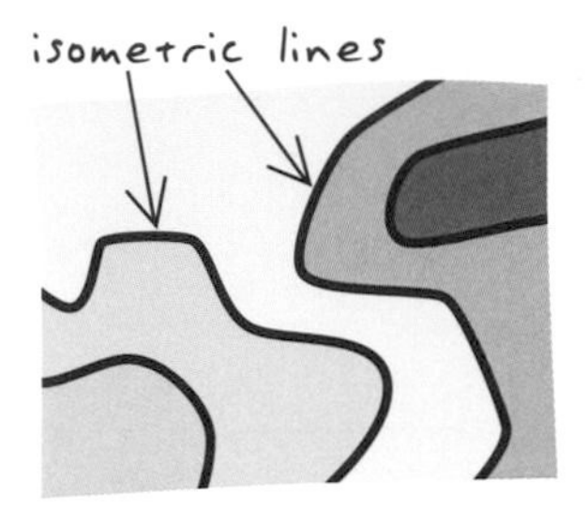

isometric line

isopleth [CARTOGRAPHY] An isoline drawn according to known values that can only be recorded for areas, not points. Examples include population per square mile or the ratio of residential land to total land for an area. *See also* isoline, isometric line.

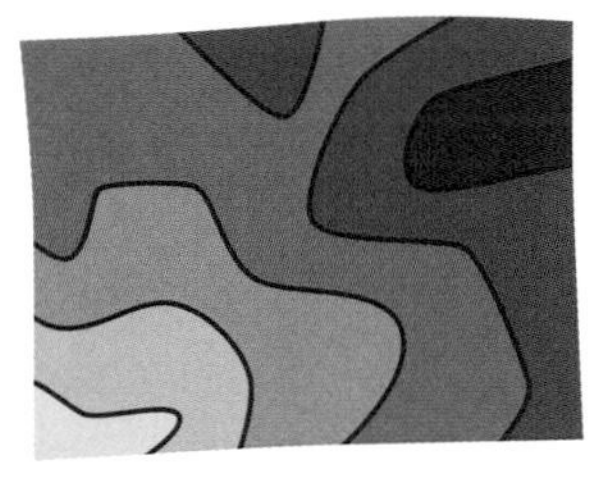

isopleth

isotherm [CARTOGRAPHY] A line on a map connecting points of equal temperature. *See also* isoline.

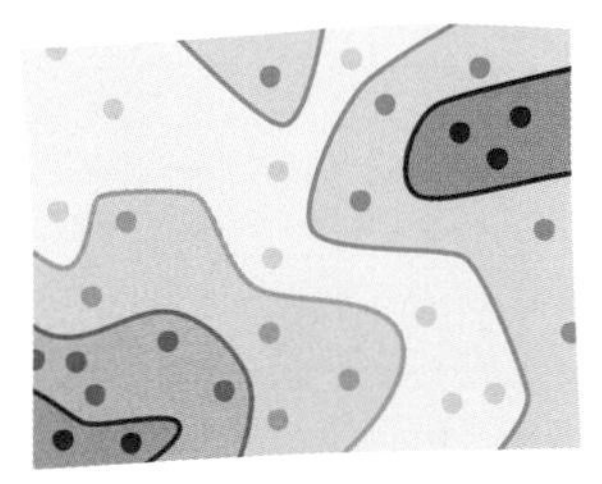

isotherm

isotropic Having uniform spatial distribution of movement or properties, usually across a surface. *See also* isotropy, anisotropic.

isotropy [STATISTICS] A property of a natural process or data where spatial dependence (autocorrelation) changes only with the distance between two locations—direction is unimportant. *See also* isotropic, anisotropy, autocorrelation.

iterative procedure [COMPUTING] A repetitive or recurring procedure.

J

jaggies *See* aliasing.

Jenks' optimization [CARTOGRAPHY] A method of statistical data classification that partitions data into classes using an algorithm that calculates groupings of data values based on the data distribution. Jenks' optimization seeks to reduce variance within groups and maximize variance between groups. *See also* classification, natural breaks classification

job [COMPUTING] A task scheduled on a computer for immediate or future processing. A job can involve a single task or a batch mode operation.

JOG *Acronym for joint operations graphic.* A 1:250,000-scale topographic map used by militaries worldwide. Joint operations graphics use a common base graphic to facilitate operations involving air, ground, and naval forces. *See also* TLM, VMap.

joining 1. Appending the fields of one table to those of another through an attribute or field common to both tables. A join is usually used to attach more attributes to the attribute table of a geographic layer. 2. Connecting two or more features from different sets of data so that they become a single feature. *See also* link, relate.

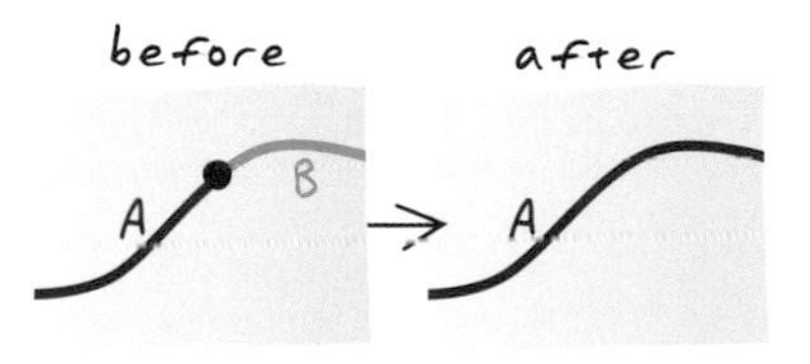

joining 2

joint operations graphic *See* JOG.

JPEG *Acronym for Joint Photographic Experts Group.* A lossy image compression format commonly used on the Internet. JPEG is well-suited for photographs or images that have graduated colors. *See also* GIF, lossy compression.

junction 1. [ESRI SOFTWARE] For network data models in a geodatabase, a point at which two or more edges meet. ▸

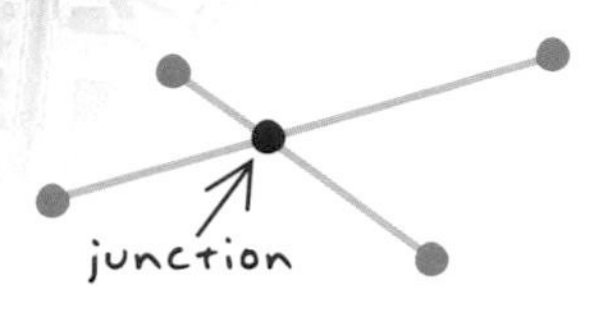

junction 1

2. [ESRI SOFTWARE] In a coverage, a node joining two or more arcs.

junction element *See* junction.

kernel *See* filter.

key An attribute or set of attributes in a database that uniquely identifies each record. *See also* alternate key, compound key, foreign key, primary key.

key attribute *See* primary key.

keyword [COMPUTING] A significant word from a document that is used to index or search content. *See also* index.

kinematic positioning [GPS] Determining the position of an antenna on a moving object such as a ship or an automobile. *See also* static positioning.

knockout In offset printing, an area that has been defined to cut through or mask specific layers of colored inks. Knockouts are used to ensure that certain ink colors are not mixed with inks that are laid down after them. *See also* overprinting, color separation.

knowledge base [COMPUTING] A database of information about a subject, used in expert systems.

known point [SURVEYING] A surveyed point that has an established x,y coordinate value. Known points are used in survey operations to extend survey computations into a project area.

Kohonen map [CARTOGRAPHY] A map that uses a neural network algorithm to classify and illustrate associations in complex datasets, and reveal multidimensional patterns. A similar set of methods produces maps referred to as self-organizing maps (SOMs). Kohonen maps are named for the Finnish engineer Teuvo Kohonen. *See also* neural network.

kriging [STATISTICS] An interpolation technique in which the surrounding measured values are weighted to derive a predicted value for an unmeasured location. Weights are based on the distance between the measured points, the prediction locations, and the overall spatial arrangement among the measured points. Kriging is unique among the interpolation methods in that it provides an easy method for characterizing the variance, or the precision, of predictions. Kriging is based on regionalized variable theory, which assumes that the spatial variation in the data being modeled is homogeneous across the surface. That is, the same pattern of variation can be observed at all locations on the surface. Kriging was named for the South African mining engineer Danie G. Krige (1919-). *See also* interpolation, block kriging, ordinary kriging, simple kriging, universal kriging.

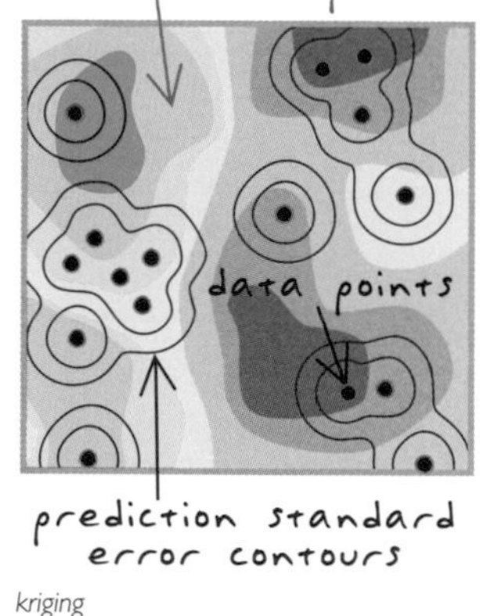

kriging

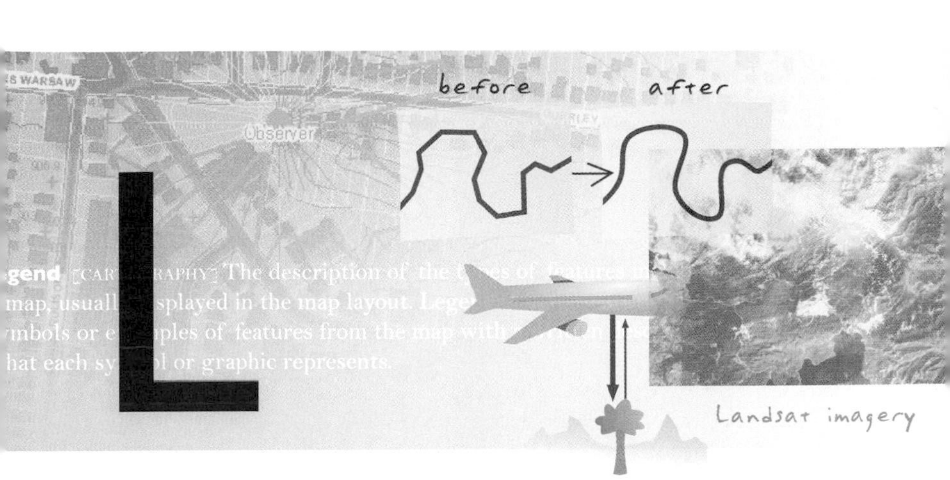

label 1. [CARTOGRAPHY] Text placed on or near a map feature that describes or identifies it.

label 1

2. [ESRI SOFTWARE] In ArcGIS, descriptive text, usually based on one or more feature attributes. Labels are placed dynamically on or near features based on user-defined rules, and in response to changes in the map display. Labels cannot be individually selected and modified by the user. Label placement rules and display properties (such as font size and color) are defined for an entire layer. *See also* annotation. **For more information about labels, see* Cartographic text in ArcGIS software: Annotation, labeling, and graphic text *on page 245.*

label orientation [ESRI SOFTWARE] The angle or direction of alignment for feature labels. Labels for features are usually placed horizontally, but they may also be oriented to an angle stored as an attribute, an angle defined by the orientation of the feature geometry, or along the graticule of the data frame. *See also* graticule, graticule alignment of labels.

lag [STATISTICS] In the creation of a semivariogram, the sample distance used to group or bin pairs of points. Using an appropriate lag distance can be helpful in revealing scale-dependent spatial correlation. *See also* bin, kriging.

LAN [COMPUTING] *Acronym for local area network.* Communications hardware and software that connect computers in a small area, such as a room or a building. Computers in a LAN can share data and peripheral devices, such as printers and plotters, but do not necessarily have a link to outside computers. *See also* WAN.

land cover [GEOGRAPHY] The classification of land according to the vegetation or material that covers most of its surface; for example, pine forest, grassland, ice, water, or sand. *See also* land use.

landform [GEOGRAPHY] Any natural feature of the land having a characteristic shape, including major forms such as plains and mountains and minor forms such as hills and valleys.

land information system A geographic information system for cadastral and land-use mapping, typically used by local governments.

landmark 1. [GEOGRAPHY] Any prominent natural or artificial object in a landscape used to determine distance, bearing, or location. 2. [GEOGRAPHY] A building or location that has historical, architectural, or cultural value.

Landsat [REMOTE SENSING] Multispectral, earth-orbiting satellites developed by NASA (National Aeronautics and Space Administration) that gather imagery for land-use inventory, geological and mineralogical exploration, crop and forestry assessment, and cartography.

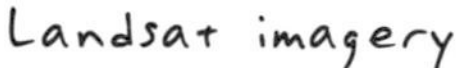

Landsat

landscape ecology The study of spatial patterns, processes, and change across biological and cultural structures within areas encompassing multiple ecosystems.

land use [GEOGRAPHY] The classification of land according to what activities take place on it or how humans occupy it; for example, agricultural, industrial, residential, urban, rural, or commercial. *See also* land cover.

large-format printer A printing device capable of producing an image on large paper or other media sized between 36 and 87 inches (91 and 220 centimeters) wide. Modern large format printers typically use inkjet printing technology to print an image on a roll of paper that is automatically cut to the desired length. Large-format printers may also be called plotters or wide-format printers. *See also* medium-format printer, plotter.

large scale [CARTOGRAPHY] Generally, a map scale that shows a small area on the ground at a high level of detail. *See also* small scale.

late binding [PROGRAMMING] A COM technique that an application uses for determining an object's properties and methods at run time, rather than when the code is compiled. Late binding is generally used by scripting languages. *See also* binding, early binding.

latitude [CARTOGRAPHY] The angular distance, usually measured in degrees north or south of the equator. Lines of latitude are also referred to as parallels. *See also* parallel, longitude.

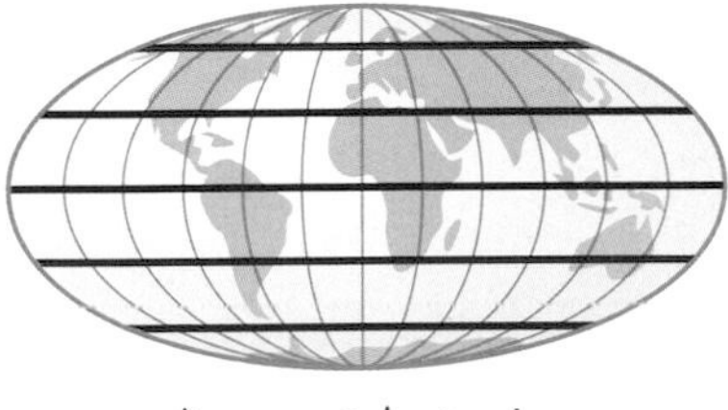

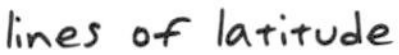

latitude

latitude–longitude [CARTOGRAPHY] A reference system used to locate positions on the earth's surface. Distances east–west are measured with lines of longitude (also called meridians), which run north–south and converge at the North and South Poles. Distance measurements begin at the prime meridian, and are measured positively 180 degrees to the east and negatively 180 degrees to the west. Distances north–south are measured with lines of latitude (also called parallels) which run east–west. Distance measurements begin at the equator and are measured positively 90 degrees to the north and negatively 90 degrees to the south. *See also* latitude, longitude.

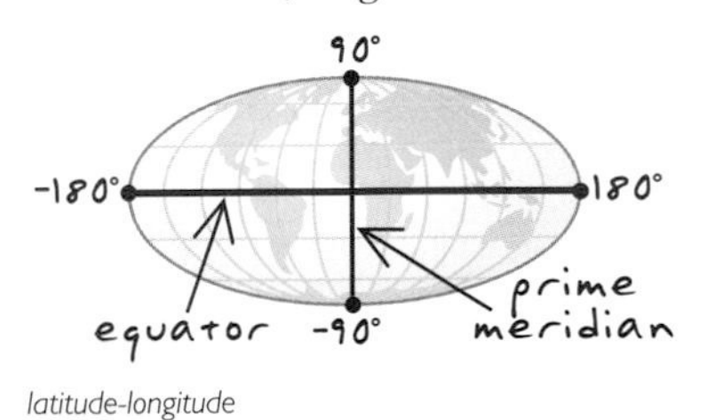

latitude-longitude

latitude of center [CARTOGRAPHY] The latitude value that defines the center, and sometimes the origin, of a projection. *See also* latitude.

latitude of origin [CARTOGRAPHY] The latitude value that defines the origin of the y-coordinate values for a projection. *See also* longitude of origin.

lattice A representation of a surface using an array of regularly spaced sample points (mesh points) that are referenced to a common origin and have a constant sampling distance in the x and y directions. Each mesh point contains the z-value at that location, which is referenced to a common base z-value, such as sea level. Z-values for locations between lattice mesh points can be approximated by interpolation based on neighboring mesh points. *See also* raster.

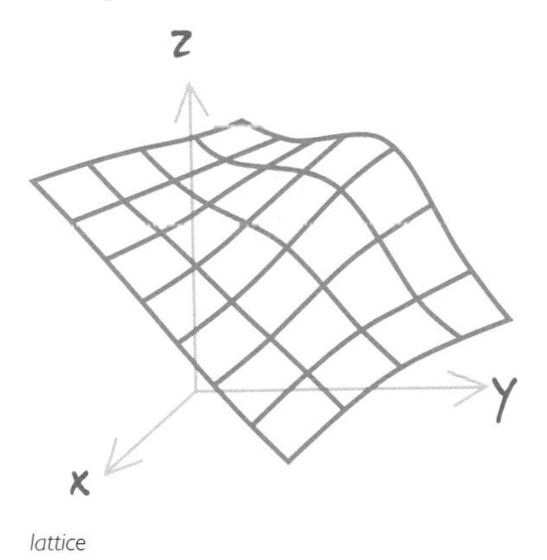

lattice

layer 1. The visual representation of a geographic dataset in any digital map environment. Conceptually, a layer is a slice or stratum of the geographic reality in a particular area, and is more or less equivalent to a legend item on a paper map. On a road map, for example, roads, national parks, political boundaries, and rivers might be considered different layers. **2.** [ESRI SOFTWARE] In ArcGIS, a reference to a data source, such as a shapefile, coverage, geodatabase feature class, or raster, that defines how the data should be symbolized on a map. Layers can also define additional properties, such as which features from the data source are included. Layers can be stored in map documents (.mxd) or saved individually as layer files (.lyr). Layers are conceptually similar to themes in ArcView 3.x. *For more information about layers, see* An overview of layers in ArcGIS *on page 255.*

L

layout [CARTOGRAPHY] The arrangement of elements on a map, possibly including a title, legend, north arrow, scale bar, and geographic data.

L-band [GPS] The group of radio frequencies that carry data from GPS satellites to GPS receivers.

LBS *See* location-based services.

least convex hull *See* convex hull.

least-cost path The path between two locations that costs the least to traverse, where cost is a function of time, distance, or some other criteria defined by the user. *See also* cost, shortest path.

least-squares adjustment [SURVEYING] A statistical method for providing a best fit for survey point locations and detecting measurement error by minimizing the sum of the squares of measurement residuals. The method allows many measurements to participate simultaneously in a single computation.

least-squares corrections [SURVEYING] The final measurement residuals of a least squares adjustment. *See also* least-squares adjustment.

legend [CARTOGRAPHY] The description of the types of features included in a map, usually displayed in the map layout. Legends often use graphics of symbols or examples of features from the map with a written description of what each symbol or graphic represents.

legend

leveling [SURVEYING] The measurement of the heights of objects and points according to a specified elevation, usually mean sea level.

level of confidence *See* confidence level.

level of significance *See* significance level.

library [PROGRAMMING] In object-oriented programming, a logical grouping of classes, usually with a header section that lists the classes in the library.

license The grant to a party of the right to use a software package or component. A license differs from a sale in that the user does not necessarily purchase the software, but is granted the legal right to use it.

lidar [REMOTE SENSING] *Acronym for light detection and ranging.* A remote-sensing technique that uses lasers to measure distances to reflective surfaces. *See also* radar, sonar.

lidar

line [MATHEMATICS] On a map, a shape defined by a connected series of unique x,y coordinate pairs. A line may be straight or curved. *See also* line feature, polyline.

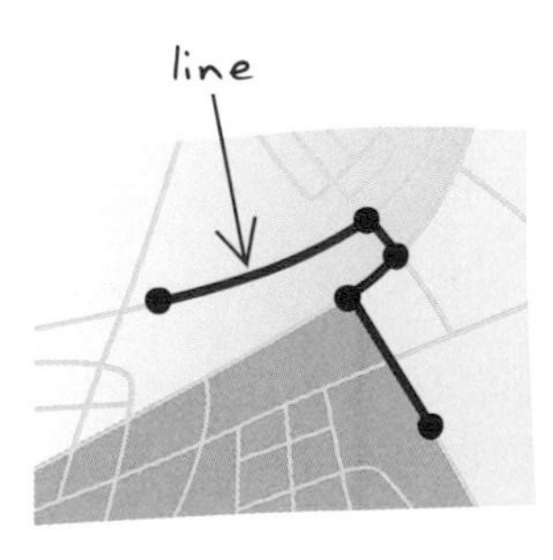

line

linear dimension [SURVEYING] A measurement of the horizontal or vertical dimension of a feature. Linear dimensions may not represent the true distance between beginning and ending dimension points because they do not take angle into account as aligned dimensions do. *See also* aligned dimension.

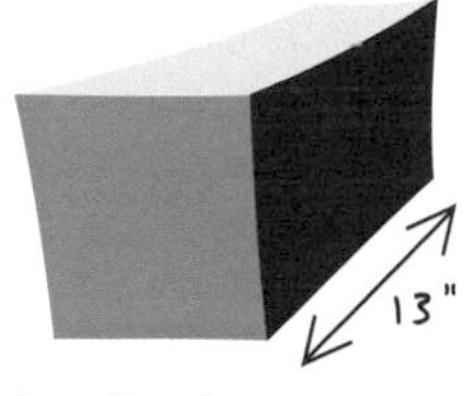

linear dimension

linear feature *See* line feature.

linear interpolation [STATISTICS] The estimation of an unknown value using the linear distance between known values.

linear referencing A method for storing geographic data by using a relative position along an already existing line feature; the ability to uniquely identify positions along lines without explicit x,y coordinates. In linear referencing, location is given in terms of a known line feature and a position, or measure, along the feature. Linear referencing is an intuitive way to associate multiple sets of attributes to portions of linear features. *See also* dynamic segmentation.

linear unit [CARTOGRAPHY] The unit of measurement on a plane or a projected coordinate system, often meters or feet. *See also* angular unit, unit of measure.

line event In linear referencing, a description of a portion of a route using a from- and to-measure value. Examples of line events include pavement quality, salmon spawning grounds, bus fares, pipe widths, and traffic volumes. *See also* event, linear referencing.

line feature [CARTOGRAPHY] A map feature that has length but not area at a given scale, such as a river on a world map or a street on a city map. *See also* feature.

line of sight 1. A line drawn between two points, an origin and a target, that is compared against a surface to show whether the target is visible from the origin and, if it is not visible, where the view is obstructed.

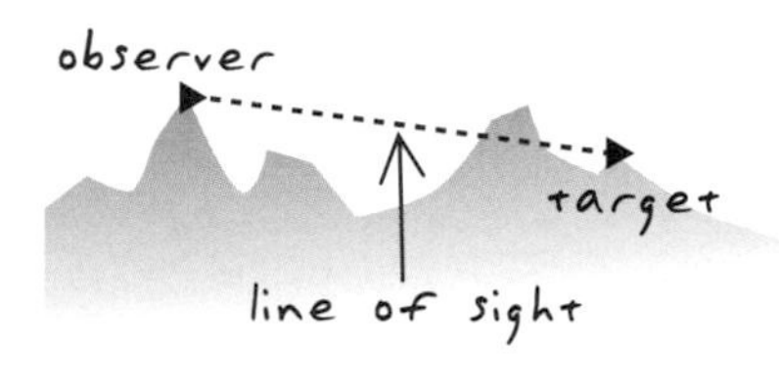

line of sight 1

2. In a perspective view, the point and direction from which a viewer looks into an image.

line simplification [CARTOGRAPHY] A generalization technique in which vertices are selectively removed from a line feature to eliminate detail while preserving the line's basic shape. *See also* simplification, generalization, line smoothing.

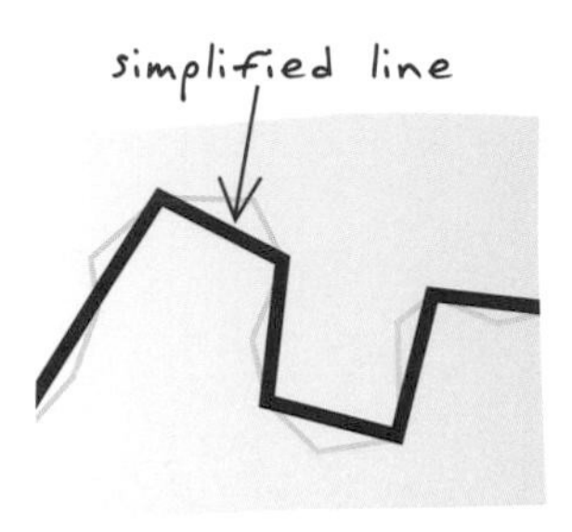

line simplification

line smoothing The process of adding extra points to lines to reduce the sharpness of angles between line segments, resulting in a smoother appearance. *See also* weeding.

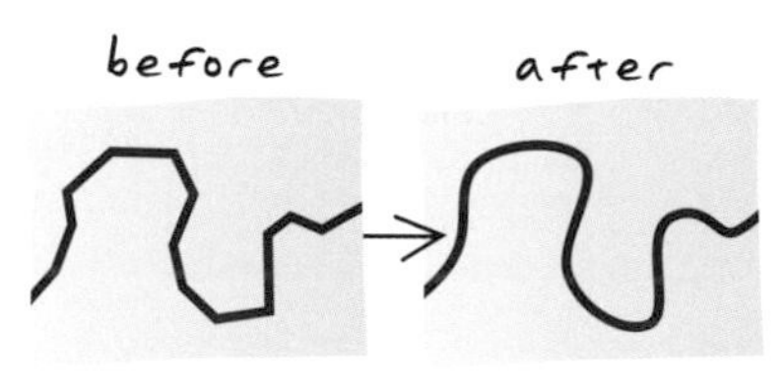

line smoothing

link 1. [REMOTE SENSING] In georeferencing, connections added between known points in a dataset being georeferenced and corresponding points in the dataset being used as a reference. 2. [COMPUTING] An operation that relates two tables using a common field, without altering either table. 3. [INTERNET] In a hyperlinked document, a graphic or piece of text that, when selected by a user, causes the display to move to another document or to another location within the same document. *See also* joining.

LIS *See* land information system.

little endian [COMPUTING] A computer hardware architecture in which, within a multibyte numeric representation, the least significant byte has the lowest address and the remaining bytes are encoded in increasing order of significance. *See also* big endian.

load balance [COMPUTING] The act of distributing application, network, and/or server resources to optimize performance.

load distribution *See* load balance.

local analysis The computation of an output raster where the output value at each location is a function of the input value at the same location. *See also* global analysis.

local area network *See* LAN.

local datum **1.** [GEODESY] A horizontal geodetic datum that serves as a basis for measurements over a limited area of the earth; that has its origin at a location on the earth's surface; that uses an ellipsoid whose dimensions conform well to its region of use; and that was originally defined for land-based surveys. A local datum in this sense stands in contrast to a geocentric datum. Examples include the North American Datum of 1927 and the Australian Geodetic Datum of 1966. **2.** [GEODESY] A horizontal or vertical datum used for measurements over a limited area of the earth, such as a nation, a supranational region, or a continent. A horizontal datum that is local in this sense may or may not be geocentric. For example, the North American Datum of 1983 and the Geocentric Datum of Australia 1994 are local in that they are applied to a particular part of the world; they are also geocentric. All vertical datums are local in that there is, at present, no global vertical datum. *See also* datum, horizontal geodetic datum, vertical geodetic datum, geocentric datum.

local functions *See* local analysis.

localization The process of adapting software to the requirements of a different language or culture, including translating user interfaces, documentation, and help systems; customizing features; and accommodating different character sets. *See also* internationalization.

location **1.** [GEOGRAPHY] An identifier assigned to a region or feature. **2.** A position defined by a coordinate value.

location-allocation The process of finding the best locations for one or more facilities that will service a given set of points and then assigning those points to the facilities, taking into account factors such as the number of facilities available, their cost, and the maximum impedance from a facility to a point. *See also* cost, impedance.

location-based services Information or a physical service delivered to multiple channels, exclusively based on the determined location of a wireless device. Some location-based applications include emergency services, information services, and tracking services.

location query *See* spatial query.

locomotion The movements of a person following a route. Locomotion is the physical component of navigation. *See also* wayfinding, navigation.

logarithm [MATHEMATICS] The power to which a fixed number (the base) must be raised to equal a given number. The three most frequently used bases for logarithms are base 10, base e, and base 2. *See also* exponent.

logical expression [MATHEMATICS] A string of numbers, constants, variables, operators, and functions that returns a value of true or false.

logical network [ESRI SOFTWARE] An abstract representation of a network, implemented as a collection of hidden tables. A logical network contains edge, junction, and turn elements, the connectivity between them, and the weights necessary for traversing the network. It does not contain information about the geometry or location of its elements; this information is one of the components of a network system. *See also* geometric network.

logical operator [MATHEMATICS] An operator used to compare logical expressions that returns a result of true or false. Examples of logical operators include less than (<), greater than (>), equal to (=), and not equal to (<>). *See also* operator, logical expression.

logical query The process of using mathematical expressions to select features from a geographic layer based on their attributes; for example, "select all polygons with an area greater than 16,000 units" or "select all street segments named Green Apple Run." *See also* query.

logical selection *See* logical query.

longitude [CARTOGRAPHY] The angular distance, usually expressed in degrees, minutes, and seconds, of the location of a point on the earth's surface east or west of an arbitrarily defined meridian (usually the Greenwich prime meridian). All lines of longitude are great circles that intersect the equator and pass through the North and South Poles. *See also* great circle, latitude.

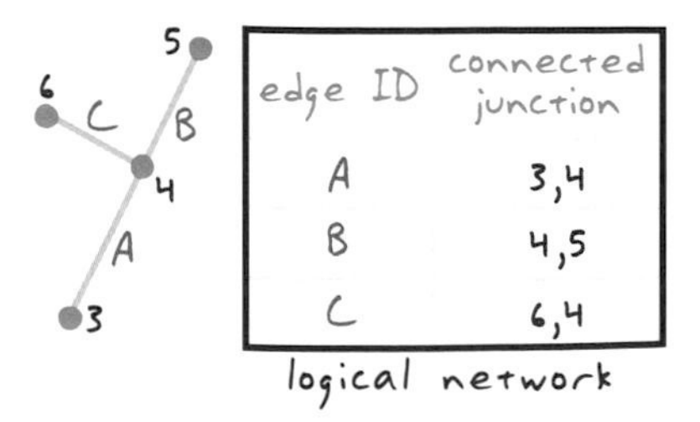

logical network

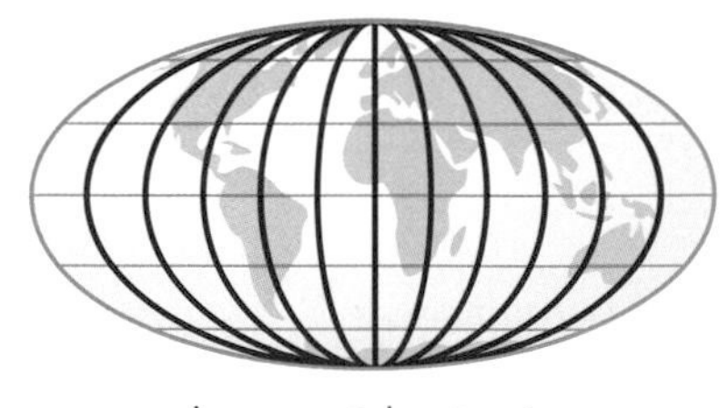

longitude

longitude of center [CARTOGRAPHY] The longitude value that defines the center, and sometimes the origin, of a projection.

longitude of origin [CARTOGRAPHY] The longitude value that defines the origin of the x-coordinate values for a projection. *See also* latitude of origin.

long-range variation [STATISTICS] In a spatial model, coarse-scale variation that is usually modeled as the trend. *See also* trend.

loop traverse *See* closed loop traverse.

loose coupling A relatively unstructured relationship between two software components or programs that work together to process data, which requires little overlap between methods, ontologies, class definitions, and so on. *See also* tight coupling.

lossless compression Data compression that has the ability to store data without changing any of the values, but is only able to compress the data at a low ratio (typically 2:1 or 3:1). In GIS, lossless compression is often used to compress raster data when the pixel values of the raster will be used for analysis or deriving other data products. *See also* lossy compression.

lossy compression Data compression that provides high compression ratios (for example 10:1 to 100:1), but does not retain all the information in the data. In GIS, lossy compression is used to compress raster datasets that will be used as background images, but is not suitable for raster datasets used for analysis or deriving other data products. *See also* lossless compression.

low-level language [PROGRAMMING] A programming language that uses keywords and statements that are little more complex than the ones and zeros of machine language. Low-level language technically includes machine language, but more commonly refers to an assembly language that uses symbols to make machine instructions easier for programmers to read and understand. Each statement in assembly language represents a single command to the processor, affording the developer only a low level of abstraction in regard to mundane functions such as memory access and register storage, meaning such operations demand the developer's close attention. *See also* high-level language.

low-pass filter [REMOTE SENSING] A spatial filter that blocks high-frequency (shortwave) radiation, resulting in a smoother image. *See also* band-pass filter, high-pass filter.

loxodrome *See* rhumb line.

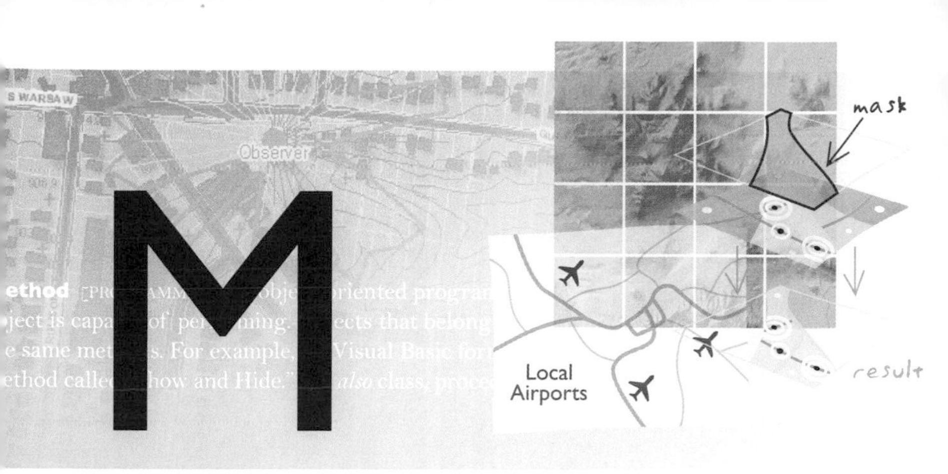

macro [PROGRAMMING] A computer program, usually a text file, containing a sequence of commands that are executed as a single command. Macros are used to perform commonly used sequences of commands or complex operations.

magnetic bearing [NAVIGATION] A bearing measured relative to magnetic north. *See also* bearing.

magnetic declination [GEODESY] The angle between magnetic north and true north observed from a point on the earth. Magnetic declination varies from place to place, and changes over time, in response to changes in the earth's magnetic field. *See also* declination.

magnetic north [GEOGRAPHY] The direction from a point on the earth's surface following a great circle toward the magnetic north pole, indicated by the north-seeking end of a compass. *See also* grid north, true north.

magnetometer [PHYSICS] An instrument used to measure variations in the strength and direction of the earth's magnetic field.

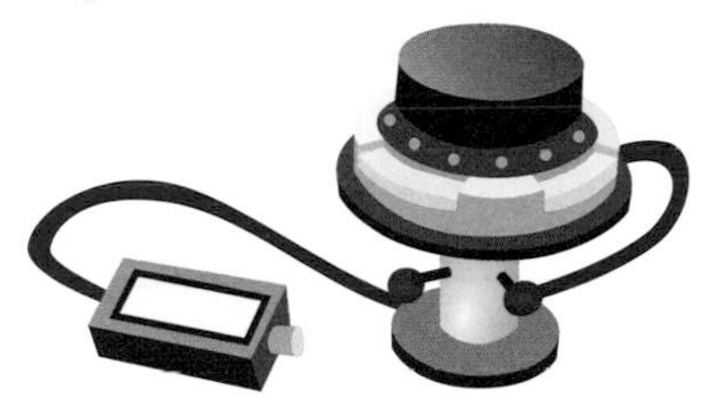

magnetometer

major axis [MATHEMATICS] The longer axis of an ellipse or spheroid. *See also* minor axis.

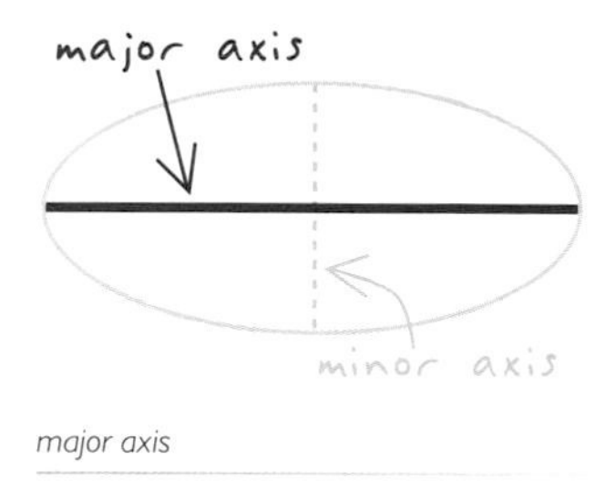

major axis

majority resampling [STATISTICS] A technique for resampling raster data in which the value of each cell in an output is calculated, most commonly using a 2 × 2 neighborhood of the input raster. Majority resampling does not create any new cell values, so it is useful for resampling categorical or integer data, such as land use, soil, or forest type. Majority resampling acts as a type of low-pass filter for discrete data, generalizing the data and filtering out anomalous data values. *See also* resampling, low-pass filter.

many-to-many relationship An association between two linked or joined tables in which one record in the first table may correspond to many records in the second table, and vice versa.

many-to-one relationship An association between two linked or joined tables in which many records in the first table may correspond to a single record in the second table. *See also* many-to-many relationship, one-to-many relationship, one-to-one relationship, joining.

map 1. [CARTOGRAPHY] A graphic representation of the spatial relationships of entities within an area.

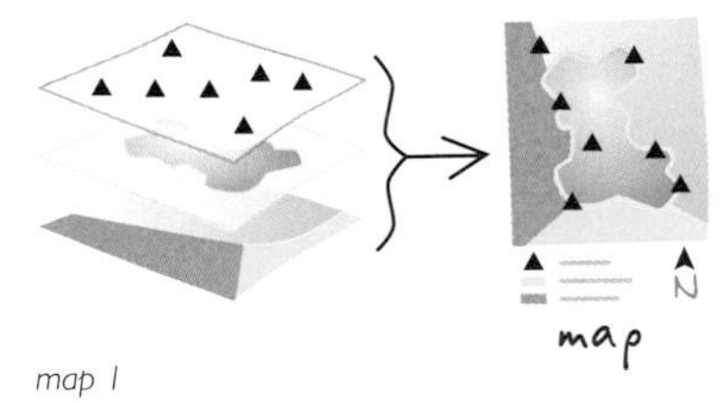

map 1

2. [CARTOGRAPHY] Any graphical representation of geographic or spatial information.

map algebra A language that defines a syntax for combining map themes by applying mathematical operations and analytical functions to create new map themes. In a map algebra expression, the operators are a combination of mathematical, logical, or Boolean operators (+, >, AND, tan, and so on), and spatial analysis functions (slope, shortest path, spline, and so on), and the operands are spatial data and numbers.

map collar *See* map surround.

map display [CARTOGRAPHY] A graphic representation of a map on a computer screen.

map document [ESRI SOFTWARE] In ArcMap, the file that contains one map, its layout, and its associated layers, tables, charts, and reports. Map documents can be printed or embedded in other documents. Map document files have a .mxd extension.

map element [CARTOGRAPHY] In digital cartography, a distinctly identifiable graphic or object in the map or page layout. For example, a map element can be a title, scale bar, legend, or other map-surround element. The map area itself can be considered a map element; or an object within the map can be referred to as a map element, such as a roads layer or a school symbol. *See also* map surround.

map extent [CARTOGRAPHY] The limit of the geographic area shown on a map, usually defined by a rectangle. In a dynamic map display, the map extent can be changed by zooming and panning. *See also* extent.

map feature *See* feature.

map generalization [CARTOGRAPHY] Decreasing the level of detail on a map so that it remains uncluttered when its scale is reduced.

map projection *See* projection.

map query *See* spatial query.

map reading [CARTOGRAPHY] The activity of viewing a map in a way that allows the viewer to make sense of or gain information from it. Map reading involves interpreting the meanings of codes and cartographic representations used on the map.

map scale *See* scale.

map series [CARTOGRAPHY] A collection of maps usually addressing a particular theme.

map service [INTERNET] A type of Web service that generates maps. *See also* Web service.

map sheet [CARTOGRAPHY] A single map or chart in a map series, such as any one of the approximately 57,000 USGS 7.5-minute topographic maps of the United States and its territories. *See also* map.

map style *See* style.

map surround [CARTOGRAPHY] Any of the supporting objects or elements that help a reader interpret a map. Typical map surround elements include the title, legend, north arrow, scale bar, border, source information and other text, and inset maps. *See also* map element.

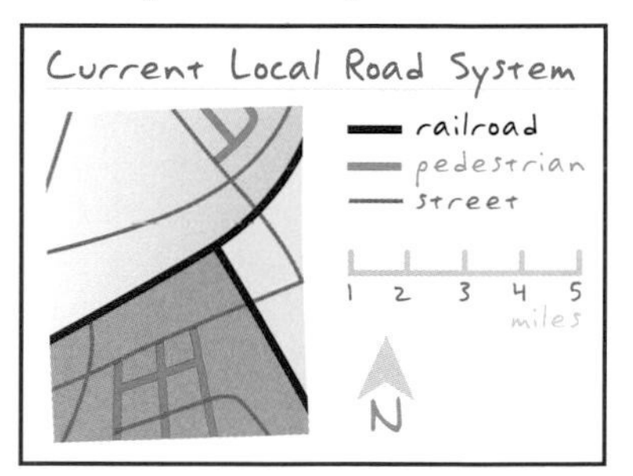

map surround

map unit The ground unit of measurement—for example, feet, miles, meters, or kilometers—in which coordinates of spatial data are stored. *See also* page unit.

marginalia *See* map surround.

marker symbol [CARTOGRAPHY] A symbol used to represent a point location on a map. *See also* fill symbol.

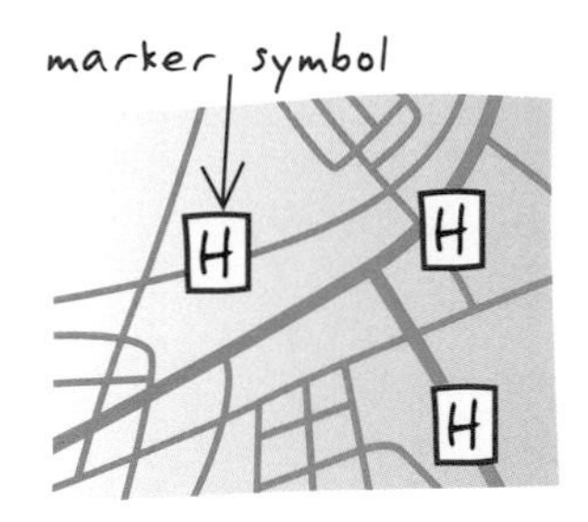

marker symbol

market area A geographic zone containing the people who are likely to purchase a firm's goods or services.

M

market penetration analysis A process that determines the percentage of a market area being reached based on the number of customers within an area divided by the total population in that area.

mask 1. [CARTOGRAPHY] In digital cartography, a means of covering or hiding features on a map to enhance cartographic representation. For example, masking is often used to cover features behind text to make the text more readable. 2. [ESRI SOFTWARE] In ArcGIS, a means of identifying areas to be included in analysis. Such a mask is often referred to as an analysis mask, and may be either a raster or feature layer.

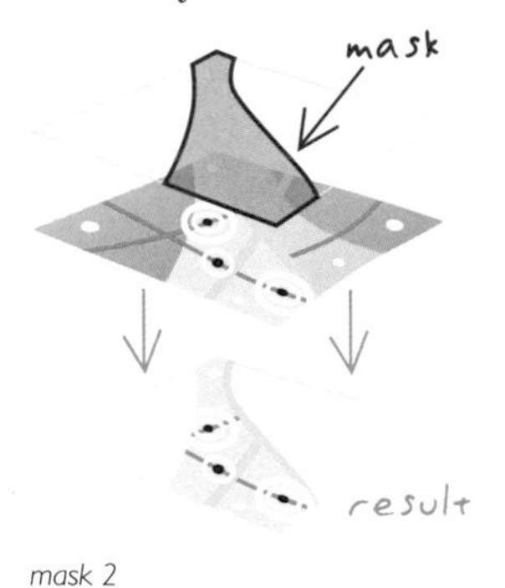

mask 2

mass point [STATISTICS] An irregularly distributed sample point, with an x-, y-, and z-value, used to build a triangulated irregular network (TIN). Ideally, mass points are chosen to capture the more important variations in the shape of the surface being modeled. *See also* TIN.

matching In geocoding, the process of linking a record, such as an address, to a set of reference data. The matched record in the reference data is used to determine the location of the input address. *See also* geocoding, reference data.

mathematical operator *See* operator.

matrix [MATHEMATICS] A rectangular arrangement of data, usually numbers, in rows and columns. In computer science, a two-dimensional array is called a matrix. In GIS, matrices are used to store raster data. *See also* array, raster.

MAUP *Acronym for modifiable areal unit problem.* A challenge that occurs during the spatial analysis of aggregated data in which the results differ when the same analysis is applied to the same data, but different aggregation schemes are used. MAUP takes two forms: the scale effect and the zone effect. The scale effect exhibits different results when the same analysis is applied to the same data, but changes the scale of the aggregation units. For example, analysis using data aggregated by county will differ from analysis using data aggregated by census tract. Often this difference in results is valid: each analysis asks a different question because each evaluates the data from a different perspective (different scale). The zone effect is observed when the scale of analysis is fixed, but the shape of the aggregation units is changed. For example, analysis using data aggregated into one-mile grid cells will differ from analysis using one-mile hexagon cells. The zone effect is a problem because it is an analysis, at least in part, of the aggregation scheme rather than the data itself.

m-coordinate *See* m-value.

mean [MATHEMATICS] The average for a set of values, computed as the sum of all values divided by the number of values in the set. *See also* median.

mean center [STATISTICS] The location of a single x,y coordinate value that represents the average x-coordinate value and the average y-coordinate value of all features in a study area.

mean sea level [GEODESY] The average height of the surface of the sea for all stages of the tide over a nineteen-year period, usually determined by averaging hourly height readings from a fixed level of reference.

mean stationarity [STATISTICS] In geostatistics, a property of a spatial process in which a spatial random variable has the same mean value at all locations. *See also* stationarity.

measure *See* route measure.

measured grid *See* grid.

measurement An observed numerical value that is an appraisal of size, extent, or amount according to a set criteria.

measurement residual [SURVEYING] The difference between a measured quantity and its theoretical true value as determined during each iteration of a least-squares adjustment. *See also* least-squares adjustment.

measure value *See* m-value.

median [MATHEMATICS] The middle value of a set of values when they are ordered by rank. Half the values in a set are higher than the median, and half are lower. When there are two middle values (if the set has an even number of elements) the median is the mean of these two values. *See also* mean.

median center **1.** [STATISTICS] A location representing the shortest total distance to all other features in a study area. **2.** [Statistics] The location of a single x,y coordinate value that represents the median x-coordinate value and the median y-coordinate value for all features in a study area. *See also* median, mean, mean center.

medium-format printer A printing device capable of producing an image on paper or other media sized between 15 and 35 inches (38 and 90 centimeters) wide. Medium-format printers typically use inkjet printing technology to print an image on a roll or sheet-fed media. While most large-format printers are large, free-standing units, most medium-format printers are small enough to fit on a desk. *See also* large-format printer, plotter.

memory leak [COMPUTING] In computer programming, the loss of computer memory that occurs when an application or component fails to free a section of computer memory when it has finished using it. During a memory leak, the section of memory allocated by one application or component may not be used by any other application.

mental map A person's perception of a place. A mental map may include the physical characteristics of a place, such as boundaries of a neighborhood, or the attributes of a place, such as a neighborhood's perceived unsafe areas. A mental map is primarily a psychological construct, although it may also be rendered as an actual map.

menu [COMPUTING] A list of available commands or operations displayed on a computer screen from which a user can make a selection.

M

mereing [SURVEYING] Establishing a boundary relative to ground features present at the time of a survey.

merging Combining features from multiple data sources of the same data type into a single, new dataset. *See also* appending.

meridian [GEODESY] A great circle on the earth that passes through the poles, often used synonymously with longitude. Meridians run north–south between the poles. By convention, meridians are labeled with positive numbers that ascend as one moves eastward from the prime meridian, and negative numbers as one moves westward from the prime meridian until the east and west hemispheres meet at the 180-degree line. Meridians can also, however, be labeled with all positive or negative numbers, including positive numbers increasing westward from the prime meridian. *See also* longitude, Greenwich meridian, prime meridian, central meridian.

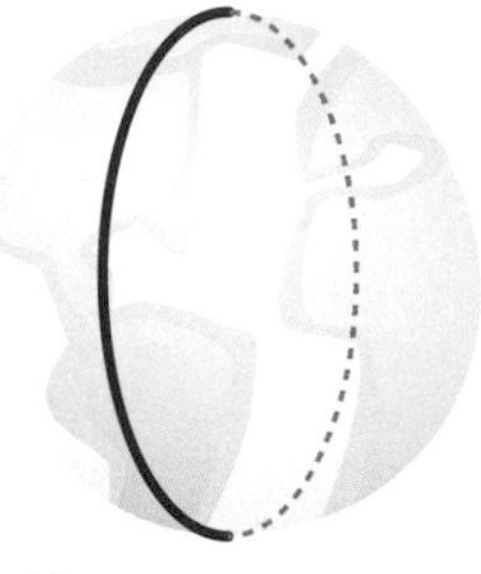

meridian

metadata Information that describes the content, quality, condition, origin, and other characteristics of data or other pieces of information. Metadata for spatial data may describe and document its subject matter; how, when, where, and by whom the data was collected; availability and distribution information; its projection, scale, resolution, and accuracy; and its reliability with regard to some standard. Metadata consists of properties and documentation. Properties are derived from the data source (for example, the coordinate system and projection of the data), while documentation is entered by a person (for example, keywords used to describe the data).

metadata element A unit of information within metadata, used to describe a particular characteristic of the data.

metadata profile A modification of an existing metadata standard to adapt to data issues, cultural issues, or both. A profile is typically a subset of a base standard that tailors the metadata elements in the base standard to better describe the data to the community that uses it. Metadata profiles allow communities to follow a metadata standard, while at the same time enhancing the standard so that it is more appropriate for a particular use or locale. *See also* metadata.

metes and bounds [SURVEYING] A surveying method in which the limits of a parcel are identified as relative distances and bearings from landmarks. Metes and bounds surveying often resulted in irregularly shaped areas.

method [PROGRAMMING] In object-oriented programming, an action that an object is capable of performing. Objects that belong to the same class all have the same methods. For example, all Visual Basic form objects can execute a method called "Show and Hide." *See also* class, procedure.

metropolitan statistical area *See* MSA.

microdensitometer [CARTOGRAPHY] A densitometer that can read densities in minute areas, used particularly for studying spectroscopic and astronomical images. *See also* densitometer.

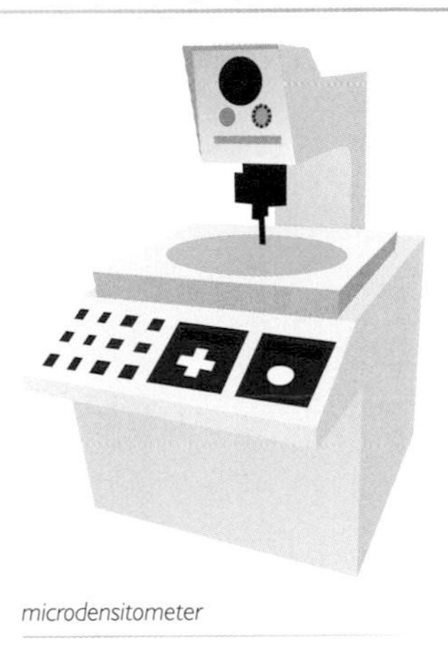

microdensitometer

micrometer [PHYSICS] An instrument for measuring minute lengths or angles.

micron [PHYSICS] One millionth of a meter, represented by the symbol µm. Microns are used to measure wavelengths in the electromagnetic spectrum.

mimetic symbol [CARTOGRAPHY] A symbol that imitates or closely resembles the thing it represents, such as an icon of a picnic table that represents a picnic area. *See also* arbitrary symbol, symbol.

mimetic symbol

M

minimum bounding rectangle A rectangle, oriented to the x- and y-axes, that bounds a geographic feature or a geographic dataset. It is specified by two coordinate pairs: xmin, ymin and xmax, ymax. *See also* extent.

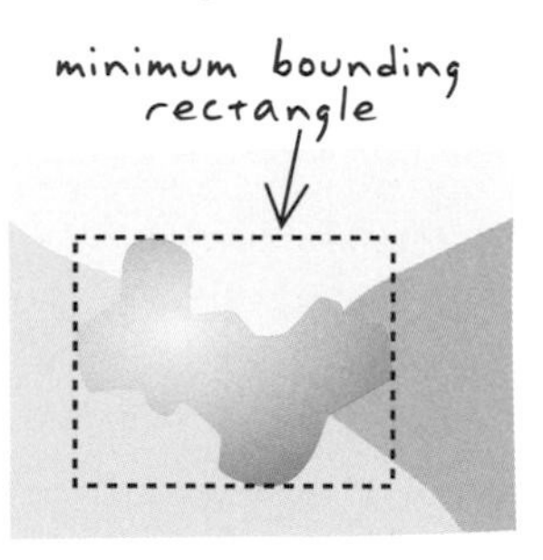

minimum bounding rectangle

minimum map unit [CARTOGRAPHY] For a given scale, the size in map units below which a narrow feature can be reasonably represented by a line and an area by a point. *See also* map unit.

minor axis [MATHEMATICS] The shorter axis of an ellipse or spheroid. *See also* major axis.

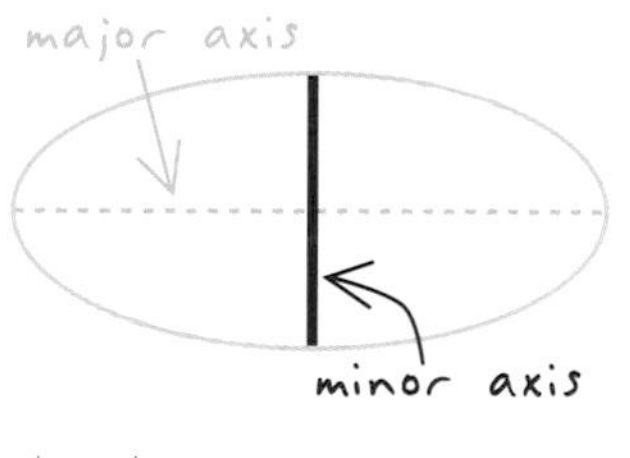

minor axis

minute 1. [GEODESY] An angle equal to 1/60 of a degree of latitude or longitude and containing sixty seconds. 2. [MATHEMATICS] An angle equal to 1/60 of a degree of arc. *See also* second.

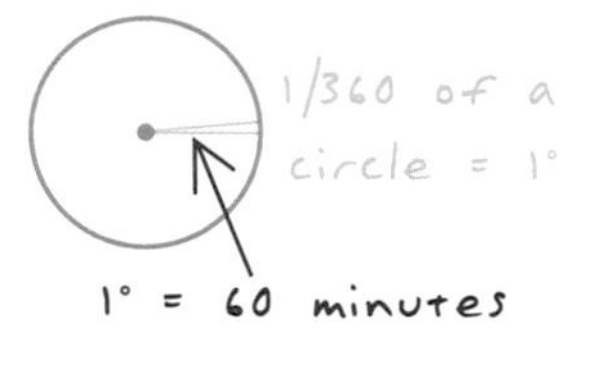

minute 2

misclosure *See* closure error.

mixed pixel [REMOTE SENSING] A pixel whose digital number represents the average of several spectral classes within the area that it covers on the ground, each emitted or reflected by a different type of material. Mixed pixels are common along the edges of features.

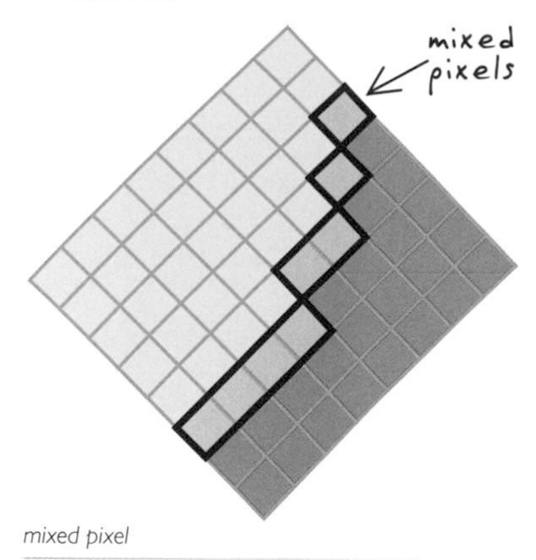

mixed pixel

model 1. An abstraction of reality used to represent objects, processes, or events. 2. A set of rules and procedures for representing a phenomenon or predicting an outcome. 3. A data representation of reality, such as the vector data model.

modifiable areal unit problem *See* MAUP.

MODIS [REMOTE SENSING] *Acronym for moderate resolution imaging spectroradiometer.* A bundle of remote-sensing equipment housed on two NASA (National Aeronautics and Space Administration) satellites, Terra and Aqua, in orbit around Earth. These two MODIS-equipped satellites constantly record multiple images of the globe in various wavelengths and resolutions, imaging the earth's entire surface in less than two days.

monochromatic 1. [PHYSICS] Related to a single wavelength or a very narrow band of wavelengths. 2. A color scheme made up of lighter and darker shades of the same color.

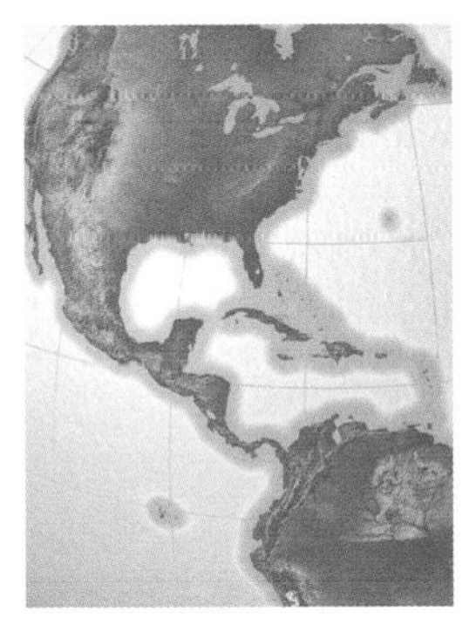

monochromatic 2

Monte Carlo method An algorithm for computing solutions to problems that contain a large number of variables by performing iterations with different sets of random numbers until the best solution is found. The Monte Carlo method is usually applied to problems too complex for analysis by anything but a computer. *See also* stochastic model.

monument *See* survey monument.

morphology The study of structure or form.

mosaic 1. [REMOTE SENSING] A raster dataset composed of two or more merged raster datasets—for example, one image created by merging several individual images or photographs of adjacent areas.

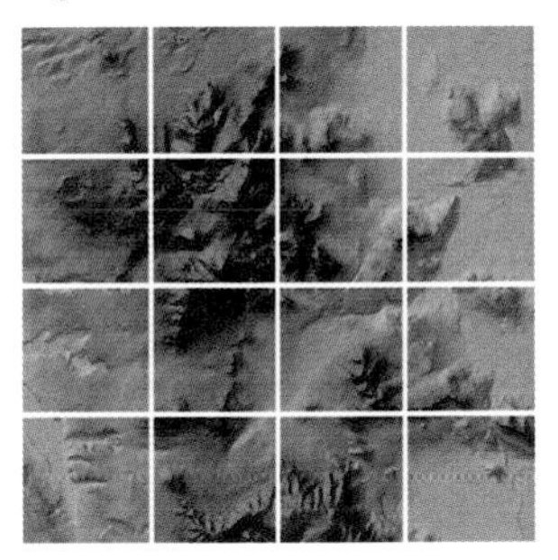

mosaic 1

2. Maps of adjacent areas with the same spatial reference and scale whose boundaries have been matched and dissolved. *See also* edgematching.

mouse mode A way of using a digitizing tablet in which the digitizer puck behaves like a mouse; the puck is used to point to interface elements rather than to trace shapes on the surface of the digitizing tablet. *See also* digitizing mode.

MSA *Acronym for metropolitan statistical area.* A geographic entity defined by the U.S. Office of

Management and Budget for use by federal statistical agencies, including the U.S. Census Bureau. An MSA is based on the concept of a core area with a large population nucleus, plus adjacent communities having a high degree of economic and social integration with that core area. According to the 1990 standards, to qualify as an MSA the area must include at least one city or urbanized area with 50,000 or more inhabitants and a total metropolitan population of at least 100,000 (75,000 in New England). The county or counties containing the largest city and surrounding densely settled territory are central counties of the MSA. Additional outlying counties qualify to be included in the MSA if they meet specified requirements. MSAs in New England are defined in terms of cities and towns rather than counties.

MSS *See* multispectral scanner.

multichannel receiver [REMOTE SENSING] A receiver that tracks several satellites at a time, using one channel for each satellite. *See also* multiplexing channel receiver.

multidimensional data Data that is comprised of multiple dimensions, such as space and time. For example, a temperature dataset could have dimensions of latitude, longitude, altitude, and time. *See also* dimension.

multipart feature [ESRI SOFTWARE] In ArcGIS, a digital representation of a place or thing that has more than one part but is defined as one feature because it references one set of attributes. In a layer of states, for example, the state of Hawaii could be considered a multipart feature because its separate geometric parts are classified as a single state. A multipart feature can be a point, line, or polygon. *See also* feature.

multipart feature

multipatch feature [ESRI SOFTWARE] In ArcGIS, a real-world geographic feature modeled using multipatch geometry.

multipatch feature

multipath error [REMOTE SENSING] Errors caused when a satellite signal reaches the receiver from two or more paths, one directly from the satellite and the others reflected from nearby buildings or other surfaces. Signals from satellites low on the horizon will produce more error.

multiple regression [STATISTICS] Regression in which the dependent variable is measured against two or more independent variables.

multiplexing channel receiver [REMOTE SENSING] A receiver that tracks several satellite signals using a single channel. *See also* multichannel receiver.

multipoint feature [ESRI SOFTWARE] In ArcGIS, a digital map feature that represents a place or thing that has neither area nor length at a given scale, and that is treated as a single object with multiple locations. For example, the entrances and exits to a prairie dog den might be represented as a multipoint feature. A multipoint feature is associated with a single record in an attribute table. *See also* feature.

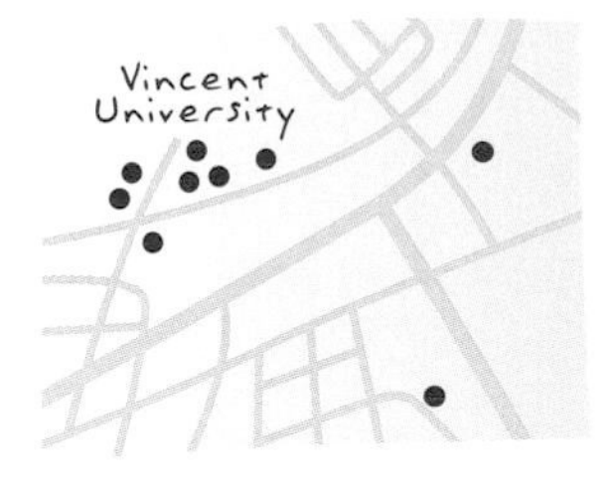

multipoint feature

multispectral [PHYSICS] Related to two or more frequencies or wavelengths in the electromagnetic spectrum.

multispectral image [REMOTE SENSING] An image created from several narrow spectral bands. *See also* multispectral, color composite.

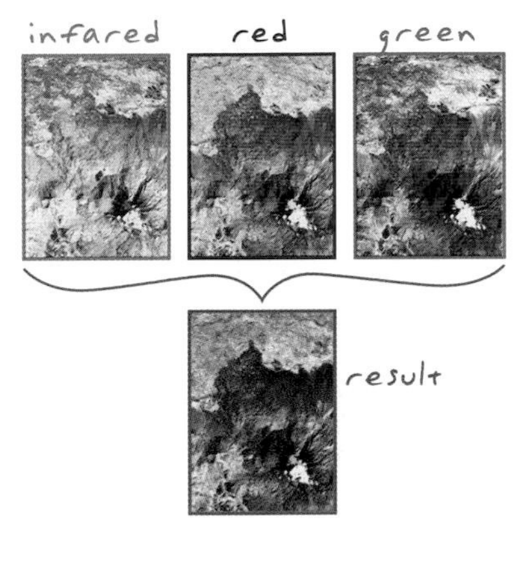

multispectral image

multispectral scanner [REMOTE SENSING] A device carried on satellites and aircraft that records energy from multiple portions of the electromagnetic spectrum. *See also* electromagnetic spectrum.

multivariate analysis [STATISTICS] Any statistical method for evaluating the relationship between two or more variables.

m-value In linear referencing, a measure value that is added to a line feature. M-values are used to measure the distance along a line feature from a vertex (a known location) to an event.

MXD *See* map document.

M

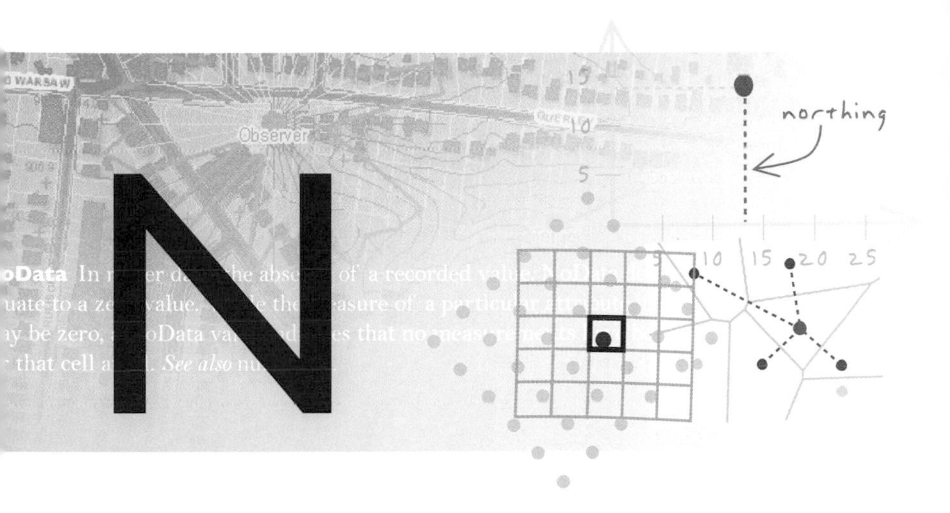

NAD 1927 [CARTOGRAPHY] *Acronym for North American Datum of 1927.* The primary local horizontal geodetic datum and geographic coordinate system used to map the United States during the middle part of the twentieth century. NAD 1927 is referenced to the Clarke spheroid of 1866 and an origin point at Meades Ranch, Kansas. Features on USGS topographic maps, including the corners of 7.5-minute quadrangle maps, are referenced to NAD27. It is gradually being replaced by the North American Datum of 1983. *See also* NAD 1983, Clarke ellipsoid of 1866.

NAD 1983 [CARTOGRAPHY] *Acronym for North American Datum of 1983.* A geocentric datum and graphic coordinate system based on the Geodetic Reference System 1980 ellipsoid (GRS80). Mainly used in North America, its measurements are obtained from both terrestrial and satellite data. *See also* NAD 1927, geocentric datum, GRS80.

nadir 1. [REMOTE SENSING] In aerial photography, the point on the ground vertically beneath the perspective center of the camera lens.

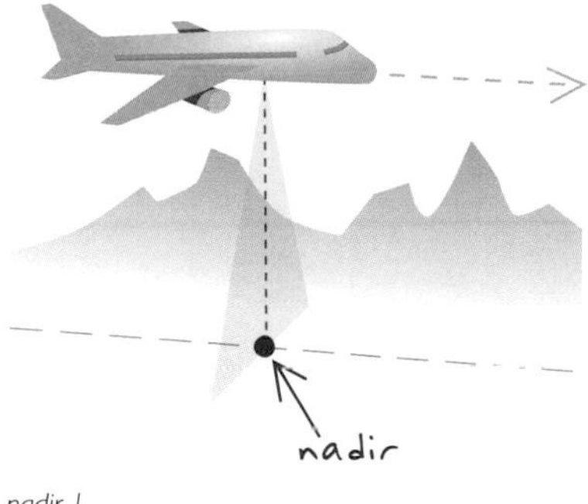

nadir 1

2. [ASTRONOMY] The point on the celestial sphere directly beneath an observer. Both the nadir and zenith lie on the observer's meridian; the nadir lies 180 degrees from the zenith and is therefore unobservable. *See also* zenith, meridian, celestial sphere.

NAICS *Acronym for North American Industry Classification System.* A system for classifying individual business locations by their types of economic activity. The statistics agencies of Canada, Mexico, and the United States collaborated on NAICS to standardize the industry statistics produced by the three countries. NAICS is used as an identification system by all federal statistical agencies, as well as many state and local agencies, trade associations, private businesses, and other organizations. NAICS replaced Standard Industrial Classification (SIC) codes in 1997. *See also* Standard Industrial Classification codes.

National Geodetic Vertical Datum of 1929 [CARTOGRAPHY] The datum established in 1929 by the U.S. Coast and Geodetic Survey as the surface against which elevation data in the United States is referenced.

National Spatial Data Infrastructure *See* NSDI.

N

natural breaks classification [CARTOGRAPHY] A method of manual data classification that seeks to partition data into classes based on natural groups in the data distribution. Natural breaks occur in the histogram at the low points of valleys. Breaks are assigned in the order of the size of the valleys, with the largest valley being assigned the first natural break. *See also* classification, Jenks' optimization.

natural neighbors [MATHEMATICS] An interpolation method for multivariate data in a Delaunay triangulation. The value for an interpolation point is estimated using weighted values of the closest surrounding points in the triangulation. These points, the natural neighbors, are the ones the interpolation point would connect to if inserted into the triangulation. *See also* interpolation, triangulation.

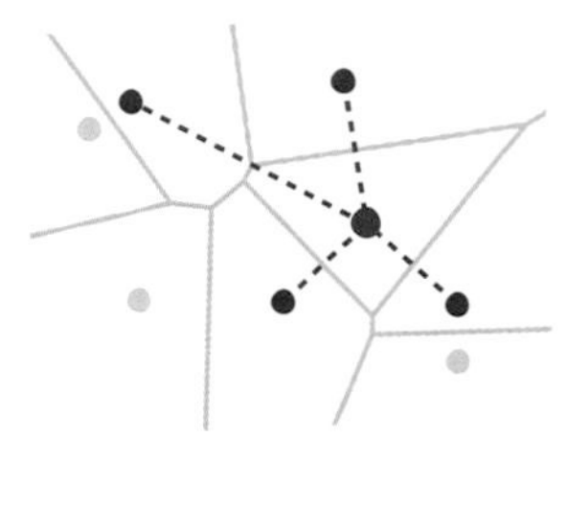

natural neighbors

navigation 1. The combined mental and physical activities involved in traveling to a destination, often a distant or unfamiliar one. Navigation comprises wayfinding and locomotion. 2. [GEOGRAPHY] The activity of guiding a ship, plane, or other vehicle to a destination, along a planned or improvised route, according to reliable methods. *See also* wayfinding, locomotion.

Navstar [GPS] The name of the U.S. Department of Defense's Global Positioning System (GPS). *See also* GPS.

nearest neighbor resampling [MATHEMATICS] A technique for resampling raster data in which the value of each cell in an output raster is calculated using the value of the nearest cell in an input raster. Nearest neighbor assignment does not change any of the values of cells from the input layer; for this reason it is often used to resample categorical or integer data (for example, land use, soil, or forest type), or radiometric values, such as those from remotely sensed images. *See also* resampling, bilinear interpolation, cubic convolution, neighborhood functions

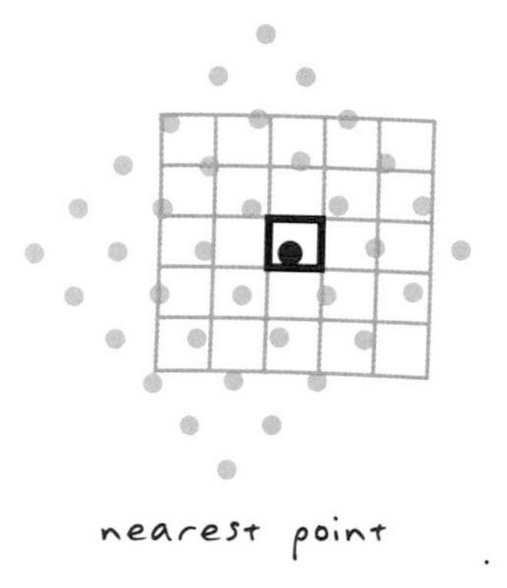

nearest neighbor resampling

neatline [CARTOGRAPHY] The border delineating and defining the extent of geographic data on a map. It demarcates map units so that, depending on the map projection, the neatline does not always have 90-degree corners. In a properly made map, it is the most accurate element of the data; other map features may be moved slightly or exaggerated for generalization or readability, but the neatline is never adjusted.

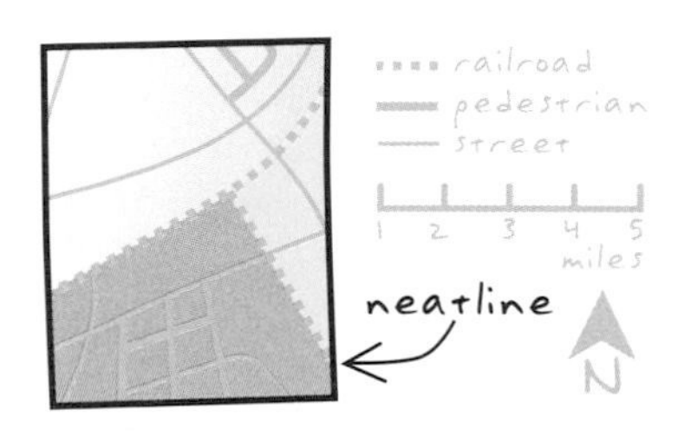

neatline

neighborhood *See* filter.

neighborhood functions [MATHEMATICS] Methods of defining new values for locations using the values of other locations within a given distance or direction. *See also* proximity analysis, nearest neighbor resampling, natural neighbors.

neighborhood statistics *See* focal analysis.

network 1. An interconnected set of points and lines that represent possible routes from one location to another. For geometric networks, this consists of edge features, junction features, and the connectivity between them. For network datasets, this consists of edge, junction, and turn elements and the connectivity between them. For example, an interconnected set of lines representing a city streets layer is a network. 2. [COMPUTING] A group of computers connected in order to share software, data, and peripheral devices, as in a local or wide area network. *See also* geometric network, logical network, network dataset.

network analysis Any method of solving network problems such as traversability, rate of flow, or capacity, using network connectivity. *See also* network.

network dataset [ESRI SOFTWARE] A collection of topologically connected network elements (edges, junctions, and turns) that are derived from network sources, typically used to represent a linear network, such as a

N

road or subway system. Each network element is associated with a collection of network attributes. Network datasets are typically used to model undirected flow systems.

network dataset

network location A geographic position in a network system.

network node [ESRI SOFTWARE] A connecting point in a geometric network, such as an intersection or interchange of a road network, confluence of streams in a hydrologic network, or switch in a power grid.

network trace A function that performs network analysis on a geometric network. Specific kinds of network tracing include finding features that are connected, finding common ancestors, finding loops, tracing upstream, and tracing downstream. *See also* geometric network.

neural network [COMPUTING] A computer architecture modeled after the human brain and designed to solve problems that human brains solve well, such as recognizing patterns and making predictions from past performance. Neural networks are composed of interconnected computer processors that calculate a number of weighted inputs to generate an output. For example, an output might be the approval or rejection of a credit application. This output would be based on several inputs, including the applicant's income, current debt, and credit history. Some of these inputs would count more than others; cumulatively, they would be compared to a threshold value that separates approvals from rejections. Neural networks "learn" to generate better outputs by adjusting the weights and thresholds applied to their inputs.

NGVD 1929 *See* National Geodetic Vertical Datum of 1929.

NGVD29 *See* National Geodetic Vertical Datum of 1929.

NoData In raster data, the absence of a recorded value. NoData does not equate to a zero value. While the measure of a particular attribute in a cell may be zero, a NoData value indicates that no measurements have been taken for that cell at all. *See also* null value.

node **1.** [ESRI SOFTWARE] In a geodatabase, the point representing the beginning or ending point of an edge, topologically linked to all the edges that meet there. **2.** [ESRI SOFTWARE] In a coverage, the beginning or ending point of an arc, topologically linked to all the arcs that meet there. **3.** In a TIN, one of the three corner points of a triangle, topologically linked to all triangles that meet there. Each sample point in a TIN becomes a node in the triangulation that may store elevation z-values and tag values.

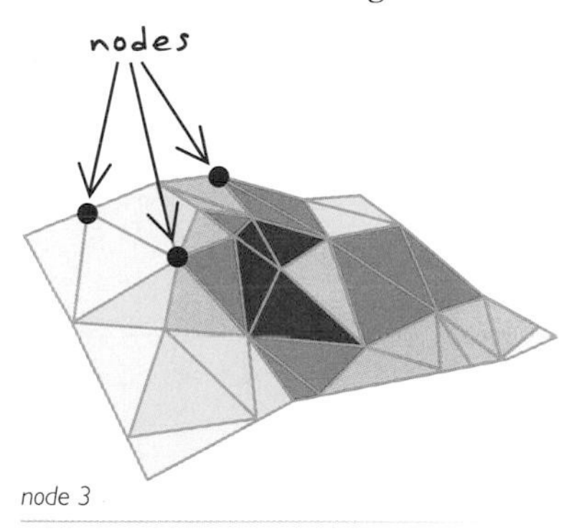

node 3

4. [MATHEMATICS] In graph theory, any vertex in a graph. **5.** [COMPUTING] The point at which a computer, or other addressable device, attaches to a communications network.

noise **1.** [REMOTE SENSING] Any disturbance in a frequency band. **2.** Any irregular, sporadic, or random oscillation in a transmission signal. **3.** Random or repetitive events that interfere with communication. **4.** In a raster, irrelevant or meaningless cells that exist due to poor scanning or imperfections in the original source document.

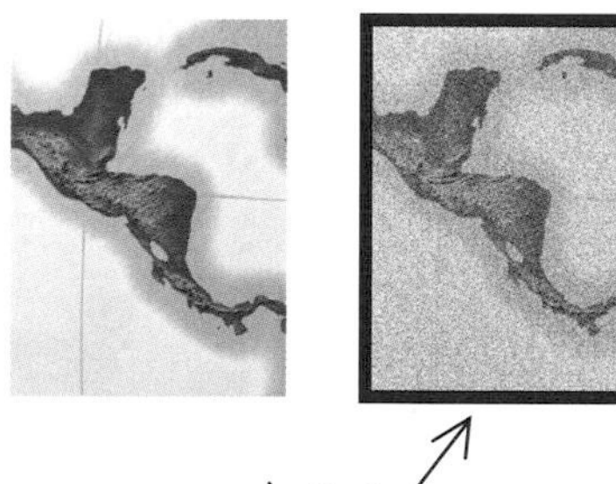

noise 4

nominal data Data divided into classes within which all elements are assumed to be equal to each other, and in which no class comes before another in sequence or importance; for example, a group of polygons colored to represent different soil types. *See also* ordinal data.

nonsimple polygon A polygon that violates topological integrity by crossing its own boundary (usually by making a small loop).

nonsimple polygon

nonsimple polygon

nonspatial data Data without inherently spatial qualities, such as attributes. *See also* attribute data.

normal distribution [STATISTICS] A theoretical frequency distribution of a dataset in which the distribution of values can be graphically represented as a symmetrical bell curve. Normal distributions are typically characterized by a clustering of values near the mean, with few values departing radically from the mean. There are as many values on the left side of the curve as on the right, so the mean and median values for the distribution are the same. Sixty-eight percent of the values are plus or minus one standard

N

deviation from the mean; 95 percent of the values are plus or minus two standard deviations; and 99 percent of the values are plus or minus three standard deviations. *See also* standard deviation.

normal form A set of guidelines for designing table and data structures in a relational database. When followed, normal form guidelines prevent data redundancy, increase database efficiency, and reduce consistency errors. A database is said to be in first normal form (1NF), second normal form (2NF), third normal form (3NF), and so on, depending on the level of normal form guidelines followed in its design. In practice, 3NF is commonly used, but higher levels are rarely used. *See also* first normal form, second normal form, third normal form.

normalization 1. The process of organizing, analyzing, and cleaning data to increase efficiency for data use and sharing. Normalization usually includes data structuring and refinement, redundancy and error elimination, and standardization.
2. [STATISTICS] The process of dividing one numeric attribute value by another to minimize differences in values based on the size of areas or the number of features in each area. For example, normalizing (dividing) total population by total area yields population per unit area, or density.

normal probability distribution *See* normal distribution.

North American Datum of 1927 *See* NAD 1927.

North American Datum of 1983 *See* NAD 1983.

North American Industry Classification System *See* NAICS.

north arrow [CARTOGRAPHY] A map symbol that shows the direction of north on the map, thereby showing how the map is oriented.

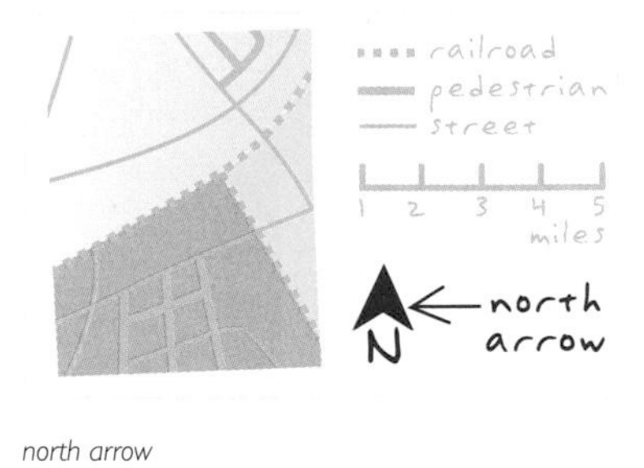

north arrow

northing 1. [CARTOGRAPHY] The distance north of the origin that a point in a Cartesian coordinate system lies, measured in that system's units.

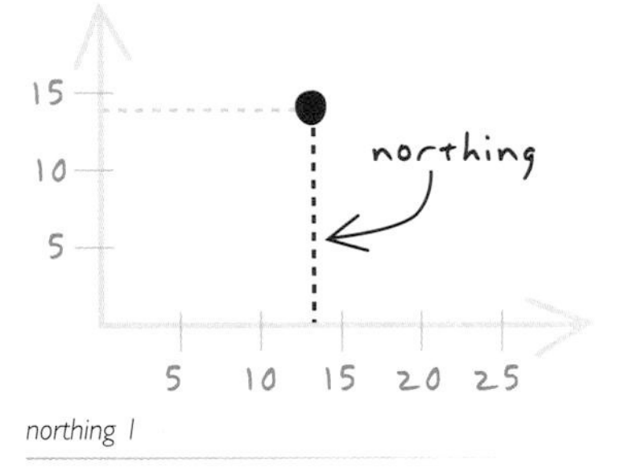

northing 1

2. [CARTOGRAPHY] The positive y-value in a rectangular coordinate system. *See also* easting.

NSDI *Acronym for National Spatial Data Infrastructure.* A federally mandated framework of spatial data that refers to U.S. locations, as well as the means of distributing and using that data effectively. Developed and coordinated by the FGDC, the NSDI encompasses policies, standards, procedures, technology, and human resources for organizations to cooperatively produce and share geographic data. The NSDI is developed by the federal governments; state, local, and tribal governments; the academic community; and the private sector. *See also* FGDC.

NSDI Clearinghouse Network A community of digital spatial data providers that maintain NSDI Clearinghouse Nodes as part of the U.S. National Spatial Data Infrastructure.

NSDI Clearinghouse Node An Internet server that hosts a collection of metadata and data maintained and stored on a computer server by a data provider. An NSDI Clearinghouse Node provides information about geographic data within the data provider's areas of responsibility. Nodes must host FGDC-compliant metadata and data and use a common access protocol. *See also* NSDI Clearinghouse Network.

nugget [STATISTICS] A parameter of a covariance or semivariogram model that represents independent error, measurement error, or microscale variation at spatial scales that are too fine to detect. The nugget effect is seen as a discontinuity at the origin of either the covariance or semivariogram model. *See also* random noise.

null hypothesis [STATISTICS] A statement that essentially outlines an expected outcome when there is no pattern, no relationship, and/or no systematic cause or process at work; any observed differences are the result of random chance alone. The null hypothesis for a spatial pattern is typically that the features are randomly distributed across the study area. Significance tests help determine whether the null hypothesis should be accepted or rejected.

null value [STATISTICS] The absence of a recorded value for a geographic feature. A null value differs from a value of zero in that zero may represent the measure of an attribute, while a null value indicates that no measurement has been taken. *See also* NoData.

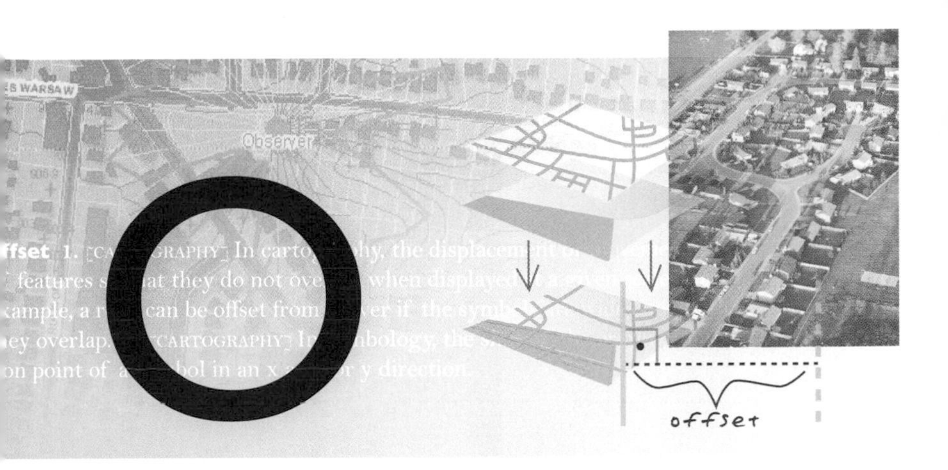

object **1.** In GIS, a digital representation of a spatial or nonspatial entity. Objects usually belong to a class of objects with common attribute values and behaviors. **2.** [PROGRAMMING] In object-oriented programming, an instance of the data structure and behavior defined by a class. **3.** [COMPUTING] A piece of software that performs a specific task and is controlled by another piece of software, called a client. For example, an object is often the interface by which an application program accesses an operating system and other services.

object class A table in a geodatabase used to store a collection of objects with similar attributes and behavior. Objects with no locational information are stored as rows or records in object classes. Spatial objects, or features, are stored as rows in feature classes, which are a specialized type of object class in which objects have an extra attribute to define their geographic location. *See also* feature class.

object-oriented database A data management structure that stores data as objects (instances of a class) instead of as rows and tables as in a relational database. *See also* RDBMS.

object-oriented programming [PROGRAMMING] A programming model in which related tasks, properties, and data structures are enclosed inside an object, and work is done when objects make requests and receive results from other objects. For example, a billing program may contain an object that maintains customer records. That object may pass information to another object that handles mailing statements, and another object that handles customer payments may ask it to update a customer record when a payment is received.

object view A philosophical view of geographic space in which space is seen as empty except where occupied by objects. In this view, every spatial location is either something (an object) or nothing. *See also* field view.

oblate ellipsoid [MATHEMATICS] An ellipsoid created by rotating an ellipse around its minor axis. The shape of the earth approximates an oblate ellipsoid with a flattening ratio of 1 to 298.257. *See also* prolate ellipsoid, ellipsoid, spheroid.

oblate ellipsoid

oblate spheroid *See* oblate ellipsoid.

oblique aspect *See* oblique projection.

oblique photograph [REMOTE SENSING] An aerial photograph taken with the axis of the camera held at an angle between the horizontal plane of the ground and the vertical plane perpendicular to the ground. A low oblique image shows only the surface of the earth; a high oblique image includes the horizon. *See also* aerial photograph.

oblique photograph

oblique projection **1.** [CARTOGRAPHY] A planar or cylindrical projection whose point of tangency is neither on the equator nor at a pole.

oblique projection 1

2. [CARTOGRAPHY] A conic projection whose axis does not line up with the polar axis of the globe. **3.** [CARTOGRAPHY] A cylindrical projection whose lines of tangency or secancy follow neither the equator nor a meridian.

off-nadir [REMOTE SENSING] Any point not directly beneath a scanner's detectors, but rather off at an angle. *See also* nadir, zenith.

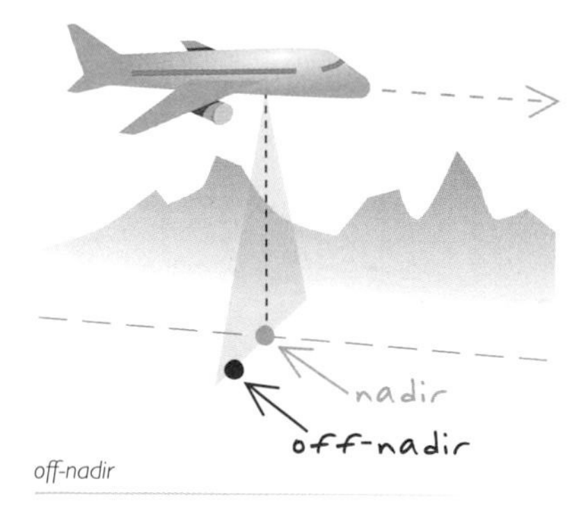

off-nadir

O

offset 1. [CARTOGRAPHY] The displacement or movement of features so that they do not overlap when displayed at a given scale. For example, a road can be offset from a river if the symbols are wide enough that they overlap. 2. [CARTOGRAPHY] In symbology, the shift of the origin or insertion point of a symbol in an x and/or y direction.

OGC *Acronym for Open Geospatial Consortium.* An international consortium of companies, government agencies, and universities participating in a consensus process to develop publicly available geospatial and location-based services. Interfaces and protocols defined by OpenGIS specifications support interoperability and seek to integrate geospatial technologies with wireless and location-based services.

OGIS *Acronym for Open Geodata Interoperability Specification.* A specification, developed by the Open Geospatial Consortium, Inc., to support interoperability of GIS in a heterogeneous computing environment. *See also* OGC.

one-to-many relationship An association between two linked or joined tables in which one record in the first table corresponds to many records in the second table. *See also* many-to-one relationship, one-to-one relationship, many-to-many relationship, joining.

one-to-one relationship An association between two linked or joined tables in which one record in the first table corresponds to only one record in the second table. *See also* many-to-many relationship, many-to-one relationship, one-to-many relationship, joining.

on-screen digitizing *See* heads-up digitizing.

on the fly [COMPUTING] Assembled, created, presented, or calculated dynamically during a transaction such as a Web page search or data display query.

ontology In computer science, a data model that represents a domain by detailing the entities that comprise it and the semantic relationships between them. Ontologies generally include individuals, classes, attributes and relations.

Open Geodata Interoperability Specification *See* OGIS.

Open Geospatial Consortium *See* OGC.

OpenGIS Consortium *See* OGC.

OpenLS *Acronym for OpenGIS Location Services.* A protocol, designed to work across the many different wireless networks and devices, that allows seamless access to multiple content repositories and service frameworks.

open traverse [SURVEYING] A traverse that does not close back upon itself or on another point of known position. As such, it does not provide a means of checking for errors. *See also* closed loop traverse.

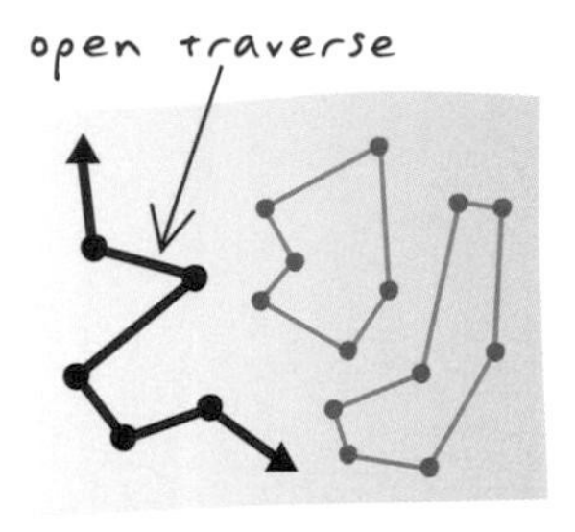

open traverse

operand [MATHEMATICS] A data value or the symbolic representation of a data value in an expression. Operands may be numbers, character strings, functions, variables, parenthetical expressions in the body of a larger expression, and so on. Symbolic representations of operands, such as variables and functions, are evaluated before they are operated upon by the operators in the expression. For example, in the expression "1 + 2", the operands are 1 and 2, and are operated upon by the + (plus) operator, which adds the operands together and returns the value 3. *See also* operator, operator precedence.

operator [MATHEMATICS] The symbolic representation of a process or operation performed against one or more operands in an expression, such as "+" (plus, or addition) and ">" (greater than). When evaluated, operators return a value as their result. If multiple operators appear in an expression, they are evaluated in order of their operator precedence. *See also* operator precedence, operand.

operator precedence [MATHEMATICS] The order in which operators are evaluated in an expression; operators with a higher precedence are evaluated before those with a lower precedence. If all operators in an expression have the same precedence, they are evaluated in the order in which they appear, from left to right. Parentheses may be used to override operator precedence; portions of an expression within parenthesis are evaluated first, and parenthetical expressions may be nested. *See also* operator, operand.

optical center *See* visual center.

optimization The process of fine-tuning data, software, or processes to increase efficiency, improve performance, and produce the best possible results.

ordinal data Data classified by comparative value; for example, a group of polygons colored lighter to darker to represent less to more densely populated areas. *See also* nominal data.

ordinary kriging [STATISTICS] A kriging method in which the weights of the values sum to unity. It uses an average of a subset of neighboring points to produce a particular interpolation point. *See also* kriging.

ordinate [MATHEMATICS] In a rectangular coordinate system, the distance of the y-coordinate along a vertical axis from the horizontal or x-axis. For example, a point with the coordinates (7,3) has an ordinate of 3. *See also* abscissa.

O

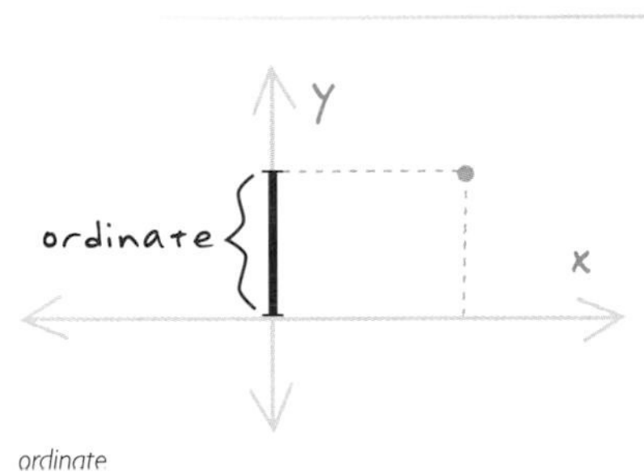

ordinate

Ordnance Survey [CARTOGRAPHY] The national mapping and cartographic agency of the United Kingdom. Now a civilian organization, the Ordnance Survey is one of the world's largest producers of maps and was the first national mapping organization in the world to complete a large-scale program of digital mapping.

orientation [GEOGRAPHY] An object's position or relationship in direction with reference to points of the compass.

origin 1. [CARTOGRAPHY] A fixed reference point in a coordinate system from which all other points are calculated, usually represented by the coordinates (0,0) in a planar coordinate system and (0,0,0) in a three-dimensional system. The center of a projection is not always its origin.

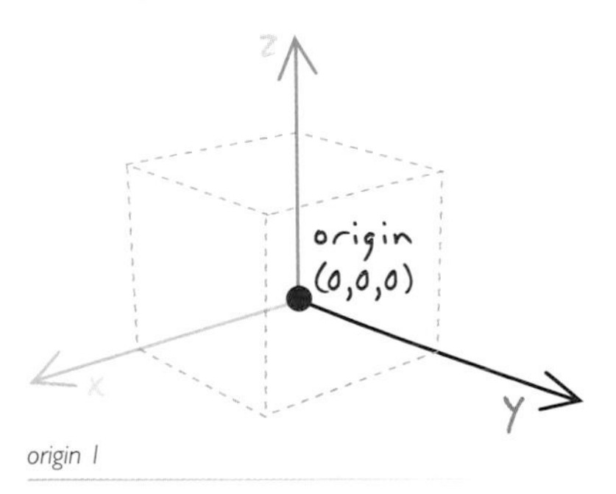

origin 1

2. [ESRI SOFTWARE] The primary object in a relationship, such as a feature class containing points where measurements are taken. The measurements are stored in another table. *See also* destination.

orthocorrection *See* orthorectification.

orthogonal [MATHEMATICS] Intersecting at right angles.

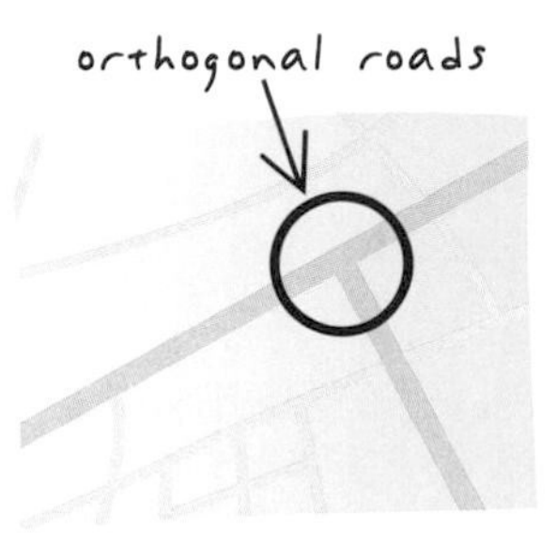

orthogonal

orthogonal offset [MATHEMATICS] A line that is perpendicular to another line at its point of tangency, often used to measure distance from a line to a separate point that does not lie along the original line.

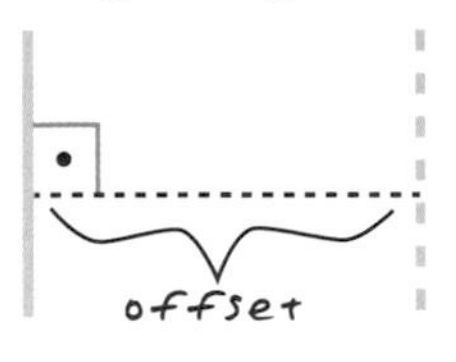

orthogonal offset

orthographic projection [CARTOGRAPHY] A planar projection, tangent to the earth at one point, that views the earth's surface from a point approaching infinity, as if from deep space. ▸

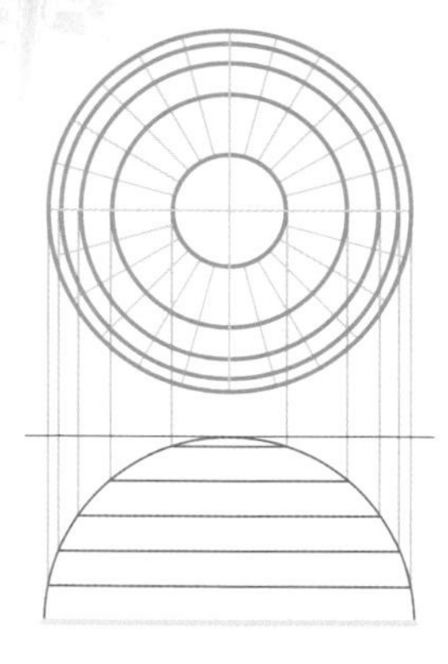

orthographic projection

orthomorphic projection *See* conformal projection.

orthophotograph [REMOTE SENSING] An aerial photograph from which distortions owing to camera tilt and ground relief have been removed. An orthophotograph has the same scale throughout and can be used as a map. *See also* aerial photograph.

orthophotograph

orthophotoquad [REMOTE SENSING] An orthophotograph that has been formatted as a USGS 1:24,000 topographic quadrangle with little or no cartographic enhancement. *See also* orthophotograph.

orthophotoscope [REMOTE SENSING] A photomechanical or optical-electronic device that creates an orthophotograph by removing geometric and relief distortion from an aerial photograph.

orthorectification [REMOTE SENSING] The process of correcting the geometry of an image so that it appears as though each pixel were acquired from directly overhead. Orthorectification uses elevation data to correct terrain distortion in aerial or satellite imagery. *See also* rectification, geometric correction.

outlier **1.** [STATISTICS] An unusual or extreme data value in a dataset. In data analysis, outliers can potentially have a strong effect on results and so must be analyzed carefully to determine if they represent valid or erroneous data. **2.** In geology, a feature that lies apart from the main body or mass to which it belongs: for example, a rock or stratum that has been separated from a formation by erosion. *See also* spike.

outline vectorization A vectorization method that generates lines along the borders of connected cells. It is typically used for vectorizing scanned land-use and vegetation maps. *See also* centerline vectorization.

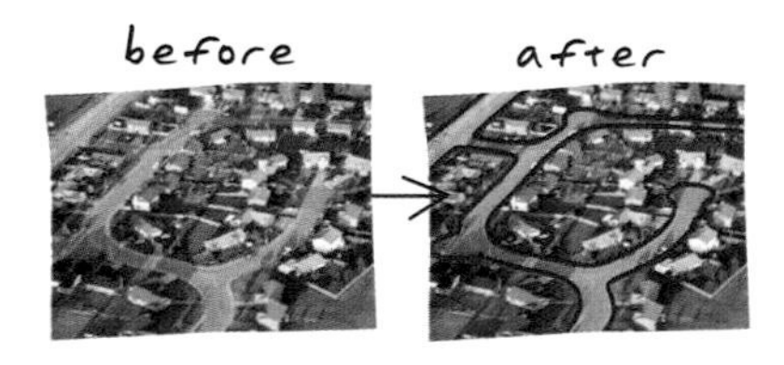

outline vectorization

O

output data Data that is the result of a computer, device, program, or process. *See also* input data, intermediate data.

overlay 1. A spatial operation in which two or more maps or layers registered to a common coordinate system are superimposed, either digitally or on a transparent material, for the purpose of showing the relationships between features that occupy the same geographic space. 2. In geoprocessing, the geometric intersection of multiple datasets to combine, erase, modify, or update features in a new output dataset. *See also* spatial overlay.

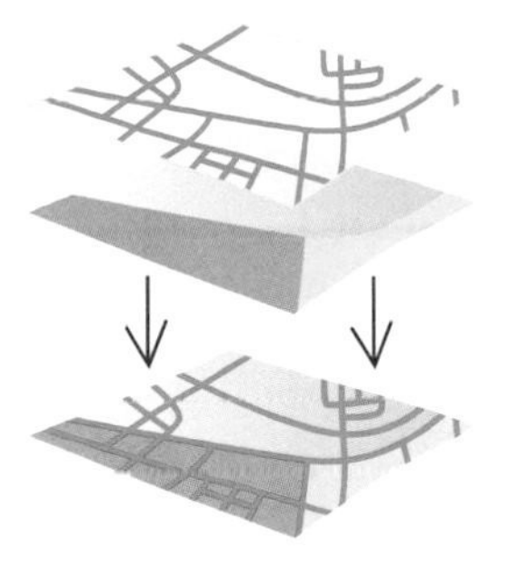

overlay 2

overprinting 1. [CARTOGRAPHY] Portraying cartographic updates on a map by printing new or modified information over the original cartography, usually in a distinctive color. 2. In offset printing, printing a color on top of areas already inked with a different color. Overprinting is often used to print thin lines, such as contours, or small intricate graphics, such as text characters, to reduce the effect of registration problems. Overprinting can also be used to create the effect of mixing two colors where they are coincident. For example, yellow printed over cyan will appear green. *See also* knockout.

overshoot The portion of an arc digitized past its intersection with another arc. *See also* dangling arc.

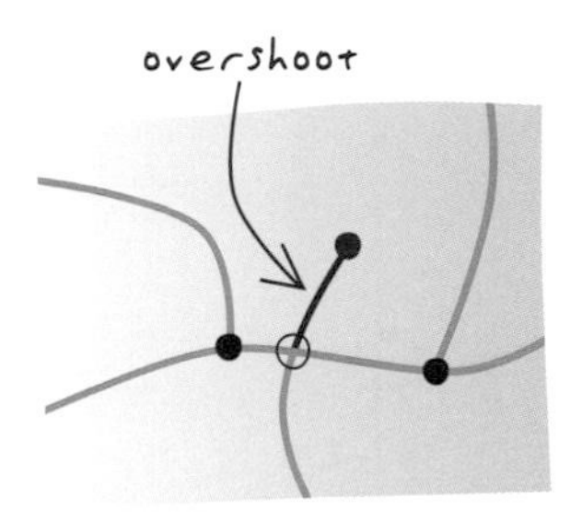

overshoot

overview map [CARTOGRAPHY] A generalized, smaller-scale map that shows the limits of another map's extent along with its surrounding area. *See also* inset map.

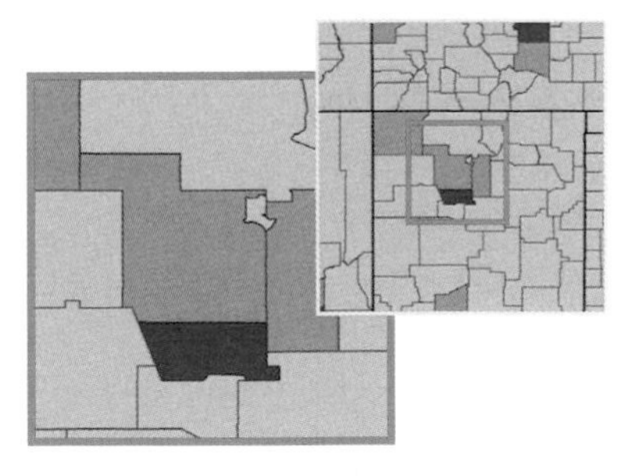

overview map

O

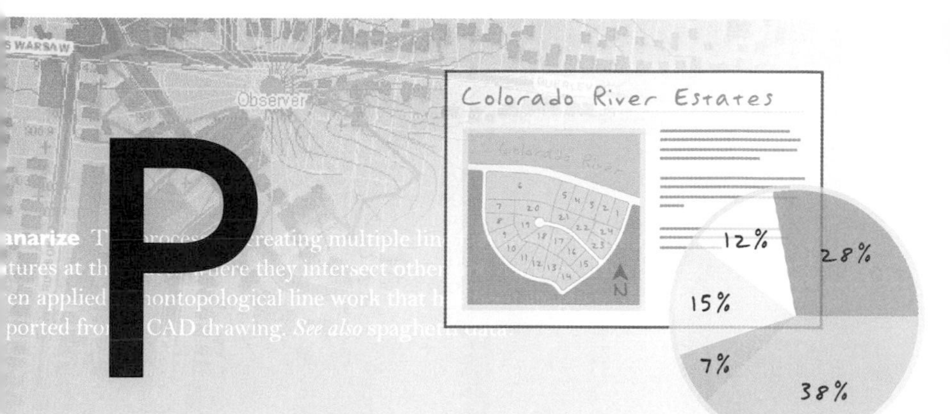

page unit The unit of measure, usually millimeters or inches, used to arrange map elements on a page for printing, as opposed to the coordinate system on the ground that the map represents. *See also* map unit.

pan To shift a map image relative to the display window without changing the viewing scale. *See also* zoom.

panchromatic [REMOTE SENSING] Sensitive to light of all wavelengths in the visible spectrum.

panchromatic sharpening [DIGITAL IMAGE PROCESSING] Sharpening a low-resolution multiband image by merging it with a high-resolution panchromatic image.

paneled map [CARTOGRAPHY] A map spliced together from smaller maps of neighboring areas.

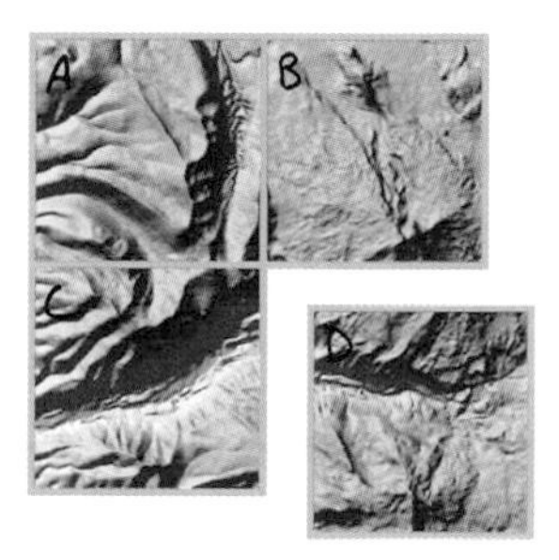

paneled map

pan sharpening *See* panchromatic sharpening.

parallax [REMOTE SENSING] The apparent shift in an object's position when it is viewed from two different angles.

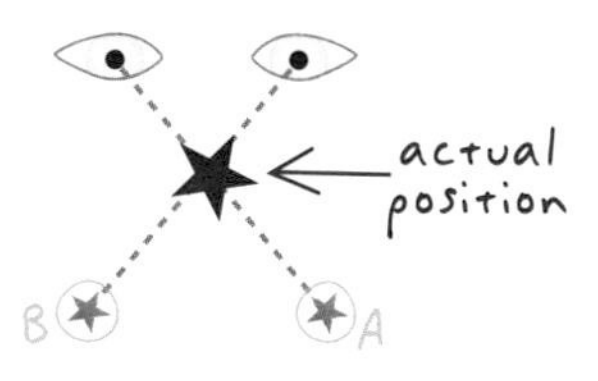

parallax

parallax bar *See* stereometer.

parallel [GEODESY] An imaginary east–west line encircling the earth, parallel to the equator and connecting all points of equal latitude. Also, the representation of this line on a globe or map. *See also* latitude.

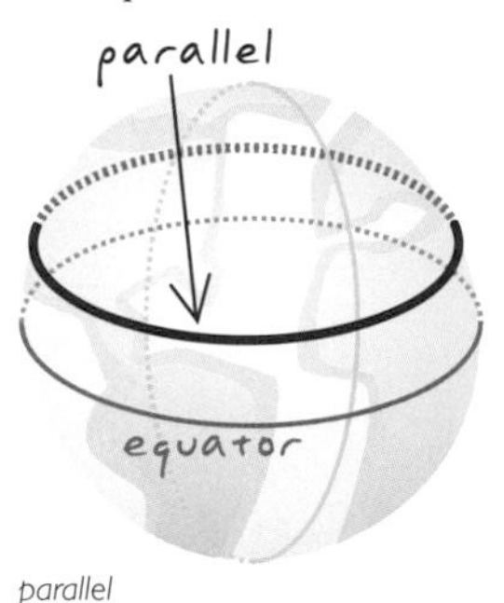

parallel

parallel processing [COMPUTING] In computer data communications, a method of storing or sending data side by side, in groups of bits. Parallel data transmission is most often used for printer ports. *See also* serialization.

parameter **1.** [CARTOGRAPHY] One of the variables that define a specific instance of a map projection or a coordinate system. Parameters differ for each projection and can include central meridian, standard parallel, scale factor, or latitude of origin. **2.** [MATHEMATICS] A variable that determines the outcome of a function or operation. **3.** [ESRI SOFTWARE] In geoprocessing in ArcGIS, a characteristic of a tool. Values set for parameters define a tool's behavior during run time.

parametric curve [MATHEMATICS] A curve that is defined mathematically rather than by a series of connected vertices. A parametric curve has only two vertices, one at each end. *See also* Bézier curve.

parcel [SURVEYING] A piece or unit of land, defined by a series of measured straight or curved lines that connect to form a polygon.

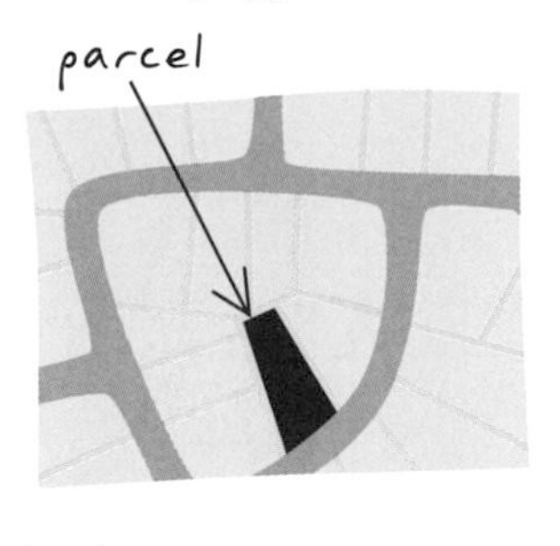

parcel

parity [MATHEMATICS] The even or odd property of an integer. In address matching, parity is used to locate a geocoded address on the correct side of the street (such as odd numbers on the left-hand side, even numbers on the right).

parse [COMPUTING] To divide a sequence of letters and numbers into parts, especially to test their agreement with a set of syntax rules.

partial sill [STATISTICS] A parameter of a covariance or semivariogram model that represents the variance of a spatially autocorrelated process without any nugget effect. In the

P

semivariogram model, the partial sill is the difference between the nugget and the sill. *See also* sill, nugget.

passive remote sensing [REMOTE SENSING] A remote-sensing system, such as an aerial photography imaging system, that only detects energy naturally reflected or emitted by an object. *See also* active remote sensing, remote sensing.

passive sensors [REMOTE SENSING] Imaging sensors that can only receive radiation, not transmit it. *See also* passive remote sensing.

password [COMPUTING] A string of characters that a user must enter to access a computer, program, database, or Web site. Passwords are a means of protecting and restricting access to information contained on networks, systems, or files.

path 1. The connecting lines, arcs, or edges that join an origin to a destination. 2. [COMPUTING] The location of a computer file, given as the drive, directories, subdirectories, and file name, in that order. *See also* segment.

pathfinding The process of calculating the optimal path between an origin and a destination point or points in a network.

pattern recognition [DIGITAL IMAGE PROCESSING] The computer-based identification, analysis, and classification of objects, features, or other meaningful regularities within an image.

P-code [GPS] The PRN code used by United States and allied military GPS receivers. *See also* PRN code, civilian code.

PDOP *See* DOP.

peak 1. [GEOGRAPHY] The highest point of a mountain or hill.

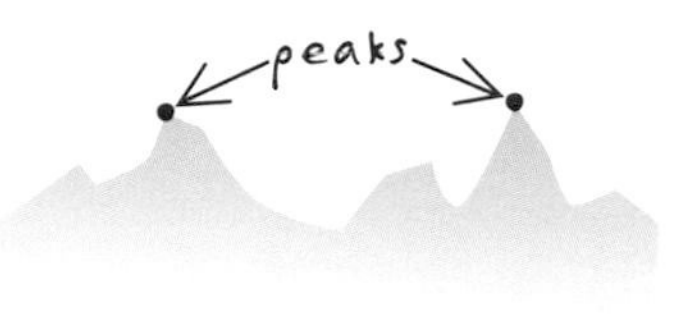

peak 1

2. In modeling, a point on a surface around which all slopes are negative. *See also* pit.

percent slope [MATHEMATICS] A measurement of the rate of change of elevation over a given horizontal distance, in which the rise is divided by the run and then multiplied by one hundred. A 45-degree slope and a 100-percent slope are the same. *See also* slope.

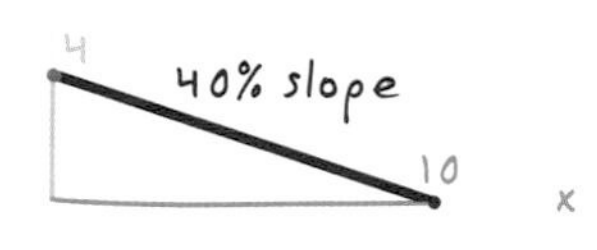

percent slope

perigee [ASTRONOMY] In an orbit path, the point at which the object in orbit is closest to the center of the body being orbited. *See also* apogee.

P

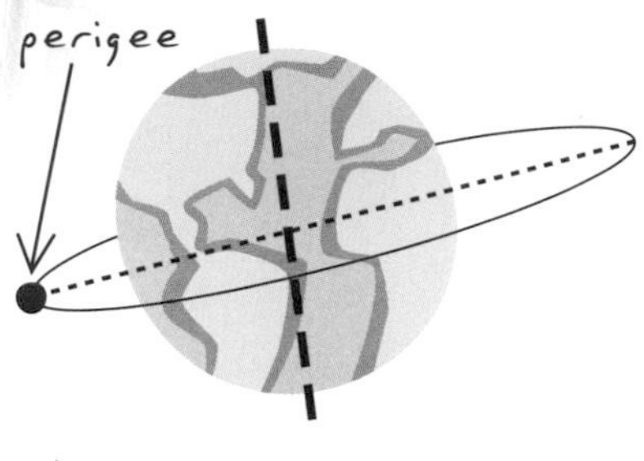

perigee

persistence [COMPUTING] The process of saving or storing data; retaining the current state of an object in a memory storage medium such as a database or file on disk.

photogeology [REMOTE SENSING] The science of interpreting and mapping geologic features from aerial photographs or remote-sensing data.

photogrammetry [REMOTE SENSING] The science of making reliable measurements of physical objects and the environment by measuring and plotting electromagnetic radiation data from aerial photographs and remote-sensing systems against land features identified in ground control surveys, generally in order to produce planimetric, topographic, and contour maps.

photomap [REMOTE SENSING] An aerial photograph or photographs, referenced to a ground control system and overprinted with map symbology.

photometer [PHYSICS] An instrument that records the intensity of light by converting incident radiation into an electrical signal and then measuring it. *See also* spectrophotometer.

physical geography [GEOGRAPHY] The field of geography concerning the natural features of the earth's surface. *See also* geography.

pie chart [STATISTICS] A chart shaped like a circle cut into wedges from a center point, that represents percentage values as proportionally sized "slices." Pie charts are used to represent the relationship between parts and the whole.

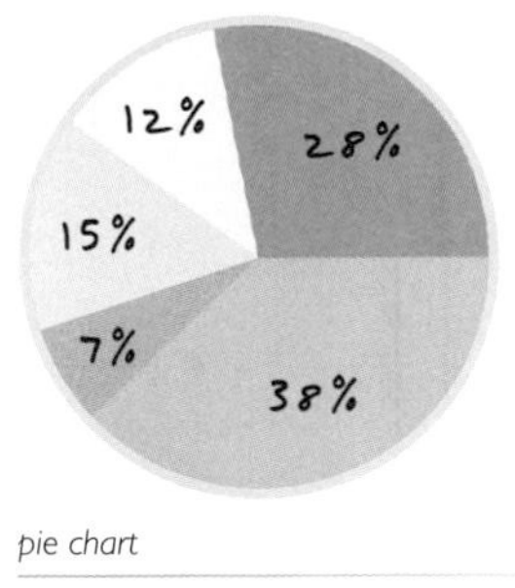

pie chart

pinch-roller *See* plotter.

pit **1.** [GEOGRAPHY] A depression in the earth's surface.

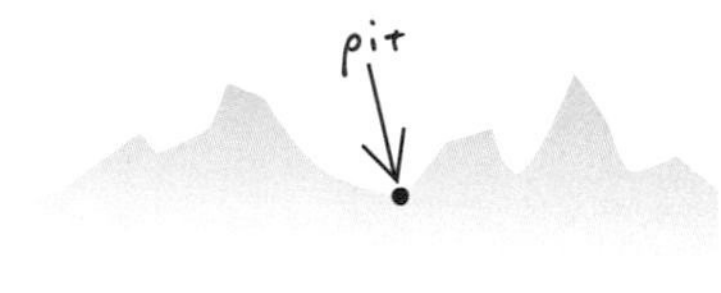

pit 1

2. In modeling, a point on a surface around which all slopes are positive. *See also* peak.

pixel **1.** The smallest unit of information in an image or raster map, usually square or rectangular. Pixel is often used synonymously with cell. **2.** [REMOTE SENSING] The fundamental unit of data collection. A pixel is represented in a remotely sensed image as a cell in an array of data values. **3.** The smallest element of a display device, such as a video monitor, that can be independently assigned attributes, such as color and intensity. Pixel is an abbreviation for picture element. *See also* cell.

pixel space The x,y coordinate space defined by the number of pixels in a computer's display area, with a pixel being a single unit of color on the screen. Most computer displays support several pixel configurations (800 x 600, 1024 x 768, 1600 x 1200, and so on). The more pixels there are, the smaller each pixel is for a given display size. Since a pixel is a piece of information, a configuration with more pixels can fit more information into a given display area. *See also* pixel, image space.

pixel type *See* data type.

planar coordinate system [CARTOGRAPHY] A two-dimensional measurement system that locates features on a plane based on their distance from an origin (0,0) along two perpendicular axes.

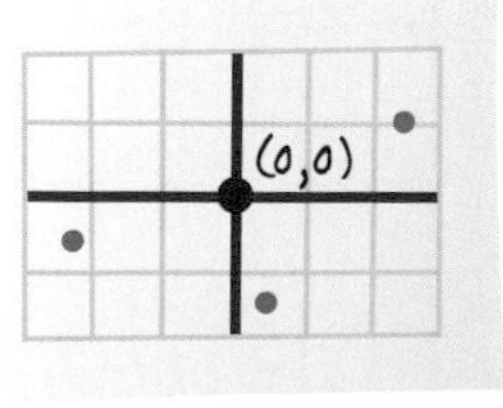

planar coordinate system

planar enforcement A set of rules used to define a consistent method of building point, line and polygon features from spaghetti-digitized data. For example, planar enforcement includes rules that polygons of differing soil types cannot overlap, and that lines must be split at intersections.

planarize The process of creating multiple line features by splitting longer features at the places where they intersect other line features. This process is often applied to nontopological line work that has been spaghetti digitized or imported from a CAD drawing. *See also* spaghetti data.

planar projection [CARTOGRAPHY] A projection that transforms points from a spheroid or sphere onto a tangent or secant plane. Because its directions are often true, the planar projection is also known as an azimuthal or zenithal projection. *See also* azimuthal projection.

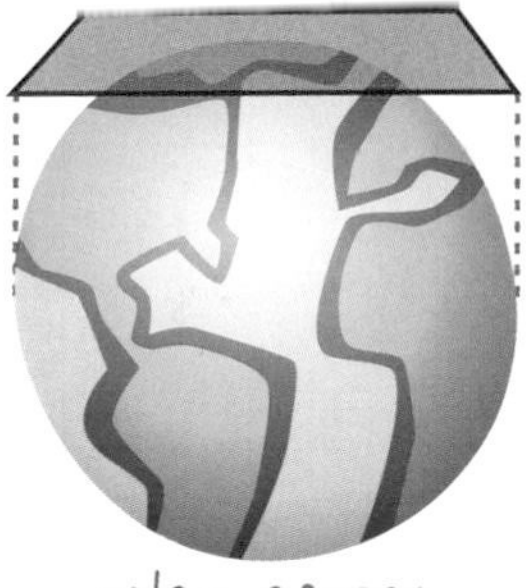

planar projection

plane survey [SURVEYING] A survey of a small area that does not take the curvature of the earth's surface into account.

planimetric [REMOTE SENSING] Two-dimensional; showing no relief.

planimetric base [REMOTE SENSING] A two-dimensional map that serves as a guide for contour mapping, usually prepared from aerial photographs.

planimetric base

planimetric map [CARTOGRAPHY] A map that displays only the x,y locations of features and represents only horizontal distances. *See also* topographic map.

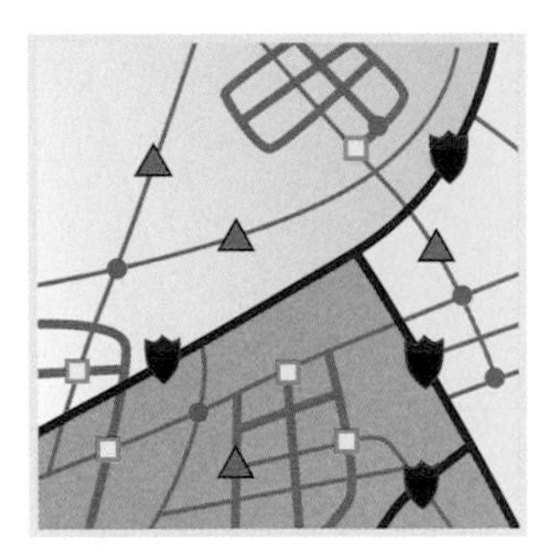

planimetric map

planimetric shift [REMOTE SENSING] Deviations in the horizontal positions of features in an aerial photograph caused by differences in elevation. Planimetric shift causes changes in scale throughout a photograph.

plat [SURVEYING] A survey diagram, drawn to scale, of the legal boundaries and divisions of a tract of land.

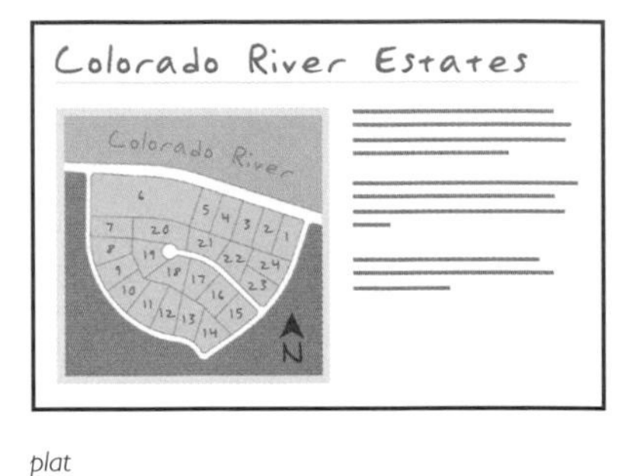

plat

platform [COMPUTING] The operating system of a machine, such as the UNIX, Linux, or Windows operating systems. Platform may also refer to a programming language or development environment, such as COM, Java, or ArcGIS 9.

plotter A printing device that draws an image onto large-size paper or transparencies. Although pen and electrostatic plotters have largely been replaced by large-format inkjet printers, the term plotter is still frequently used to refer to all large print devices. *See also* large-format printer, medium-format printer.

PLSS [SURVEYING] *Acronym for Public Land Survey System.* The description of the location of land in the United States using a survey system established by the federal government in 1785. The system is based on the concept of a township, a square parcel of land measuring 6 miles on each side. The township's position is described as a number of

6-mile units east of a north–south line (called the meridian) and north or south of an east–west line (called the baseline). Each township is divided into 36 sections, each of which is 1 square mile. A section is divided into quarters equal to 160 acres. The quarter section may be further divided into four 40-acre parcels. The PLSS is also called the rectangular survey.

plug-in [INTERNET] A small software application that extends the functionality of a Web browser.

plumb line [SURVEYING] A line that corresponds to the direction of gravity at a point on the earth's surface; the line along which an object will fall when dropped.

point [MATHEMATICS] A geometric element defined by a pair of x,y coordinates. *See also* point feature.

point digitizing *See* point mode digitizing.

point event In linear referencing, a feature that occurs at a precise point location along a route and uses a single measure value. Examples include accident locations along highways, signals along rail lines, bus stops along bus routes, and pumping stations along pipelines.

point feature [ESRI SOFTWARE] A map feature that has neither length nor area at a given scale, such as a city on a world map or a building on a city map. *See also* feature.

point-in-polygon overlay A spatial operation in which points from one feature dataset are overlaid on the polygons of another to determine which points are contained within the polygons.

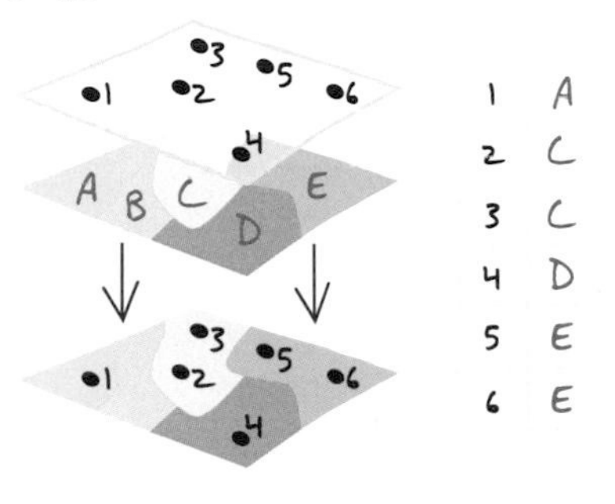

point-in-polygon overlay

point mode digitizing A method of digitizing in which the digitizer selects particular points, or vertices, to encode. *See also* stream mode digitizing.

point size [CARTOGRAPHY] A unit of measure for fonts, nearly equal to 1/72 of an inch. *See also* font.

polar aspect [CARTOGRAPHY] A planar projection with its central point located at either the north or south pole.

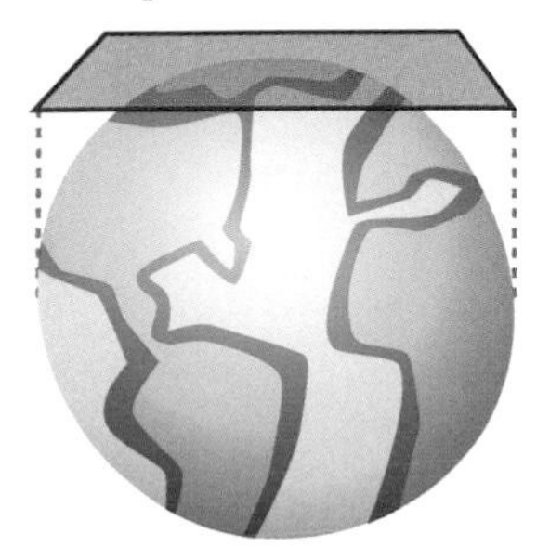

polar aspect

polar flattening *See* flattening.

polar orbit [REMOTE SENSING] A satellite orbit with an inclination near 90 degrees that passes over each polar region.

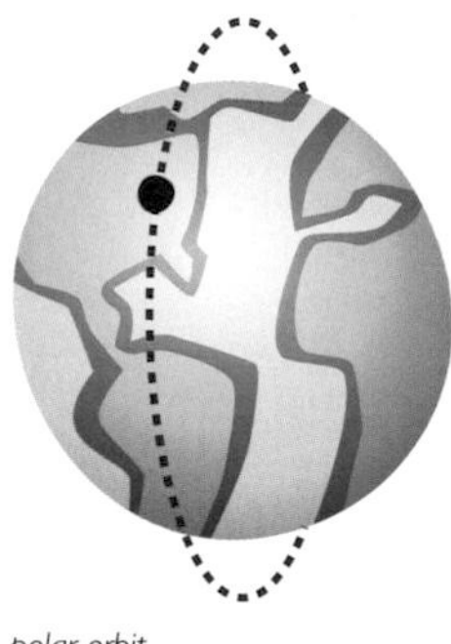
polar orbit

polar radius [GEODESY] The distance from the earth's geometric center to either pole.

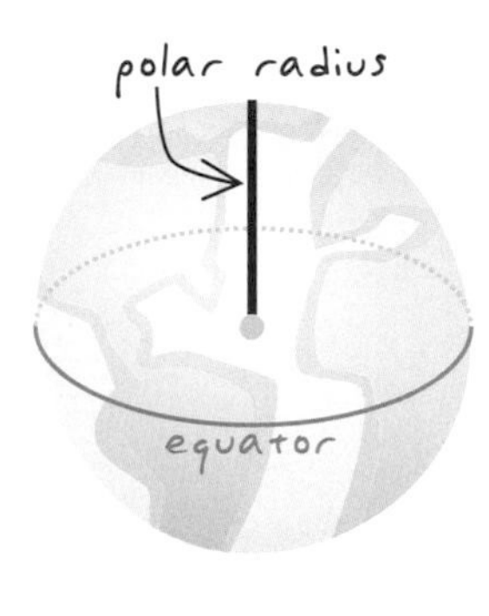

polar radius

polygon 1. On a map, a closed shape defined by a connected sequence of x,y coordinate pairs, where the first and last coordinate pair are the same and all other pairs are unique.

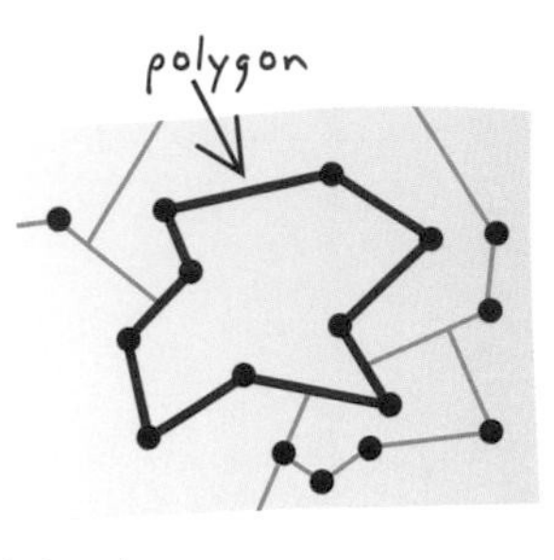

polygon 1

2. [ESRI SOFTWARE] In ArcGIS, a shape defined by one or more rings, where a ring is a path that starts and ends at the same point. If a polygon has more than one ring, the rings may be separate from one another or they may nest inside one another, but they may not overlap. *See also* polygon feature, ring.

polygon feature A map feature that bounds an area at a given scale, such as a country on a world map or a district on a city map. *See also* feature.

polygon overlay The process of superimposing two or more geographic polygon layers and their attributes to produce a new polygon layer. *See also* overlay.

polyhedron [MATHEMATICS] A three-dimensional object or volume defined by a number of plane faces or polygons.

polyhedron

polyline [ESRI SOFTWARE] In ArcGIS, a shape defined by one or more paths, in which a path is a series of connected segments. If a polyline has more than one path (a multipart polyline), the paths may either branch or be discontinuous. *See also* polyline feature, path.

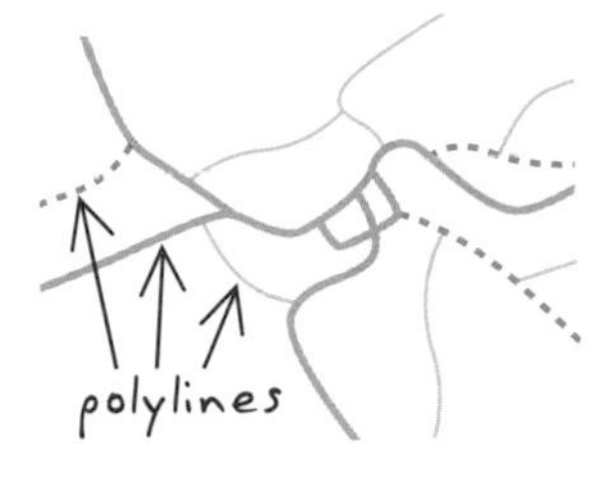

polyline

polyline feature [ESRI SOFTWARE] In ArcGIS, a digital map feature that represents a place or thing that has length but not area at a given scale. A polyline feature may have one or more parts. For example, a stream is typically a polyline feature with one part. If the stream goes underground and later reemerges, it might be represented as a multipart polyline feature with discontinuous parts. If the stream diverges around an island and then rejoins itself, it might be represented as a multipart polyline feature with branching parts. A multipart polyline feature is associated with a single record in an attribute table. *See also* feature.

portal [INTERNET] A Web resource that provides access to a broad array of related resources and services.

portlet [INTERNET] A Web component that processes requests and generates dynamic content. Portlets are used in portals as pluggable user interfaces to add specialized content, such as weather information, news, or maps, to Web pages. Users can customize the content, appearance, and position of a portlet.

position [SURVEYING] The latitude, longitude, and altitude (x,y,z coordinates) of a point, often accompanied by an estimate of error. Position may refer to an object's orientation (facing east, for example) without referring to its location.

postal code A series of letters or numbers, or both, in a specific format, used by the postal service of a country to divide geographic areas into zones in order to simplify delivery of mail. *See also* ZIP Code.

precise code *See* P-code.

precision **1.** The closeness of a repeated set of observations of the same quantity to one another. Precision is a measure of the control over random error. For example, assessment of the quality of a surveyor's work is based in part on the precision of their measured values. **2.** The number of significant digits used to store numbers, particularly coordinate values. Precision is important for accurate feature representation, analysis, and mapping. *See also* accuracy, single precision, double precision, dataset precision.

precision code *See* P-code.

P

prediction [STATISTICS] In spatial modeling, the process of forming a statistic from observed data to assign values to random variables at locations where data has not been collected. *See also* estimation.

prediction standard error [STATISTICS] A value quantifying the uncertainty of a prediction; mathematically, the square-root of the prediction variance. (The prediction variance is the variation associated with the difference between the true and predicted value.) As a rule, 95 percent of the time the true value will lie within the predicted value plus or minus two times the prediction standard error if data is normally distributed. *See also* kriging.

prediction standard error

primary colors Colors from which all other colors are derived in a particular color system. On a display monitor, these colors are red, green, and blue. In printing, they are cyan, magenta, and yellow. In traditional pigments, they are red, blue, and yellow. *See also* RGB, CMYK.

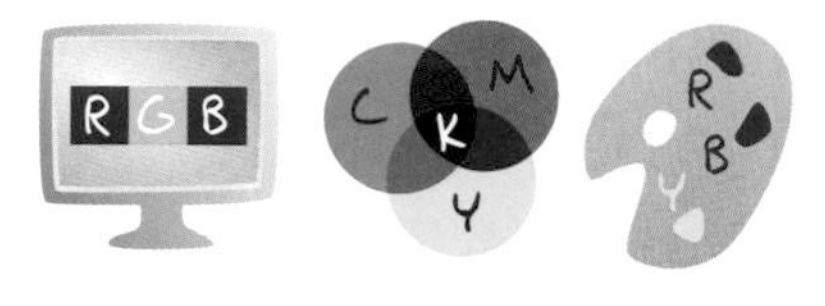

primary colors

primary key An attribute or set of attributes in a database that uniquely identifies each record. A primary key allows no duplicate values and cannot be null. *See also* key.

primary reference data In geocoding, the most basic reference material used in an address locator, usually consisting of the geometry of features in a region and an associated address attribute table.

primary table In geocoding, the attribute table associated with the primary reference data. Based on the address locator style selected, certain address elements must be present in the primary table.

prime meridian [CARTOGRAPHY] In a coordinate system, any line of longitude designated as 0 degrees east and west, to which all other meridians are referenced. The Greenwich meridian is internationally recognized as the prime meridian for most official purposes, such as civil timekeeping. *See also* Greenwich meridian.

prime vertical [GEODESY] In astronomy and geodesy, the vertical circle that passes through an observer's zenith and through the east and west points of the horizon.

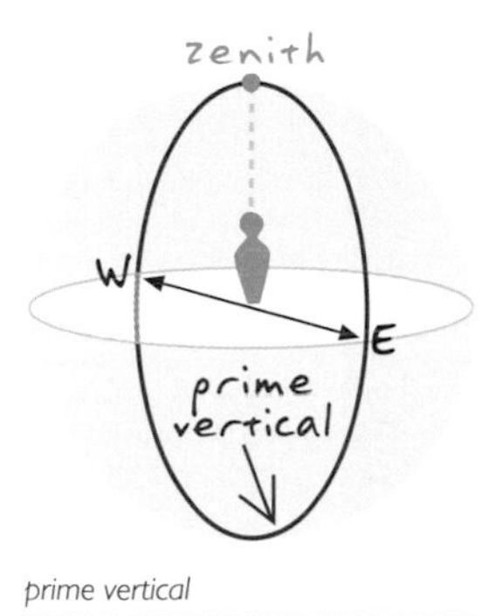

prime vertical

PRN 1. [GPS] *Acronym for pseudo-random noise.* A signal carrying a code that can be reproduced exactly, but appears to be randomly distributed like noise. Each NAVSTAR satellite has a unique PRN code. **2.** [GPS] *Acronym for pseudo-random number.* A number representing a unique GPS satellite ID or code.

PRN code [GPS] *Acronym for pseudo-random noise code.* A repeating radio signal broadcast by each GPS satellite and generated by each GPS receiver. In a given cycle, the satellite and the receiver start generating their codes at the same moment, and the receiver measures how much later the satellite's broadcast reaches it. By multiplying that time by the speed of radio waves, the receiver can compute the distance between the satellite's antenna and its own. *See also* civilian code, P-code.

probability [STATISTICS] A measure of the likelihood that a particular outcome, such as a spatial pattern or event, will occur given a set of possible outcomes. Probability values range from 0 for impossible outcomes to 1 for completely certain outcomes. The probability that a tossed coin will land heads-up, for example, is 0.5, since landing heads-up is one of two possible outcomes.

probability map [STATISTICS] A surface that gives the probability that the variable of interest is above or below a specified threshold value. *See also* probability.

procedure [PROGRAMMING] In software, a block of code that performs some task. Procedures are commonly used to organize code into reusable units. In object-oriented programming, a procedure that is specific to an object or class is called a method. *See also* method.

process [ESRI SOFTWARE] In geoprocessing in ArcGIS, a tool and its parameter values. One process, or multiple processes connected together, creates a model.

profile graph [CARTOGRAPHY] A graph of the elevation of a surface along a specified line.

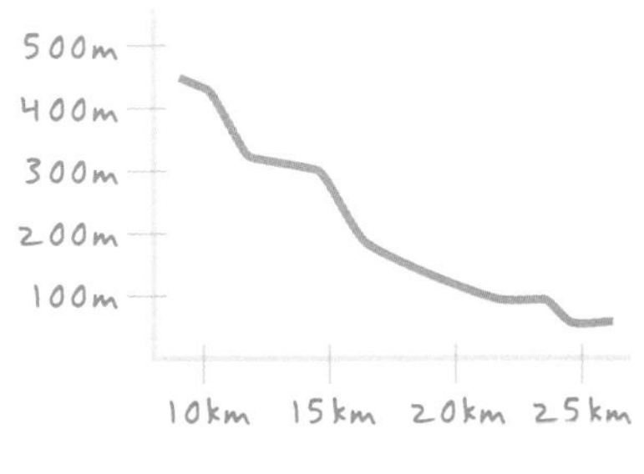

profile graph

P

project data Any data in a process that existed before the process existed. *See also* input data.

projected coordinates [CARTOGRAPHY] A measurement of locations on the earth's surface expressed in a two-dimensional system that locates features based on their distance from an origin (0,0) along two axes, a horizontal x-axis representing east–west and a vertical y-axis representing north–south. Projected coordinates are transformed from latitude and longitude to x,y coordinates using a map projection. *See also* geographic coordinates.

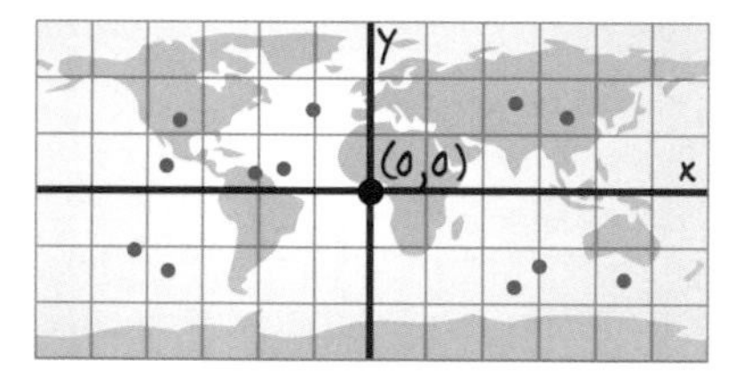

projected coordinates

projected coordinate system [CARTOGRAPHY] A reference system used to locate x, y, and z positions of point, line, and area features in two or three dimensions. A projected coordinate system is defined by a geographic coordinate system, a map projection, any parameters needed by the map projection, and a linear unit of measure. *See also* projected coordinates.

projection [CARTOGRAPHY] A method by which the curved surface of the earth is portrayed on a flat surface. This generally requires a systematic mathematical transformation of the earth's graticule of lines of longitude and latitude onto a plane. Some projections can be visualized as a transparent globe with a light bulb at its center (though not all projections emanate from the globe's center) casting lines of latitude and longitude onto a sheet of paper. Generally, the paper is either flat and placed tangent to the globe (a planar or azimuthal projection) or formed into a cone or cylinder and placed over the globe (cylindrical and conical projections). Every map projection distorts distance, area, shape, direction, or some combination thereof. *See also* cylindrical projection, conformal projection, tangent projection, secant projection, conic projection, azimuthal projection, oblique projection, equidistant projection, azimuthal projection, stereographic projection, gnomonic projection. *For more information about projections, see Projected and geographic coordinate systems: What is the difference? on page 259.

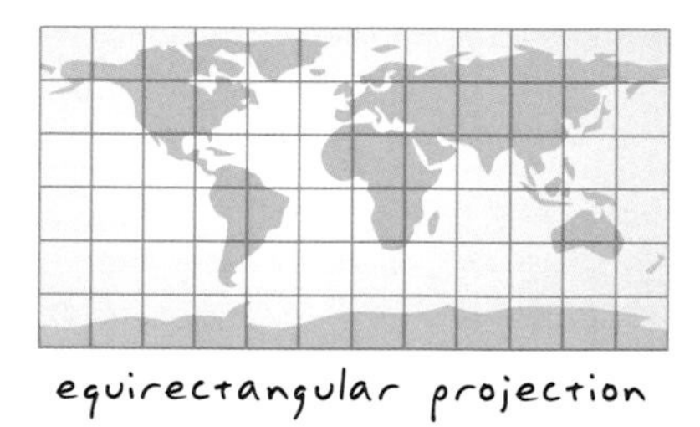

projection

projection transformation The mathematical conversion of a map from one projected coordinate system to another, generally used to integrate maps from two or more projected coordinate systems into a GIS. *See also* projected coordinate system.

projective transformation A transformation used only to transform coordinates digitized directly from high-altitude aerial photographs of relatively flat terrain, assuming there is no systematic distortion in the photographs. *See also* transformation.

prolate ellipsoid [MATHEMATICS] An ellipsoid created by rotating an ellipse around its major axis. *See also* oblate ellipsoid, ellipsoid, spheroid.

property [PROGRAMMING] An attribute of an object defining one of its characteristics or an aspect of its behavior.

proportional symbol [CARTOGRAPHY] A symbol whose size differs in relation to the phenomenon being mapped.

protected code *See* P-code.

proximity analysis A type of analysis in which geographic features (points, lines, polygons, or raster cells) are selected based on their distance from other features or cells.

proximity query A form of spatial query in which geographic features within a specified distance of a particular feature are selected.

pseudo node **1.** [ESRI SOFTWARE] In a geodatabase topology, a temporary feature marking the location where an edge has been split during an edit session. This type of pseudo node becomes a vertex when the edit is saved. **2.** [ESRI SOFTWARE] In a geodatabase topology or ArcInfo coverage, a node connecting only two edges or arcs, or the endpoint of an edge or arc that connects to itself.

pseudo-random noise *See* PRN.

pseudo-random noise code *See* PRN code.

pseudo-random number *See* PRN.

Public Land Survey System *See* PLSS.

public participation The active involvement of stakeholders outside an organization in the decision-making or planning processes of that organization. Public participation in GIS processes may include making GIS tools and data accessible, at an appropriate technical level, to stakeholders, or it may result in knowledge gained from stakeholders being incorporated into GIS analyses.

puck The handheld device used with a digitizer to record positions from the tablet surface. *See also* digitizing.

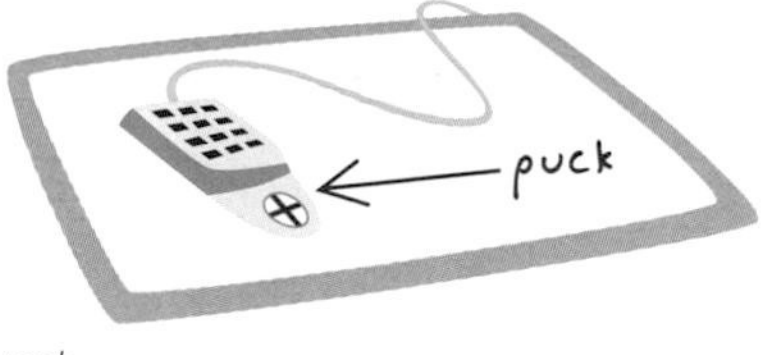

puck

push broom scanner *See* along-track scanner.

pyramid In raster datasets, a reduced resolution layer that copies the original data in decreasing levels of resolution to enhance performance.

pyramid

The coarsest level of resolution is used to quickly draw the entire dataset. As the display zooms in, layers with finer resolutions are drawn; drawing speed is maintained because fewer pixels are needed to represent the successively smaller areas.

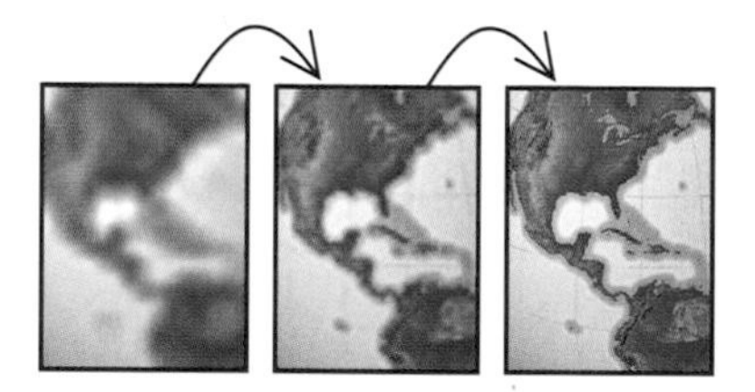

pyramid

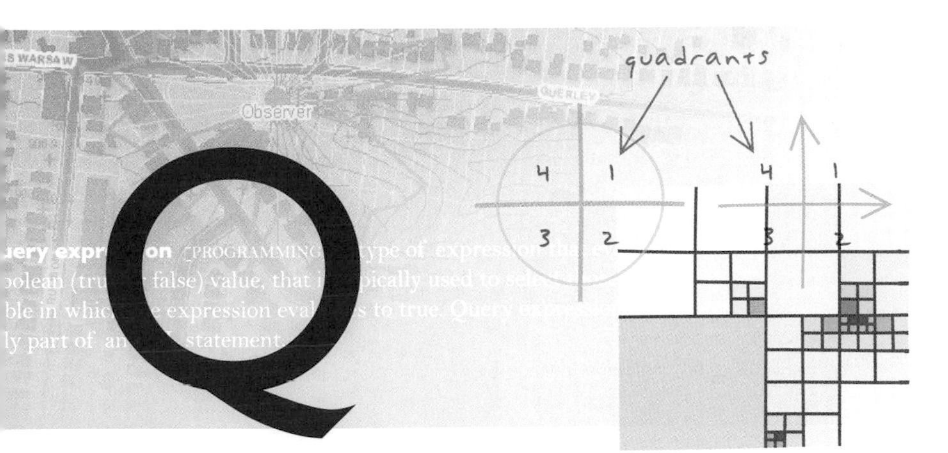

QQ plot [STATISTICS] A scatter chart in which the quantiles of two distributions are plotted against each other. *See also* quantile, scatter chart.

quadrangle [CARTOGRAPHY] A rectangular map bounded by lines of latitude and longitude, often a map sheet in either the 7.5-minute or 15-minute series published by the U.S. Geological Survey. Quadrangles are also called topo sheets.

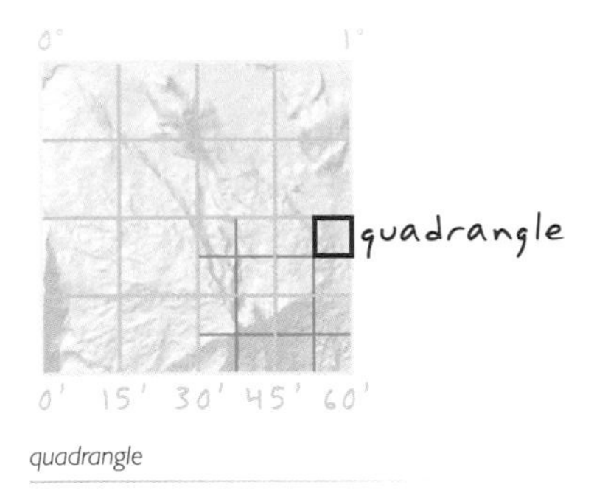

quadrangle

quadrant 1. [CARTOGRAPHY] In a rectangular coordinate system, any of the quarters formed by the central intersection of x and y axes that divide a plane into four equal parts.

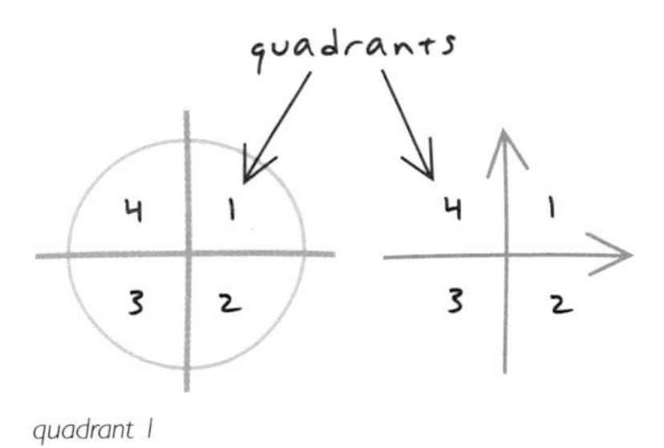

quadrant 1

2. [MATHEMATICS] One quarter of a circle measured from the center, having an arc of 90 degrees.

quadrat [STATISTICS] In spatial sampling, one of a set of identically-sized areas, often rectangular, within which the members of a population are counted. The size, number, and location of quadrats within a study area are chosen by the sampler. Population counts in each quadrat are compared to determine distribution patterns. *See also* quadrat analysis.

quadrat analysis [STATISTICS] Comparison of the statistically expected and actual counts of objects within spatial sampling areas (quadrats) to test for distribution patterns such as randomness and clustering. *See also* quadrat.

quad sheet *See* quadrangle.

quadtree A method for encoding raster data that reduces storage requirements and improves access speeds by storing values only for homogeneous regions rather than for every pixel. The raster is recursively subdivided into quadrants until all regions are homogeneous or until some specified level has been reached. *See also* tree data structure.

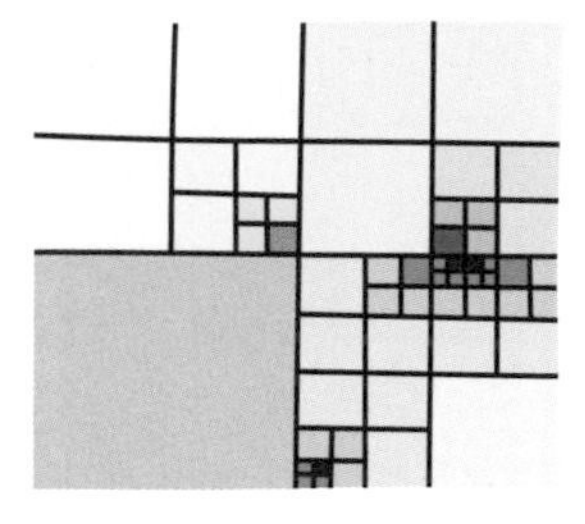

quadtree

Q

qualitative data Data classified or shown by category, rather than by amount or rank, such as soil by type or animals by species. *See also* quantitative data.

quality assurance A process used to verify the quality of a product after its production. *See also* quality control.

quality control A process used during production of a product to ensure its quality. *See also* quality assurance.

quantile [STATISTICS] In a data distribution, a value representing a class break, where classes contain approximately equal numbers of observations. The p-th quantile, where p is between 0 and 1, is that value that has a proportion p of the data below the value. For theoretical distributions, the p-th quantile is the value that has p probability below the value.

quantile classification A data classification method that distributes a set of values into groups that contain an equal number of values. *See also* classification.

quantitative data Data grouped or shown by measurements of number or amount, such as population per unit area. *See also* qualitative data.

quantitative geography [GEOGRAPHY] The application of mathematical and statistical concepts and methods to the study of geography.

query [PROGRAMMING] A request to select features or records from a database. A query is often written as a statement or logical expression.

query expression [PROGRAMMING] A type of expression that evaluates to a Boolean (true or false) value, that is typically used to select those rows in a table in which the expression evaluates to true. Query expressions are generally part of an SQL statement.

query language [COMPUTING] A language for storing, retrieving, and editing data in a database. *See also* SQL.

radar [PHYSICS] *Acronym for radio detection and ranging.* A device or system that detects surface features on the earth by bouncing radio waves off them and measuring the energy reflected back. *See also* lidar, sonar.

radar

radar altimeter [PHYSICS] An instrument that determines elevation, usually from mean sea level, by measuring the amount of time an electromagnetic pulse takes to travel from an aircraft to the ground and back again.

radar interferometry [REMOTE SENSING] The analysis of interferograms that have been created by IFSAR, or artificially. Radar interferometry involves the comparison of two or more images of the same area taken from different positions and calibrated with surveyed ground points to generate three-dimensional digital elevation models (DEMs), or models demonstrating slight movements of surface features. *See also* IFSAR, interferogram.

radian [MATHEMATICS] The angle subtended by an arc of a circle that is the same length as the radius of the circle, approximately 57 degrees, 17 minutes, and 44.6 seconds. There are 2π radians in one complete rotation.

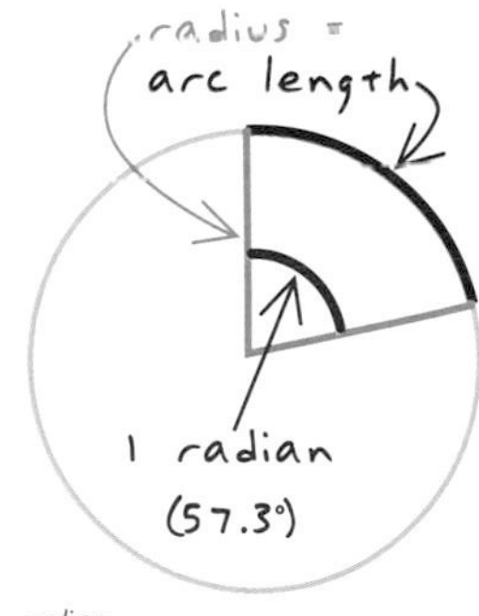

radian

radiation [PHYSICS] The emission and propagation of energy through space in the form of waves. Electromagnetic energy and sound are examples of radiation.

radiometer [PHYSICS] An instrument that measures the intensity of radiation in a particular band of wavelengths in the electromagnetic spectrum, such as infrared or microwave.

radiometer

radiometric correction [REMOTE SENSING] Procedures that correct or calibrate aberrations in data values due to specific distortions from such things as atmosphere effects (such as haze) or instrumentation errors (such as striping) in remotely sensed data. *See also* radiometric resolution.

radiometric resolution [PHYSICS] The sensitivity of a sensor to incoming reflectance. Radiometric resolution refers to the number of divisions of bit depth (for example, 255 for 8-bit, 65,536 for 16-bit, and so on) in data collected by a sensor.

radius [MATHEMATICS] The distance from the center to a point on the outer edge of a circle, circular curve, or sphere.

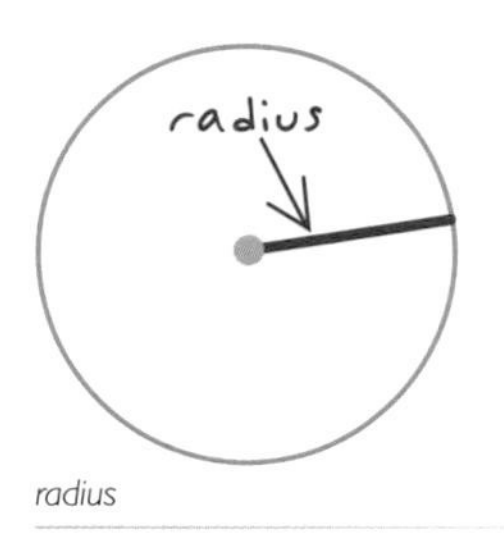

radius

random noise [STATISTICS] In a spatial model, variation in the value of a variable that cannot be described by a mathematical function and is not spatially correlated: it includes measurement error and microscale variation (variation at a finer scale than that at which the data has been sampled). Random noise is one of the three main components—along with drift and spatially correlated variation—that contribute to the change in value of a variable over a surface. In a semivariogram, random noise is represented by the nugget. Random noise is sometimes called white noise. *See also* kriging, drift, correlation, nugget.

range [STATISTICS] A parameter of a variogram or semivariogram model that represents a distance beyond which there is little or no autocorrelation among variables. *See also* autocorrelation.

range domain A type of attribute domain that defines the range of permissible values for a numeric attribute. For example, the permissible range of values for a pipe diameter could be between 1 and 32 inches. *See also* domain.

raster A spatial data model that defines space as an array of equally sized cells arranged in rows and columns, and comprised of single or multiple bands. Each cell contains an attribute value and location coordinates. Unlike a vector structure, which stores coordinates explicitly, raster coordinates are contained in the ordering of the matrix. Groups of cells that share the same value represent the same type of geographic feature. *See also* vector, lattice.

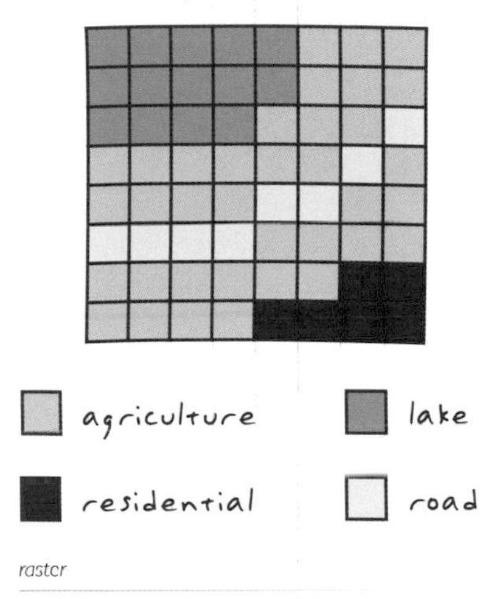

raster

raster band *See* raster dataset band.

raster cell *See* cell.

raster data model A representation of the world as a surface divided into a regular grid of cells. Raster models are useful for storing data that varies continuously, as in an aerial photograph, a satellite image, a surface of chemical concentrations, or an elevation surface. *See also* vector data model.

raster dataset band [REMOTE SENSING] One layer in a raster dataset that represents data values for a specific range in the electromagnetic spectrum (such as ultraviolet, blue, green, red, and infrared), or radar, or other values derived by manipulating the original image bands. A raster dataset can contain more than one band. For example, satellite imagery commonly has multiple bands representing different wavelengths of energy from along the electromagnetic spectrum. *See also* band.

rasterization The conversion of points, lines, and polygons into cell data. *See also* vectorization.

raster model *See* raster data model.

raster preprocessing Simple raster editing that prepares images for viewing and analysis. Preprocessing includes georeferencing, clipping, positioning, resizing, enhancing, and mosaicking.

raster snapping *See* snapping.

raster tracing An interactive vectorization process that involves drawing along the boundary of contiguous raster cells to create vector features. *See also* vectorization.

ratioing [DIGITAL IMAGE PROCESSING] Enhancing the contrast between features in an image by dividing the values of pixels in one image by the values of corresponding pixels in a second image.

ray tracing A technique that traces imaginary rays of light from a viewer's eye to the objects in a three-dimensional scene to determine which parts of the scene should be displayed from that perspective.

ray tracing

RDBMS *Acronym for relational database management system.* A type of database in which data is organized across one or more tables. Tables are associated with each other through common fields called keys. In contrast to other database structures, an RDBMS requires few assumptions about how data is related or how it will be extracted from the database. *See also* database.

real-time data Data that is displayed immediately, as it is collected. Real-time data is often used for navigation or tracking.

reclassification The process of taking input cell values and replacing them with new output cell values. Reclassification is often used to simplify or change the interpretation of raster data by changing a single value to a new value, or grouping ranges of values into single values—for example, assigning a value of 1 to cells that have values of 1 to 50, 2 to cells that range from 51 to 100, and so on.

reconcile In concurrency management, to merge all modified data in the current database edit session with a second version of the data.

record **1.** A set of related data fields, often a row in a database, containing all the attribute values for a single feature. For example, in an address database, the fields that together provide the address for a specific individual comprise one record. In the SQL query language, a record is analogous to a tuple.

FID	name	shape
1	road	line
2	market	point
3	lake	polygon

record

record 1

2. A row in a table. *See also* tuple.

rectangular survey *See* PLSS.

rectification The process of applying a mathematical transformation to an image so that the result is a planimetric image. *See also* transformation, orthorectification, geometric correction.

rectilinear **1.** [MATHEMATICS] Characterized by straight lines, usually parallel to orthogonal axes.

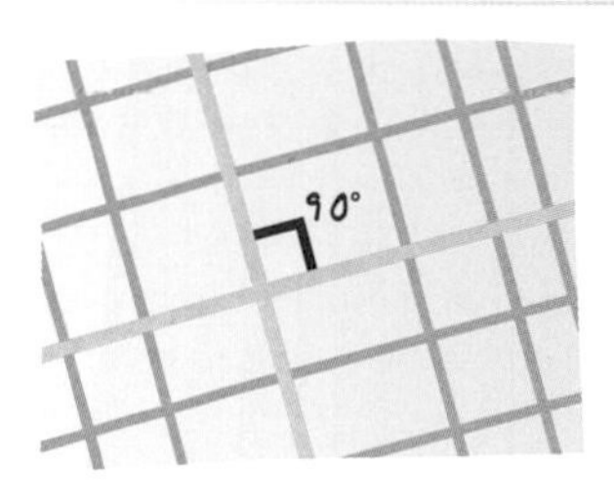

rectilinear 1

2. [CARTOGRAPHY] A map or image whose horizontal and vertical scales are identical. *See also* orthogonal.

redistricting The process of revising the boundaries of administrative, legislative, or election districts.

reference data In geocoding, material containing the location and address information of specific features. Reference data consists of the spatial representation of the data and the related attribute table.

reference datum [GEODESY] Any datum, plane, or surface from which other quantities are measured. *See also* datum.

reference ellipsoid [GEODESY] An ellipsoid associated with a geodetic reference system or geodetic datum.

reference grid *See* alphanumeric grid.

reference level *See* datum level.

reference map [CARTOGRAPHY] A map designed to show where geographic features are in relation to each other.

reference scale [CARTOGRAPHY] The scale at which symbols appear on a digital page at their true size, specified in page units. As the extent is changed, text and symbols will change scale along with the display. Without a reference scale, symbols will look the same at all map scales.

reference spheroid *See* reference ellipsoid.

reference system [CARTOGRAPHY] A method for identifying positions on the globe. This is often constructed with a grid that either refers to the earth's latitude and longitude (graticule), or a planar equivalent that divides grid lines by a fixed length from a predefined point of origin.

referential integrity A mechanism for ensuring that data remains accurate and consistent as a database changes. When changes are made to a table related to another table by a common key, the changes are automatically reflected in both tables.

reflectance [PHYSICS] The proportion of incident radiant energy that is reflected by a surface. Reflectance varies according to the wavelengths of the incident radiant energy and the color and composition of the surface.

region **1.** [GEOGRAPHY] In geography, an area usually distinguished by common cultural or physical characteristics, such as Southern California, Western Europe, or Southeast Asia. ▸

region 1

2. A set of contiguous cells with the same value. **3.** [ESRI SOFTWARE] In the coverage data structure, a polygon feature made up of multiple polygons that may be separate, overlapping, nested, or adjacent. The polygons that compose a region are stored in a polygon feature class, while the region is stored in a subclass of this feature class. A region has its own attributes but no shape geometry; its shape is defined by the shape geometry of the polygons that compose it. *See also* coverage, feature, feature class, polygon feature.

register 1. To align two or more maps or images so that equivalent geographic coordinates coincide. **2.** To link map coordinates to ground control points. **3.** [COMPUTING] To add information about a software component to the system registry, generally performed using the RegSvr32 utility. Programs search the system registry to locate software components. Registering is commonly used to make dynamic link libraries (DLLs) available to other programs. The Unregister command may be used to remove the component.

regression [STATISTICS] A statistical method for evaluating the relationship between a single dependent variable and one or more independent variables thought to influence the dependent variable. Regression is used to predict the value of the dependent variable or to determine whether an independent variable in fact influences the dependent variable, and to what extent.

rehydrate [PROGRAMMING] In programming, to reinstantiate an object and its state from persisted storage.

relate An operation that establishes a temporary connection between records in two tables using a key common to both. *See also* joining, key.

relational database A data structure in which collections of tables are logically associated with each other by shared fields. *See also* RDBMS, relate, primary key, joining.

relational database management system *See* RDBMS.

relational join *See* joining.

relational operator An expression used to compare values associated with data: greater than, less than, maximum, minimum, contains, and so forth.

relationship An association or link between two objects in a database. Relationships can exist between spatial objects (features), between nonspatial objects (rows in a table), or between spatial and nonspatial objects.

relative accuracy A measure of positional consistency between a data point and other, near data points. Relative accuracy compares the scaled distance of objects on a map with the same measured distance on the ground.

relative bearing [NAVIGATION] A bearing measured relative to a vessel or aircraft's heading. *See also* bearing.

relative coordinates [CARTOGRAPHY] Coordinates identifying the position of a point with respect to another point. *See also* absolute coordinates.

relative mode *See* mouse mode.

relative path [COMPUTING] The location of a computer file given in relation to the current working directory. *See also* path.

relief [CARTOGRAPHY] Elevations and depressions of the earth's surface, including those of the ocean floor. Relief can be represented on maps by contours, shading, hypsometric tints, digital terrain modeling, or spot elevations. *See also* TLM.

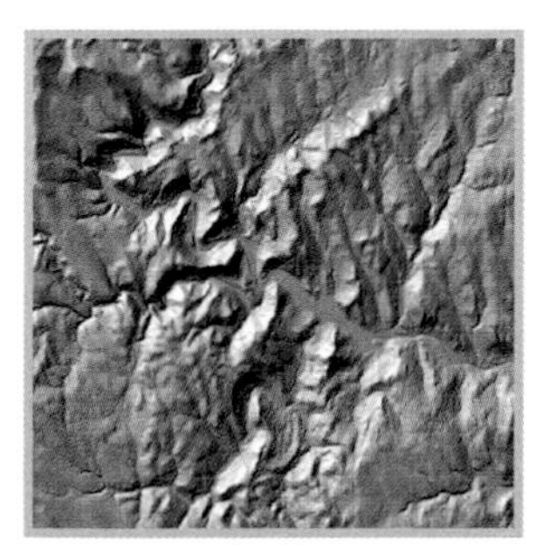

relief

relief shading *See* hillshading.

remote sensing [REMOTE SENSING] Collecting and interpreting information about the environment and the surface of the earth from a distance, primarily by sensing radiation that is naturally emitted or reflected by the earth's surface or from the atmosphere, or by sensing signals transmitted from a device and reflected back to it. Examples of remote-sensing methods include aerial photography, radar, and satellite imaging. *See also* active remote sensing, passive remote sensing. **For more information about remote sensing, see* Remote sensing *on page 261.*

remote-sensing imagery [REMOTE SENSING] Imagery acquired from satellites and aircraft, including panchromatic, radar, microwave, and multispectral satellite imagery.

multispectral satellite imagery

remote-sensing imagery

renderer A mechanism that defines how data appears when displayed. For example, the hillshade renderer for raster data in ArcMap calculates and applies shading based on existing data values for slope and aspect.

rendering The process of drawing to a display; the conversion of the geometry, coloring, texturing, lighting, and other characteristics of an object into a display image.

representation 1. [CARTOGRAPHY] A method of illustrating data so it can be viewed and understood. In cartography, representation is used to depict likenesses of real-world features in such a way that the depictions symbolize or correspond to the real features. Representation is used to present information in a format that is viewable, storable, and transferable. 2. [CARTOGRAPHY] A visual likeness or depiction of an entity that acts as a substitute for the actual entity.

representative fraction [CARTOGRAPHY] The ratio of a distance on a map to the equivalent distance measured in the same units on the ground. A scale of 1:50,000 means that one inch on the map equals 50,000 inches on the ground. *See also* scale.

resampling [MATHEMATICS] The process of interpolating new cell values when transforming rasters to a new coordinate space or cell size. *See also* nearest neighbor resampling, majority resampling, bilinear interpolation, cubic convolution.

residuals [STATISTICS] In a spatial model, values formed by subtracting the trend surface from the original data values. *See also* trend.

resolution 1. [CARTOGRAPHY] The detail with which a map depicts the location and shape of geographic features. The larger the map scale, the higher the possible resolution. As scale decreases, resolution diminishes and feature boundaries must be smoothed, simplified, or not shown at all; for example, small areas may have to be represented as points.

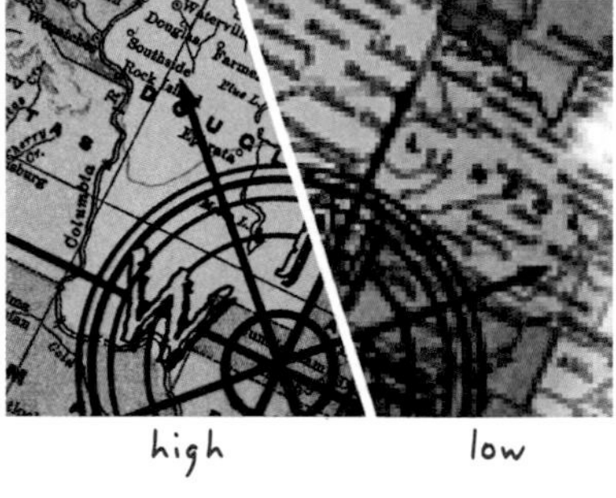

resolution 1

2. The dimensions represented by each cell or pixel in a raster. 3. The smallest spacing between two display elements, expressed as dots per inch, pixels per line, or lines per millimeter. 4. [ESRI SOFTWARE] In ArcGIS, the smallest allowable separation between two coordinate values in a feature class. A spatial reference can include x, y, z, and m resolution values. The inverse of a resolution value is called a precision or scale value.

resolution merging *See* panchromatic sharpening.

R

restriction A Boolean network element attribute used for limiting traversal through a network dataset. "One way street," "no trucks allowed," and "buses only" are examples of restrictions. *See also* cost.

reverse geocoding The process of finding a street address from a point on a map. *See also* geocoding.

RF *See* representative fraction.

RGB A color model that uses red, green, and blue, the primary additive colors used to display images on a monitor. RGB colors are produced by emitting light, rather than by absorbing it as is the case with ink on paper. Adding 100 percent of all three colors results in white. *See also* CMYK, color model, primary colors.

RGB

rhumb line [GEODESY] A complex curve on the earth's surface that crosses every meridian at the same oblique angle. A rhumb line path follows a single compass bearing; it is a straight line on a Mercator projection, or a logarithmic spiral on a polar projection. A rhumb line is not the shortest distance between two points on a sphere. *See also* great circle, sphere.

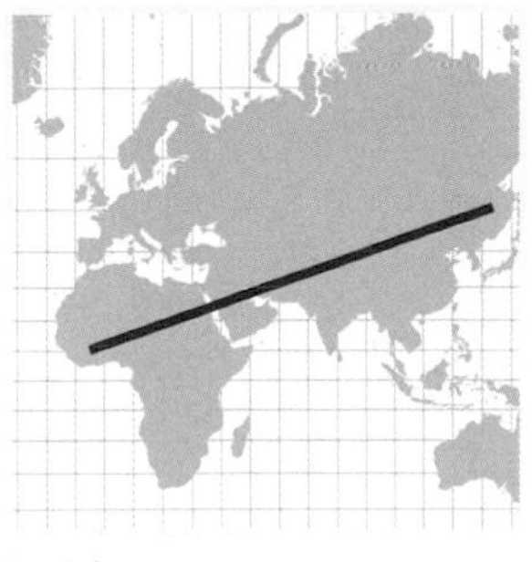

rhumb line

ring [ESRI SOFTWARE] In ArcGIS, a geometric element from which polygons are constructed. A ring is a closed path (one that begins and ends at the same point). *See also* path.

ring study The simplest and most widely used type of market-area analysis, in which a circle is generated around an area on a map; then the underlying demographics are extracted from the area delineated by the circle. Generally, a ring study is used to generate a rough visualization of the market area around a point. *See also* market area.

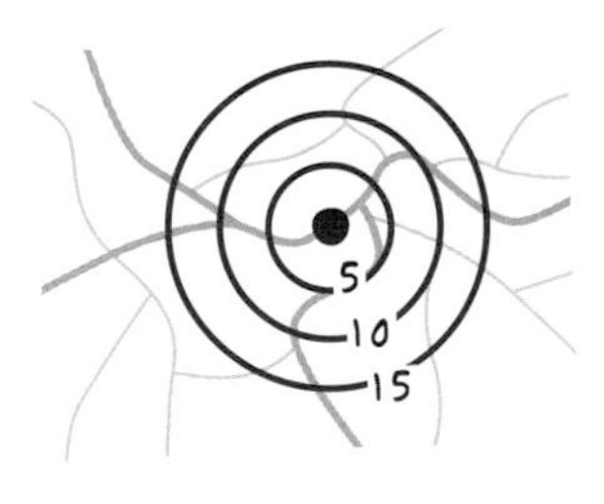

ring study

river addressing In hydrology applications, another name for linear referencing. River addressing allows

R

objects such as gauging stations to be located by their relative positions along a line feature. *See also* linear referencing.

RMSE *See* RMS error.

RMS error [STATISTICS] *Acronym for root mean square error.* A measure of the difference between locations that are known and locations that have been interpolated or digitized. RMS error is derived by squaring the differences between known and unknown points, adding those together, dividing that by the number of test points, and then taking the square root of that result.

roamer [NAVIGATION] A transparent gauge that represents easting and northing distances at a given map scale, used to locate positions on a map.

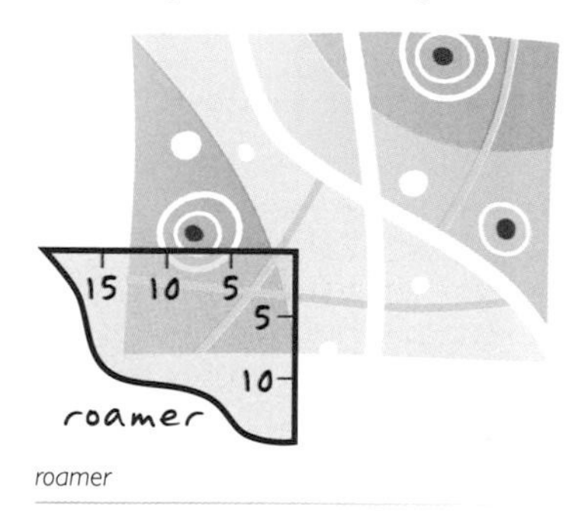

roamer

roller-feed scanner A type of scanner that moves a document through a roller assembly over camera sensors that capture a digital image. *See also* scanner, drum scanner, flatbed scanner.

root mean square error *See* RMS error.

route **1.** Any line feature, such as a street, highway, river, or pipe, that has a unique identifier. **2.** A path through a network.

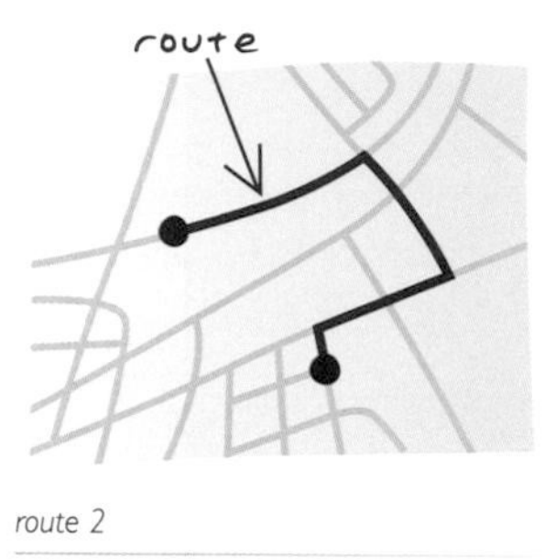

route 2

route feature class *See* route reference.

route location In linear referencing, a discrete location along a route (point) or a portion of a route (line). A point route location uses only a single measure value to describe a discrete location along a route. A line route location uses both a from- and to-measure value to describe a portion of a route.

route measure In linear referencing, a value stored along a linear feature that represents a location relative to the beginning of the feature, or some point along it, rather than as an x,y coordinate. Measures are used to map events such as distance, time, or addresses along line features. *See also* m-value, route, linear referencing, dynamic segmentation.

route reference In linear referencing, a collection of routes with a common system of measurement stored in a

R

single feature class (for example, a set of all highways in a county).

rover [GPS] A portable GPS receiver used to collect data in the field. The rover's position can be computed relative to a second, stationary GPS receiver.

rover

roving window *See* filter.

row **1.** A record in a table. **2.** The horizontal dimension of a table composed of a set of columns containing one data item each. **3.** A horizontal group of cells in a raster, or pixels in an image.

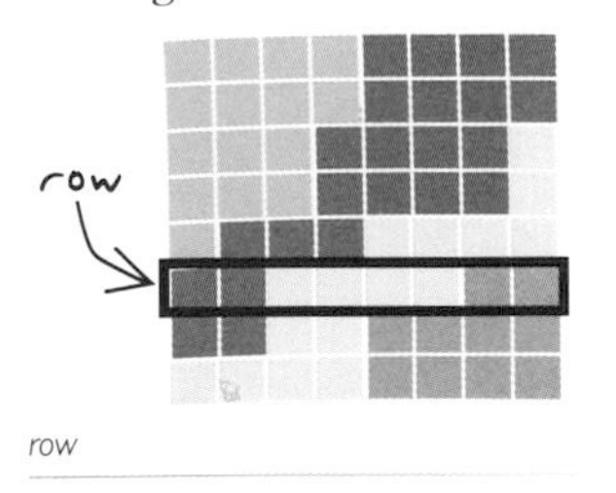

row

row standardization [STATISTICS] A technique for adjusting the weights in a spatial weights matrix. When weights are row standardized, each weight is divided by its row sum. The row sum is the sum of weights for a feature's neighbors. *See also* spatial weights matrix, weight.

R-tree A tree data structure, similar to a B-tree, used for indexing spatial data within a database. In an R-tree structure, data is sorted into a set of hierarchical nodes that may overlap. Each node has a variable number of entries, each of which includes an identifier for child nodes or actual data elements and a bounding box for all entries within the child node or the data elements. Searching algorithms check the bounding boxes before searching within a child node, thus avoiding extensive searches. *See also* B-tree, tree data structure, hierarchical database.

rubber banding *See* rubber sheeting.

rubber sheeting A procedure for adjusting the coordinates of all the data points in a dataset to allow a more accurate match between known locations and a few data points within the dataset. Rubber sheeting preserves the interconnectivity between points and objects through stretching, shrinking, or reorienting their interconnecting lines. *See also* edgematching.

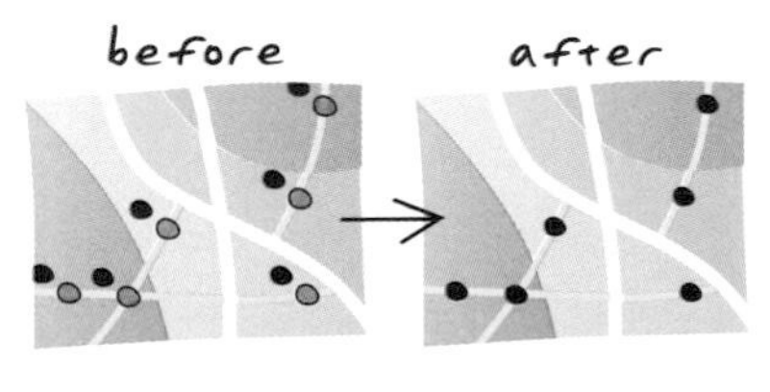

rubber sheeting

R

run-length encoding [COMPUTING]
A data compression technique for storing raster data. Run-length encoding stores data by row. If two or more adjacent cells in a row have the same value, the database stores that value once instead of recording a separate value for each cell. The more adjacent cells there are with the same value, the greater the compression.

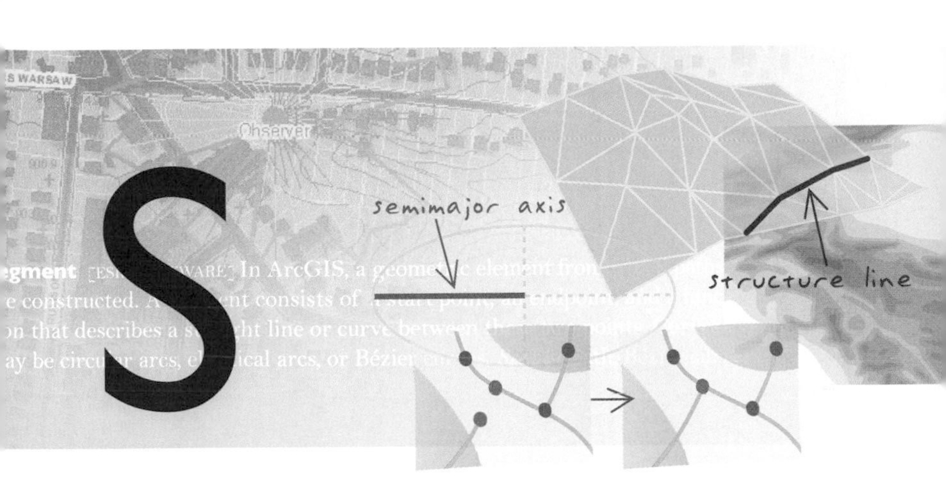

S/A *See* selective availability.

satellite constellation **1.** [REMOTE SENSING] The arrangement of a set of satellites in space.

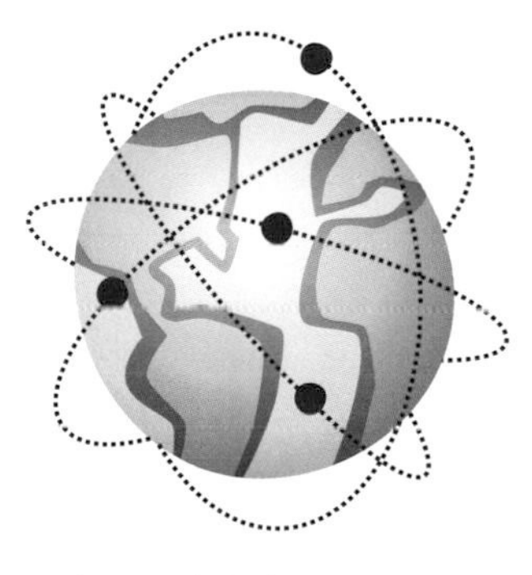

satellite constellation 1

2. [GPS] All the satellites visible to GPS receiver at one time. **3.** [GPS] The set of satellites that a GPS receiver uses to calculate positions.

satellite imagery *See* remote-sensing imagery.

saturation **1.** [CARTOGRAPHY] The intensity or purity of a color; the perceived amount of white in a hue relative to its brightness, or how free it is of gray of the same value.

saturation 1

2. [PHYSICS] The point at which energy flux exceeds the sensitivity range of a detector. *See also* intensity, chroma, value, hue.

scalable [COMPUTING] A system that does not show negative effects when its size or complexity grows greater.

scale **1.** [CARTOGRAPHY] The ratio or relationship between a distance or area on a map and the corresponding distance or area on the ground, commonly expressed as a fraction or ratio. A map scale of 1/100,000 or 1:100,000 means that one unit of measure on the map equals 100,000 of the same unit on the earth. **2.** In reference to double precision, the number of digits to the right of the decimal point in a number. For example, the number 56.78 has a scale of 2. *See also* representative fraction, small scale, large scale, scale bar.

scale bar [CARTOGRAPHY] A map element used to graphically represent the scale of a map. A scale bar is typically a line marked like a ruler in units proportional to the map's scale.

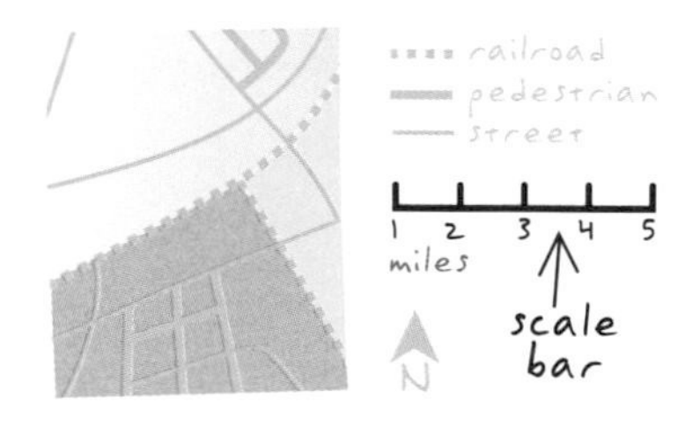

scale bar

scale factor **1.** [CARTOGRAPHY] The reciprocal of the ratio used to specify scale on a map. For example, if the scale of a map is given as 1:50,000, the scale factor is 50,000. **2.** [CARTOGRAPHY] In a coordinate system, a value (usually less than one) that converts a tangent projection to a secant projection, represented by "k_0" or "k." If a projected coordinate system doesn't support a scale factor, the standard lines of the projection have a scale factor of 1.0. Other points on the map have scale factors greater or less than 1.0. If a projected coordinate system supports a scale factor, the defining parameters no longer have a scale factor of 1.0. *See also* representative fraction.

scale range The scales at which a layer is visible on a map. Scale ranges are commonly used to prevent detailed layers from displaying at small scales (zoomed out) and to prevent general layers from displaying at large scales (zoomed in).

scanner **1.** A device that captures a print or hard-copy image, such as a text document or map, and records the information in digital format. **2.** [REMOTE SENSING] A device that records the radiation reflected or emitted by the earth's surface. *See also* drum scanner, flatbed scanner, roller-feed scanner.

scanning The process of capturing data from hard-copy maps or images in digital format using a device called a scanner. *See also* scanner.

scatter chart [STATISTICS] A chart in which each data point is marked against perpendicular x- and y-axes. Scatter charts are frequently used in analysis to find data trends.

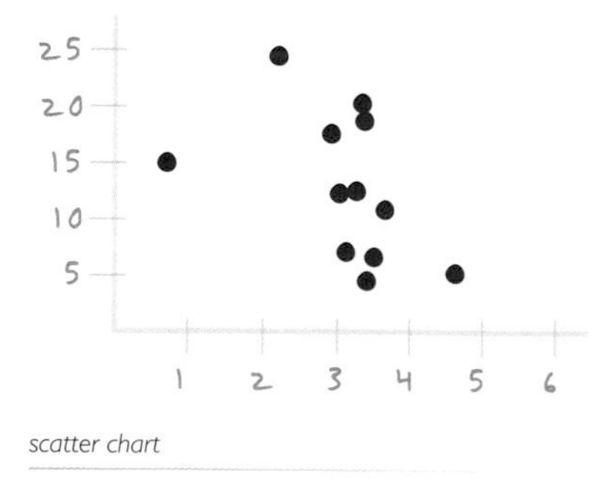

scatter chart

S

schema 1. [COMPUTING] The structure or design of a database or database object, such as a table. In a relational database, the schema defines the tables, the fields in each table, and the relationships between fields and tables. Schemas are generally documented in a data dictionary. 2. [COMPUTING] A set of rules, stored in a file, that describe the structure of an XML document. The number, type, and order of elements allowed in an XML document are described in the schema. An XML parser can compare XML documents against the schema. An XML document that uses open and close tags properly is said to be well formed; if it also follows the rules of its designated schema, it is said to be valid. *See also* data dictionary.

S-code *See* civilian code.

scratch file [COMPUTING] A file, created by either a software user or an operating system, that holds temporary data or results during an operation. When the operation is complete, the file is deleted.

script [PROGRAMMING] A set of computing instructions, usually stored in a file and interpreted at run time.

scrubbing 1. Checking the accuracy of data before it is converted into a different format. 2. Improving the appearance of data by closing open polygons, fixing overshoots and undershoots, refining thick lines, and so forth.

SDI *Acronym for spatial data infrastructure.* A framework of technologies, policies, standards, and human resources necessary to acquire, process, store, distribute, and improve the use of geospatial data across multiple public and private organizations. *See also* NSDI, GSDI.

SDTS *Acronym for Spatial Data Transfer Standard.* A data exchange format for transferring different databases between dissimilar computing systems, preserving meaning and minimizing the amount of external information needed to describe the data. All federal agencies are required to make their digital map data available in SDTS format upon request, and the standard is widely used in other sectors.

searching neighborhood [STATISTICS] In spatial interpolation, a polygon that forms a subset of data around the prediction location. Only data within the searching neighborhood is used for interpolation. *See also* interpolation.

seat [COMPUTING] In software licensing, the number of simultaneous instances of software that can be used at one time. Most often, seats represent software users at individual computers. Seats may, however, also represent the simultaneous number of servers or connections in use.

secant [MATHEMATICS] A straight line that intersects a curve or surface at two or more points.

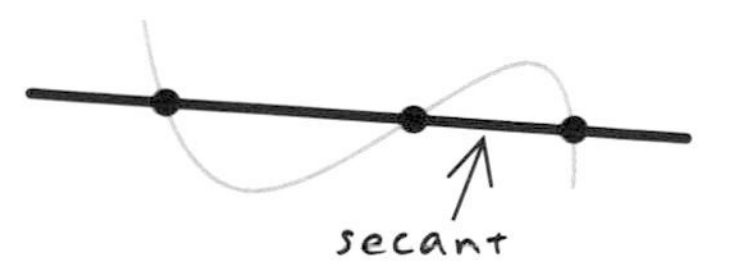

secant

S

secant projection [CARTOGRAPHY] A projection whose surface intersects the surface of a globe. A secant conic or cylindrical projection, for example, is recessed into a globe, intersecting it at two circles. At the lines of intersection, the projection is free from distortion. *See also* tangent projection.

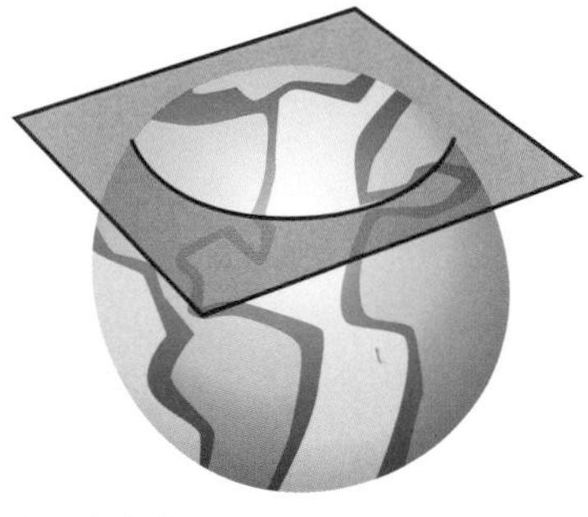

secant projection

second 1. [GEODESY] An angle equal to one sixtieth of a minute of latitude or longitude. 2. [MATHEMATICS] An angle equal to one sixtieth of a minute of arc. *See also* minute.

second 2

second normal form The second level of guidelines for designing table and data structures in a relational database. The second-normal-form guideline incorporates the guidelines of first normal form; in addition, it recommends removing data that applies to multiple rows in a table into its own table and using a foreign key to create a relationship to the original table. A database that follows these guidelines is said to be in second normal form. *See also* normal form.

second-order stationarity [STATISTICS] In geostatistics, the assumption that a set of data comes from a random process with a constant mean, and spatial covariance that depends only on the distance and direction separating any two locations. *See also* stationarity.

section 1. The arcs or portions of arcs used to define a route. 2. [SURVEYING] One thirty-sixth of a township, bounded by parallels and meridians, equal to one square mile and containing 640 acres. *See also* township.

segment [ESRI SOFTWARE] In ArcGIS, a geometric element from which paths are constructed. A segment consists of a start point, an endpoint, and a function that describes a straight line or curve between these two points. Curves may be circular arcs, elliptical arcs, or Bézier curves. *See also* path, Bézier curve.

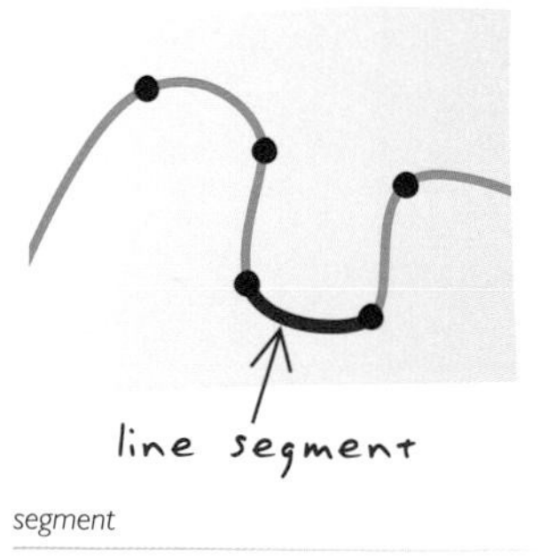

segment

select To choose from a number or group of features or records; to create a separate set or subset.

selected set A subset of features in a layer, or records in a table, that is chosen by the software user.

selective availability [GPS] The intentional degradation by the U.S. Department of Defense of the GPS signal for civilian receivers, which could cause errors in position of up to 100 meters. Selective availability (S/A) was removed from the civilian signal in May 2000. Since the lifting of S/A restrictions, position accuracy levels have improved to 20 meters or less. *See also* GPS.

self-organizing map *See* Kohonen map.

semantics The definition of the meaning of concepts within a data model by their relationships to other concepts.

semimajor axis [GEODESY] The equatorial radius of a spheroid, often referred to as "a."

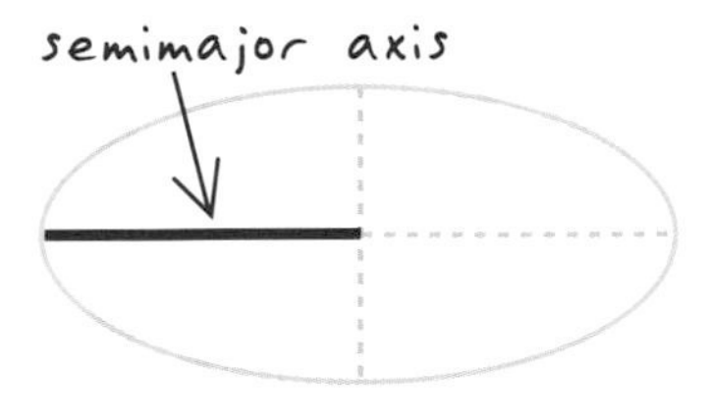

semimajor axis

semiminor axis [GEODESY] The polar radius of a spheroid, often referred to as "b."

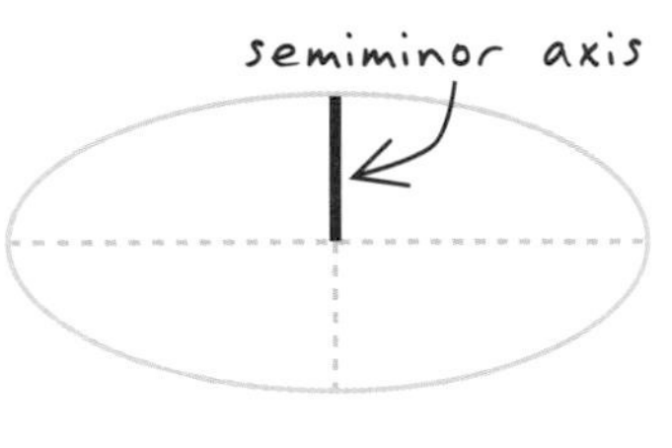

semiminor axis

semivariogram [STATISTICS] The variogram divided by two. *See also* variogram.

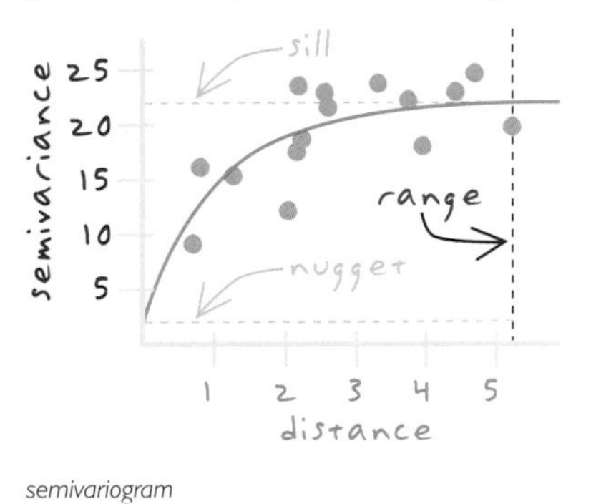

semivariogram

sense of place One's perception of the essential character of a place in which one resides or has resided, stemming from a personal response to the environment. Sense of place usually refers to perceptions of a neighborhood or city, but can also describe feelings about a larger region, state, or country.

sensitivity analysis [STATISTICS] Analysis designed to test the robustness of model and analytical results to ensure that small changes in model parameters or data structure do not exhibit large changes in the results.

sensor [REMOTE SENSING] An electronic device for detecting energy, whether emitted or radiated, and converting it into a signal that can be recorded and displayed as numbers or as an image.

sequential analysis [STATISTICS] Analysis based on a sample of an unfixed size in which testing continues only until a trend is observed with a predefined level of certainty.

serialization A method of data conversion in which data is encoded as a sequence, stored in a file, memory buffer, or other medium, and transmitted across a network connection. Data is often serialized for transmission over phone lines or networks that require data to be sent one bit at a time. *See also* persistence, parallel.

server 1. [COMPUTING] A computer that manages shared resources, such as disks, printers, and databases, on a network. 2. [COMPUTING] Software that provides services or functionality to client software. For example, a Web server is software that sends Web pages to browsers. *See also* client.

service [COMPUTING] A persistent software process that provides data or computing resources for client applications. *See also* Web service.

servlet [PROGRAMMING] A Java platform technology for extending Web servers that provides a component-based, platform-independent method for building Web-based applications.

session state [PROGRAMMING] The process by which a Web application maintains information across a sequence of requests by the same client to the same Web application.

sextant [NAVIGATION] A handheld navigational instrument that measures, from its point of observation, the angle between a celestial body and the horizon or between two objects. The angle is measured on a graduated arc that covers one sixth of a circle (60 degrees).

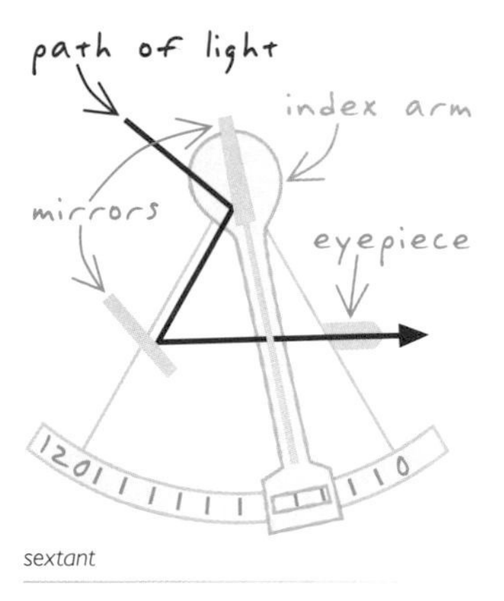

sextant

SGML [PROGRAMMING] *Acronym for Standard Generalized Markup Language.* A markup language with a predefined vocabulary and structure. SGML is used to structure information in a DTD, or Document Type Definition format, for exchanging information between different programs and machines. SGML uses a tag structure, and was standardized by ISO in 1986. *See also* XML.

shaded relief image A raster image that shows changes in elevation using light and shadows on

S

terrain from a given angle and altitude of the sun.

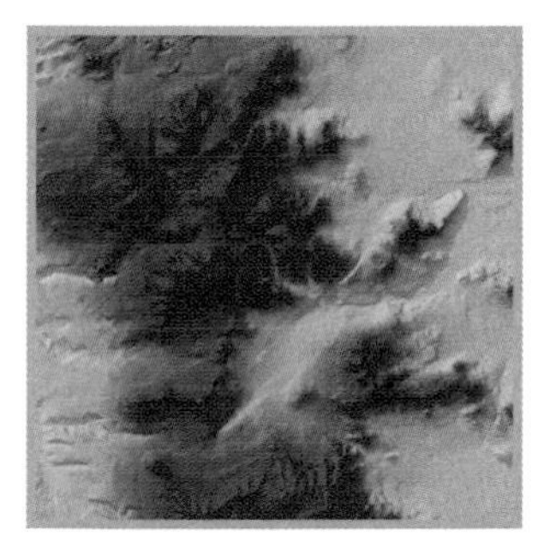

shaded relief image

shade symbol *See* fill symbol.

shading [CARTOGRAPHY] Graphic patterns such as cross hatching, lines, or color or grayscale tones that distinguish one area from another on a map.

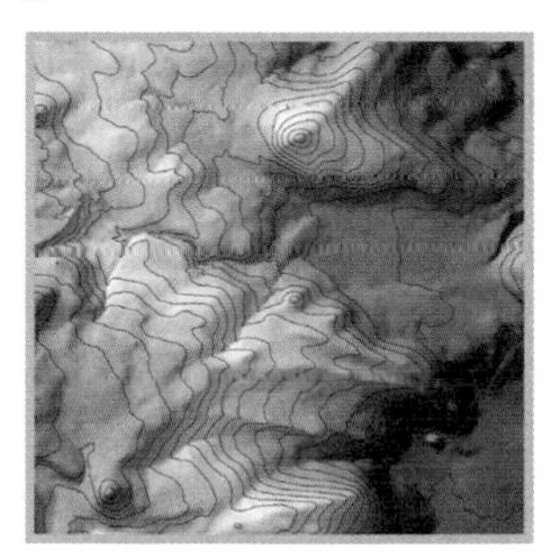

shading

shape The characteristic appearance or visible form of a geographic object as represented on a map. A GIS uses points, lines, and polygons to represent the shapes of geographic objects.

shapefile [ESRI SOFTWARE] A vector data storage format for storing the location, shape, and attributes of geographic features. A shapefile is stored in a set of related files and contains one feature class.

shared boundary A boundary common to two features. For example, in a parcel database, adjacent parcels share a boundary. Another example is a parcel that shares a boundary on one side with a river. The segment of the river that coincides with the parcel boundary shares the same coordinates as the parcel boundary.

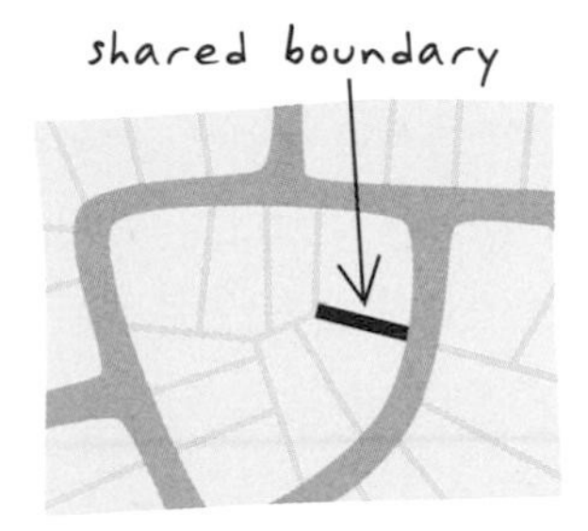

shared boundary

shared vertex A vertex common to multiple features. For example, in a parcel database, adjacent parcels share a vertex at the common corner.

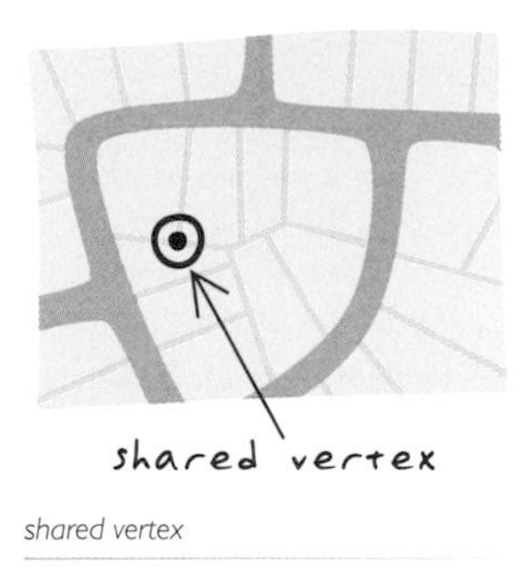

shared vertex

sheet-fed scanner *See* roller-feed scanner.

S

shield [CARTOGRAPHY] A map symbol that serves as a route marker. Shields come in many varieties, but the most common shields in the United States are for interstate highways, U.S. routes, state routes, and county routes. A uniform standard exists for interstate highways, U.S. routes, and most county routes across the United States, while shields for state routes vary by state.

shift *See* translation.

shortest path The best route or the route of least impedance between two or more points, taking into account connectivity and travel restrictions such as one-way streets and rush-hour traffic. *See also* least-cost path.

short-range variation [STATISTICS] In a spatial model, fine-scale variation that is usually modeled as spatially-dependent random variation.

SIC codes *See* Standard Industrial Classification codes.

side offset An adjustable value that dictates how far away from either the left or right side of a line feature an address location should be placed. A side offset prevents a point feature from being placed directly over a line feature. *See also* end offset.

signal **1.** [REMOTE SENSING] Information conveyed via an electric current or electromagnetic wave. **2.** [PHYSICS] The modulation of an electric current, electromagnetic wave, or other type of flow in order to convey information.

signal-to-noise ratio [REMOTE SENSING] The ratio of the information content of a signal to its noninformation content (noise).

signature *See* spectral signature.

significance level [STATISTICS] In statistical testing, the probability of an incorrect rejection of the null hypotheses. *See also* confidence level.

sill [STATISTICS] A parameter of a variogram or semivariogram model that represents a value that the variogram tends toward when distances become large. Under second-order stationarity, variables become uncorrelated at large distances, so the sill of the semivariogram is equal to the variance of the random variable.

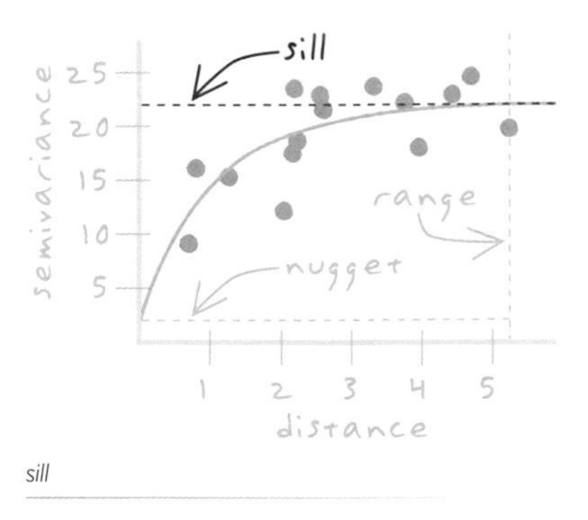

sill

simple kriging [STATISTICS] A kriging method in which the weights of the values do not sum to unity. Simple kriging uses the average of the entire dataset, which is less accurate than ordinary kriging but produces a smoother result. *See also* kriging.

simple market area An area defined by a generalized boundary drawn around the most distant set of customer points (a convex hull) that total to some value. The calculation may be unweighted (in which case every point has the same value) or weighted by a value in the underlying database, such as sales. *See also* complex market area.

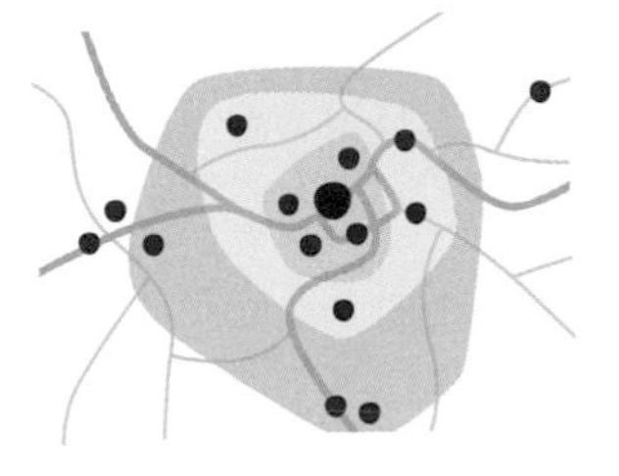

simple market area

simple relationship A link or association between data sources that exist independently of each other. *See also* composite relationship.

simplification [CARTOGRAPHY] A type of cartographic generalization in which the important characteristics of features are determined and unwanted detail is eliminated to retain clarity on a map whose scale has been reduced. *See also* generalization.

simultaneous conveyance [SURVEYING] A means of defining multiple units of land in a single survey document in such a way that all their boundaries have equal legal status. A common example of simultaneous conveyance is the modern subdivision.

single-coordinate precision *See* single precision.

single precision A level of coordinate exactness based on the number of significant digits that can be stored for each coordinate. Single precision numbers store up to seven significant digits for each coordinate, retaining a precision of plus or minus 5 meters in an extent of 1,000,000 meters. Datasets can be stored in either single or double precision coordinates. *See also* double precision.

single use [COMPUTING] In software licensing, a software product that can be used on only one machine.

sink **1.** The location or group of locations used as the endpoint for distance analysis.

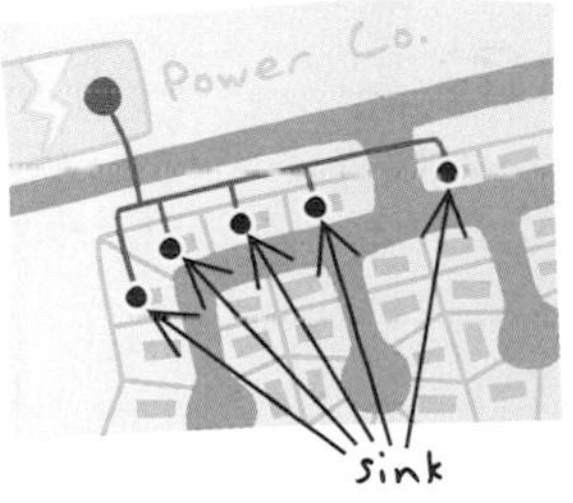

sink 1

2. A junction feature that pulls flow toward itself through the edges of a geometric network. For example, in a river network, the mouth of a river can be modeled as a sink, since gravity pulls all water toward it. *See also* source.

site prospecting The process of evaluating demographic data

S

surrounding potential locations for a business, based on a user-defined trade area or areas.

sliver polygon A small, narrow, polygon feature that appears along the borders of polygons following the overlay of two or more geographic datasets. Sliver polygons may indicate topology problems with the source polygon features, or they may be a legitimate result of the overlay.

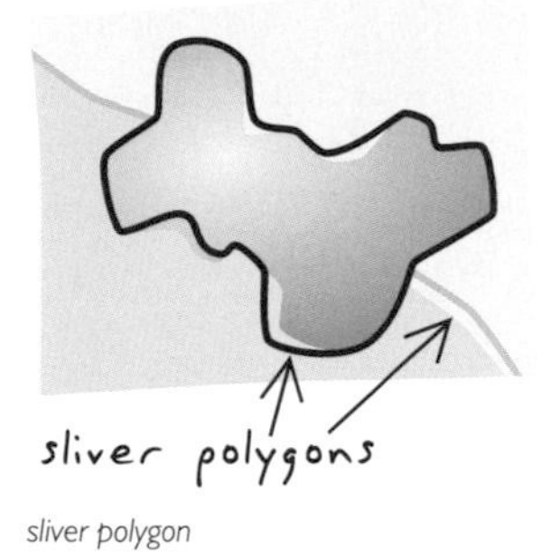

sliver polygon

sliver removal The act of deleting unwanted sliver polygons.

slope [MATHEMATICS] The incline, or steepness, of a surface. Slope can be measured in degrees from horizontal (0–90), or percent slope (which is the rise divided by the run, multiplied by 100). A slope of 45 degrees equals 100 percent slope. As slope angle approaches vertical (90 degrees), the percent slope approaches infinity. The slope of a TIN face is the steepest downhill slope of a plane defined by the face. The slope for a cell in a raster is the steepest slope of a plane defined by the cell and its eight surrounding neighbors. *See also* aspect.

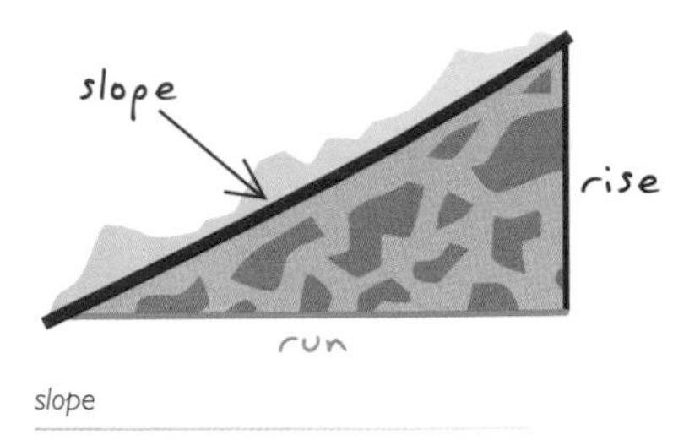

slope

slope image *See* shaded relief image.

small circle [GEODESY] The circle made when a flat plane intersects a sphere anywhere but through its center. Parallels of latitude other than the equator are small circles. *See also* great circle

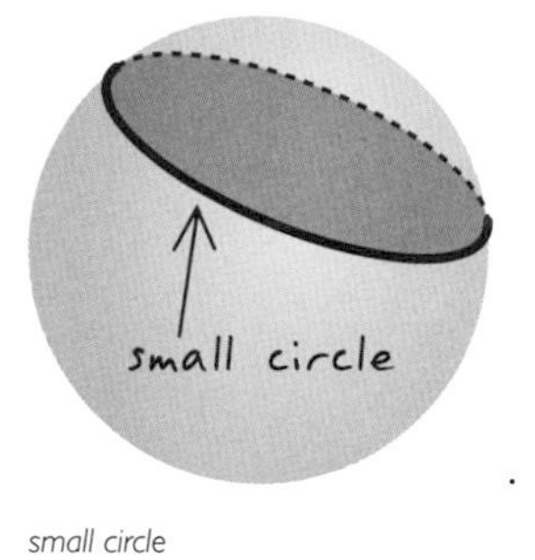

small circle

small scale [CARTOGRAPHY] Generally, a map scale that shows a relatively large area on the ground with a low level of detail. *See also* large scale.

smoothing **1.** [CARTOGRAPHY] Reducing or removing small variations in a line or other feature to improve its appearance or simplify the feature's representation. **2.** [DIGITAL IMAGE PROCESSING] Reducing or removing small variations in an image to reveal the global pattern or trend,

S

either through interpolation or by passing a filter over the image.

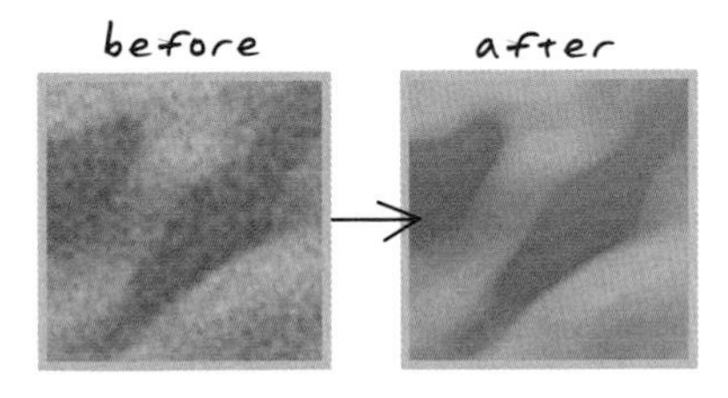

smoothing

snapping An automatic editing operation in which points or features within a specified distance (tolerance) of other points or features are moved to match or coincide exactly with each others' coordinates.

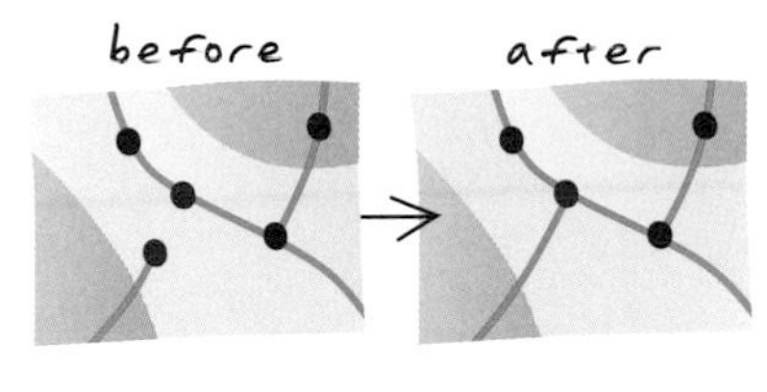

snapping

snapping tolerance A specified distance within which points or features within are moved to match or coincide exactly with each others' coordinates.

SNR *See* signal-to-noise ratio.

SOAP [COMPUTING] An XML-based protocol developed by Microsoft, SAP, and IBM for exchanging information between peers in a decentralized, distributed environment. SOAP allows programs on different computers to communicate independently of an operating system or platform by using the World Wide Web's Hypertext Transfer Protocol (HTTP) and XML as the basis of information exchange. SOAP is used in Web services, and is now a W3C specification. SOAP was originally an acronym for *Simple Object Access Protocol*, but the acronym has fallen out of use. *See also* TCP/IP, XML, W3C.

sonar [REMOTE SENSING] *Acronym for sound navigation and ranging.* A system or device that measures the time lapse between emitting a sound and receiving a returned echo to determine the location, depth and shape of objects under water. Certain types of sonar consist only of a listening device that picks up sound emitted by underwater objects, such as submarines. *See also* radar, lidar.

soundex [COMPUTING] A method of phonetic spelling used for searches and address matching. Soundex uses an algorithm to represent letters and numbers with similar phonetic equivalents to facilitate searching.

source 1. The location or group of locations used as the starting point for distance analysis.

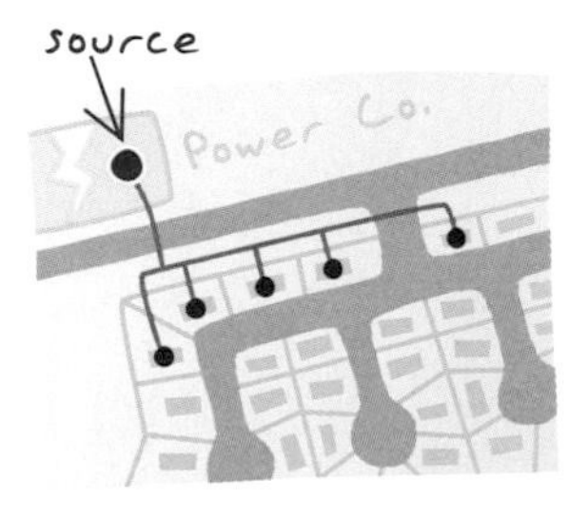

source 1

S

2. A junction feature that pushes flow away from itself through the edges of a geometric network. For example, in a water distribution network, pump stations can be modeled as sources, since they drive the water through the pipes away from the pump stations. *See also* sink.

space coordinate system [CARTOGRAPHY] A three-dimensional, rectangular, Cartesian coordinate system that has not been adjusted for the earth's curvature. In a space coordinate system, the x- and y-axes lie in a plane tangent to the earth's surface, and the z-axis points upward.

spaghetti data Vector data composed of simple lines with no topology and usually no attributes. Spaghetti lines may cross, but no intersections are created at those crossings.

spaghetti digitizing Digitizing that does not identify intersections as it records lines.

spatial Related to or existing within space. *See also* geographic.

spatial analysis The process of examining the locations, attributes, and relationships of features in spatial data through overlay and other analytical techniques in order to address a question or gain useful knowledge. Spatial analysis extracts or creates new information from spatial data. *See also* spatial modeling.

spatial autocorrelation [STATISTICS] A measure of the degree to which a set of spatial features and their associated data values tend to be clustered together in space (positive spatial autocorrelation) or dispersed (negative spatial autocorrelation). *See also* correlation, autocorrelation, Tobler's First Law of Geography.

spatial cognition The mental processes involved in gaining and using knowledge and beliefs about spatial environments. Spatial cognition includes issues of perception, memory, language, learning, and problem solving, and is an object of study in humans, nonhuman animals, and machines.

spatial data 1. Information about the locations and shapes of geographic features and the relationships between them, usually stored as coordinates and topology. **2.** Any data that can be mapped. *See also* nonspatial data, thematic data, temporal data.

spatial database A structured collection of spatial data and its related attribute data, organized for efficient storage and retrieval.

spatial data infrastructure *See* SDI.

Spatial Data Transfer Standard *See* SDTS.

spatial dependence *See* spatial autocorrelation.

spatial domain For a spatial dataset, the defined precision and allowable range for x- and y-coordinates and for m- and z-values, if present.

spatialization The transformation of complex, multivariate, nonspatial data into a spatial representation located in an information space. The relative positioning of data elements within the spatial representation shows relationships between them. Spatialization is used to allow exploration of nonspatial data using spatial metaphors and spatial analysis.

spatial join A type of table join operation in which fields from one layer's attribute table are appended to another layer's attribute table based on the relative locations of the features in the two layers.

spatial modeling A methodology or set of analytical procedures used to derive information about spatial relationships between geographic phenomena. *See also* spatial analysis.

spatial overlay The process of superimposing layers of geographic data that cover the same area to study the relationships between them. *See also* overlay.

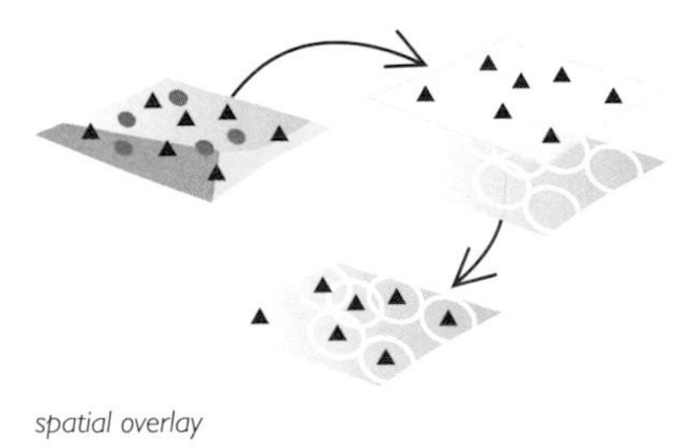

spatial overlay

spatial query A statement or logical expression that selects geographic features based on location or spatial relationship. For example, a spatial query might find which points are contained within a polygon or set of polygons, find features within a specified distance of a feature, or find features that are adjacent to each other.

spatial reference [CARTOGRAPHY] The coordinate system used to store a spatial dataset. For feature classes and feature datasets within an ESRI geodatabase, the spatial reference also includes the spatial domain.

spatial statistics [STATISTICS] The field of study concerning statistical methods that use space and spatial relationships (such as distance, area, volume, length, height, orientation, centrality and/or other spatial characteristics of data) directly in their mathematical computations. Spatial statistics are used for a variety of different types of analyses, including pattern analysis, shape analysis, surface modeling and surface prediction, spatial regression, statistical comparisons of spatial datasets, statistical modeling and prediction of spatial interaction, and more. The many types of spatial statistics include descriptive, inferential, exploratory, geostatistical, and econometric statistics. *See also* geostatistics.

spatial weights matrix [STATISTICS] A file that quantifies spatial relationships among a set of features. Typical examples of such relationships are inverse distance, contiguity, travel time, and fixed distance.

spectral resolution [REMOTE SENSING] The range of wavelengths

S

that an imaging system can detect. *See also* band.

spectral signature [PHYSICS] The pattern of electromagnetic radiation that identifies a chemical or compound. Materials can be distinguished from one another by examining which portions of the spectrum they reflect and absorb.

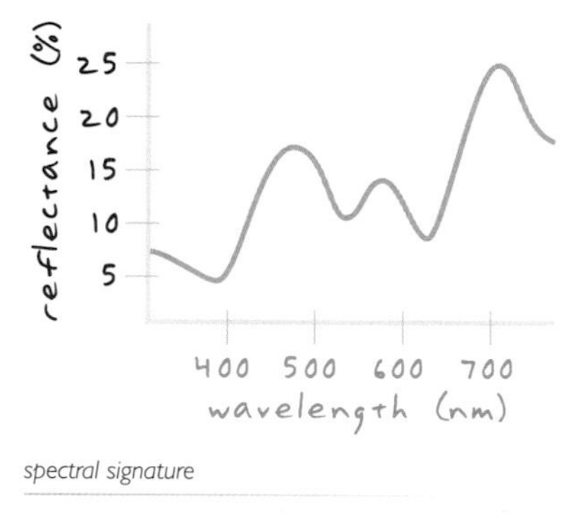

spectral signature

spectrometer *See* spectrophotometer.

spectrophotometer [PHYSICS] A photometer that measures the intensity of electromagnetic radiation as a function of its frequency. Spectrophotometers are usually used for measuring the visible portion of the spectrum.

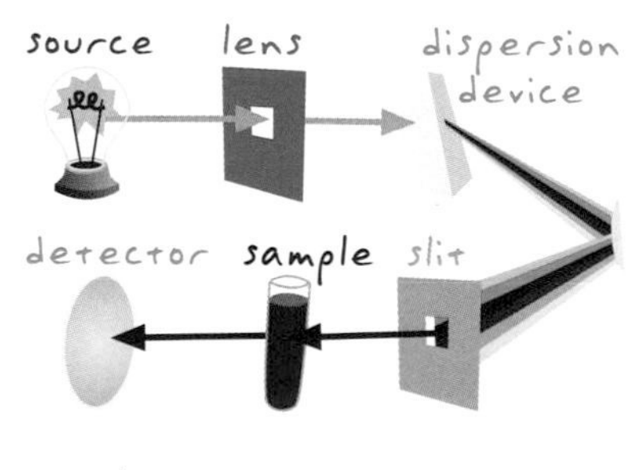

spectrophotometer

spectroscopy [PHYSICS] The scientific study of how different chemicals and other substances absorb and reflect different parts of the electromagnetic spectrum.

spectrum *See* electromagnetic spectrum.

sphere [MATHEMATICS] A three-dimensional shape whose center is equidistant from every point on its surface, made by revolving a circle around its diameter.

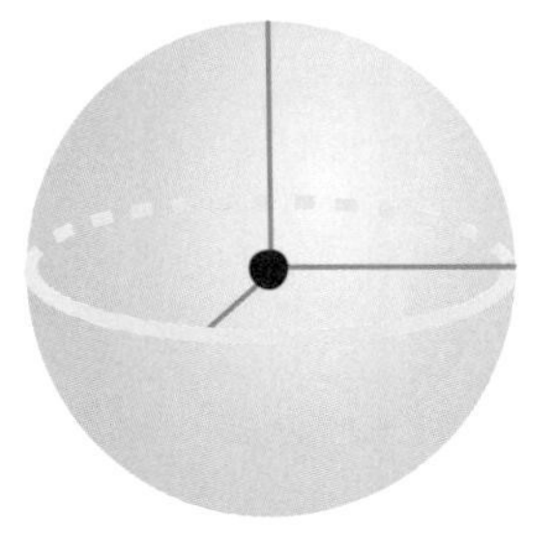

sphere

spherical coordinate system [CARTOGRAPHY] A reference system using positions of latitude and longitude to define the locations of points on the surface of a sphere or spheroid. *See also* geographic coordinate system.

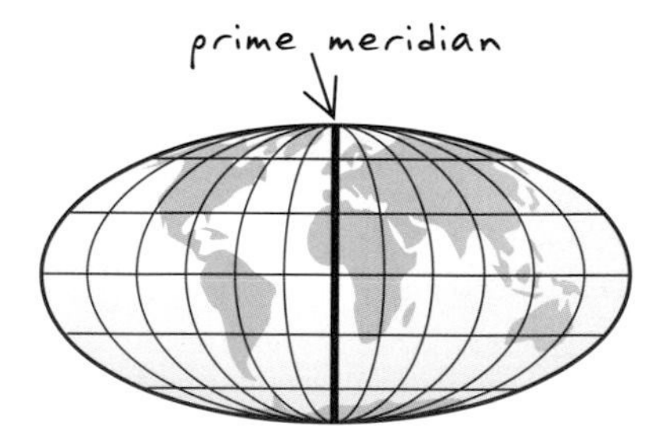

spherical coordinate system

S

spheroid 1. [MATHEMATICS] A three-dimensional shape obtained by rotating an ellipse about its minor axis, resulting in an oblate spheroid, or about its major axis, resulting in a prolate spheroid. 2. [GEODESY] When used to represent the earth, a three-dimensional shape obtained by rotating an ellipse about its minor axis, with dimensions that either approximate the earth as a whole, or with a part that approximates the corresponding portion of the geoid. *See also* ellipsoid, geoid.

spider diagram *See* desire-line analysis.

spike 1. [STATISTICS] An anomalous data point that protrudes above or below an interpolated surface.

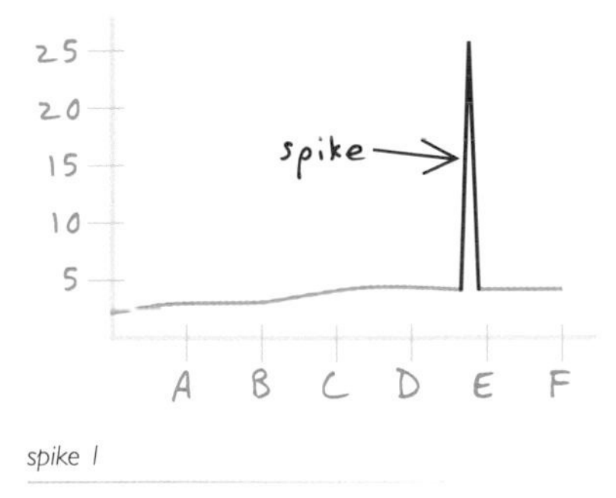

spike 1

2. An overshoot line created erroneously by a scanner and its rasterizing software. *See also* outlier.

spline [MATHEMATICS] A piecewise polynomial function used to approximate a smooth curve in a line or surface. *See also* spline interpolation, densify.

spline

spline interpolation [STATISTICS] An interpolation method in which cell values are estimated using a mathematical function that minimizes overall surface curvature, resulting in a smooth surface that passes exactly through the input points. *See also* spline, interpolation.

SPOT [REMOTE SENSING] *Acronym for Satellite Pour l'Observation de la Terre.* Earth observation satellites developed by Centre National d'Etudes Spatiales (CNES), the space agency of France. The SPOT satellites gather high-resolution imagery used in natural resource management, climatology, oceanography, environmental monitoring, and the monitoring of human activities.

spot elevation [SURVEYING] An elevation measurement taken at a single location. *See also* elevation, contour line.

spot height *See* spot elevation.

spurious polygon *See* sliver polygon.

SQL [PROGRAMMING] *Acronym for Structured Query Language.* A syntax for retrieving and manipulating data from a relational database. SQL has become an industry standard query language in most relational database management systems.

stable base [CARTOGRAPHY] Any material such as a Mylar sheet or film that is more durable than paper and less likely to shrink or stretch.

stack [COMPUTING] A data storage structure that operates on last in, first out (LIFO) protocol. As with a stack of dishes, the item placed on top of the stack last must be removed before the others may be manipulated.

standard deviation [STATISTICS] A statistical measure of the spread of values from their mean, calculated as the square root of the sum of the squared deviations from the mean value, divided by the number of elements minus one. The standard deviation for a distribution is the square root of the variance.

standard deviation classification [STATISTICS] A data classification method that finds the mean value, then places class breaks above and below the mean at intervals of either .25, .5, or 1 standard deviation until all the data values are contained within the classes. Values that are beyond three standard deviations from the mean are aggregated into two classes, greater than three standard deviations above the mean and less than three standard deviations below the mean. *See also* classification.

standard distance [STATISTICS] A measure of the compactness of a spatial distribution of features around its mean center. Standard distance (or standard distance deviation) is usually represented as a circle where the radius of the circle is the standard distance. *See also* mean center.

Standard Generalized Markup Language *See* SGML.

Standard Industrial Classification codes The federal U.S. standard for classifying establishments by their primary type of business activity. Standard Industrial Classification codes (SIC codes) are used as an identification system in business directories, publications, and statistical sources. The classification system was officially replaced by NAICS in 1997, but it is still used by some organizations outside the federal government. *See also* NAICS.

standard line [CARTOGRAPHY] A line on a sphere or spheroid that has no length compression or expansion after being projected; usually a standard parallel or central meridian.

standard parallel [CARTOGRAPHY] The line of latitude in a conic or cylindrical projection in normal aspect where the projection surface touches the globe. A tangent conic or cylindrical projection has one standard parallel, while a secant conic or cylindrical projection has two. At the standard parallel, the projection shows no distortion. *See also* parallel.

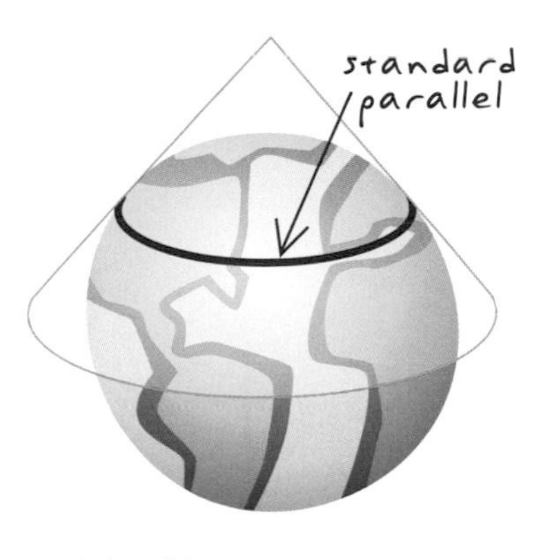

standard parallel

star diagram [CARTOGRAPHY] A type of diagram that consists essentially of a central point from which lines radiate outward. The central point usually represents a geographic location while the length of each line represents an attribute value or ratio. The direction of the line may represent a compass direction, a period of time, or some other attribute classification. A wind rose is a common example of a star diagram.

state [PROGRAMMING] The current data contained by an object.

stateful operation [PROGRAMMING] An operation that makes changes to an object or one of its associated objects, such as removing a layer from a map. *See also* stateless operation.

stateless operation [PROGRAMMING] An operation that does not make changes to an object, such as drawing a map. *See also* stateful operation.

state plane coordinate system [CARTOGRAPHY] A group of planar coordinate systems based on the division of the United States into more than 130 zones to minimize distortion caused by map projections. Each zone has its own map projection and parameters and uses either the NAD27 or NAD83 horizontal datum. The Lambert conformal conic projection is used for states that extend mostly east–west, while transverse Mercator is used for those that extend mostly north–south. The oblique Mercator projection is used for the panhandle of Alaska.

static binding *See* early binding.

static positioning [GPS] Determining a position on the earth by averaging the readings taken by a stationary antenna over a period of time. *See also* kinematic positioning.

stationarity [STATISTICS] In geostatistics, a property of a spatial process in which all statistical properties of an attribute depend only on the relative locations of attribute values. *See also* second-order stationarity, intrinsic stationarity.

stationing In the pipeline industry, another name for linear referencing. Stationing allows any point along a line feature representing a pipeline to be uniquely identified by its relative position along the line feature. *See also* linear referencing.

statistical surface [STATISTICS] Ordinal, interval, or ratio data represented as a surface in which the height of each area is proportional to a numerical value.

statistic F *See* F statistic.

steepest path A line that follows the steepest downhill direction on a surface. Paths terminate at the surface perimeter or in surface concavities or pits.

steradian [MATHEMATICS] The solid (conical) angle subtended at the center of a sphere of radius *r* by a bounded region on the surface of the sphere having an area *r* squared. There are 4π steradians in a sphere. *See also* radian.

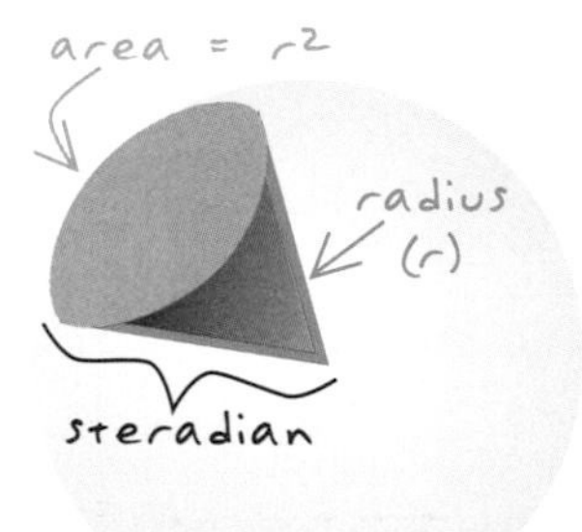

steradian

stereocompilation [CARTOGRAPHY] A map produced with a stereoscopic plotter using aerial photographs and geodetic control data.

stereogrammatic organization *See* visual hierarchy.

stereographic projection 1. [CARTOGRAPHY] A tangent planar projection that views the earth's surface from a point on the globe opposite the tangent point. 2. [CARTOGRAPHY] A secant planar projection that views the earth from a point on the globe opposite the center of the projection. *See also* projection.

stereometer [REMOTE SENSING] A stereoscope containing a micrometer for measuring the effects of parallax in a stereoscopic image.

stereomodel [REMOTE SENSING] The three-dimensional image formed where rays from points in the images of a stereoscopic pair intersect.

stereopair [REMOTE SENSING] Two aerial photographs of the same area taken from slightly different angles that when viewed together through a stereoscope produce a three-dimensional image. *See also* stereoscope.

stereopair

stereoplotter [REMOTE SENSING] An instrument that projects a stereoscopic image from aerial photographs, converts the locations of objects and landforms on the image to x-, y-, and z-coordinates, and plots these coordinates as a drawing or map.

stereoscope [REMOTE SENSING] A binocular device that produces the impression of a three-dimensional image from two overlapping images of the same area. *See also* stereopair.

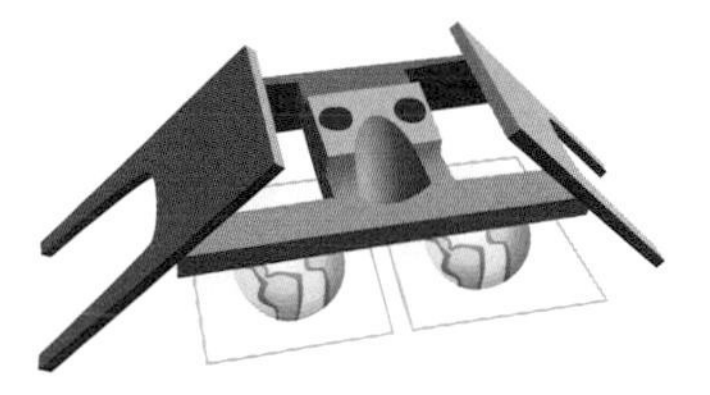

stereoscope

stereoscopic pair *See* stereopair.

stochastic model A model that includes a random component. The random component can be a model variable, or it can be added to existing input data or model parameters. *See also* model, deterministic model, Monte Carlo method.

stop impedance In network analysis, the time it takes for a stop to occur, used to compute the impedance of a path or tour. For example, when a school bus drops children off or picks them up at their homes, the stop impedance might be 2 minutes at each stop. *See also* impedance.

store market analysis A type of business analysis that uses mostly data about a store or stores, rather than about customers. Examples include ring studies and analyses of equal competition areas and drive-time areas.

store prospecting A type of business analysis that assesses the potential of a site by performing simple ring or drive-time analysis. *See also* site prospecting.

stream digitizing *See* stream mode digitizing.

streaming A technique for transferring data, usually over the Internet, in a real-time flow as opposed to storing it in a local file first. Streaming allows large multimedia files to be viewed before the entire file has been downloaded to a client's computer. When received by the client (local computer) the data is uncompressed and displayed using software designed to interpret and display the data rapidly.

stream mode digitizing A method of digitizing in which, as the cursor is moved, points are recorded automatically at preset intervals of either distance or time. *See also* point mode digitizing.

stream tolerance During stream mode digitizing, the minimum interval between vertices. Stream tolerance is measured in map units.

street-based mapping A form of digital mapping that links information to geographic locations and displays address locations as point features on a map.

street network A system of interconnecting lines and points that represent a system of roads for a given area. A street network provides the foundation for network analysis; for example, finding the best route or creating service areas.

stretch A display technique applied to the histogram of raster datasets, most often used to increase the visual contrast between cells. *See also* histogram.

stretch

string 1. A set of coordinates that defines a group of linked line segments. 2. [PROGRAMMING] A sequence of letters or numbers, or both, sometimes with a fixed length.

structure *See* drift.

Structured Query Language
See SQL.

structure line A line feature enforced in a TIN. There are two types of structure lines: hard and soft. Hard structure lines, also known as breaklines, represent interruptions in the slope of the surface. Soft structure lines are used to add information about the surface without implying a change in the surface behavior across the line. *See also* breakline.

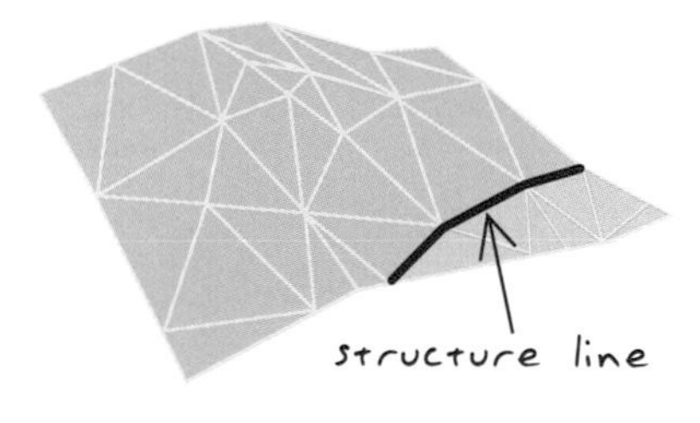

structure line

study area The geographic area treated in an analysis.

style [CARTOGRAPHY] An organized collection of predefined colors, symbols, properties of symbols, and map elements. Styles promote standardization and consistency in mapping products.

style sheet [INTERNET] A file or form that provides style and layout information, such as margins, fonts, and alignment, for tagged content within an XML or HTML document. Style sheets are frequently used to simplify XML and HTML document design, since one style sheet may be applied to several documents. Transformational style sheets may also contain code to transform the structure of an XML document and write its content into another document.

subtractive primary colors In printing, the three primary colors—cyan, magenta, and yellow—that when used as filters for white light remove blue, green, and red light, respectively. *See also* CMYK.

suitability model A model that weights locations relative to each other based on given criteria. Suitability models might aid in finding a favorable location for a new facility, road, or habitat for a species of bird.

supplemental contour [CARTOGRAPHY] A contour line placed between regularly spaced contours, used when the terrain change is not large enough to be depicted with consistent contour intervals. *See also* contour line.

surface A geographic phenomenon represented as a set of continuous data (such as elevation, geological boundaries, or air pollution); a spatial distribution which associates a single value with each position in a plane, usually associated with continuous attributes.

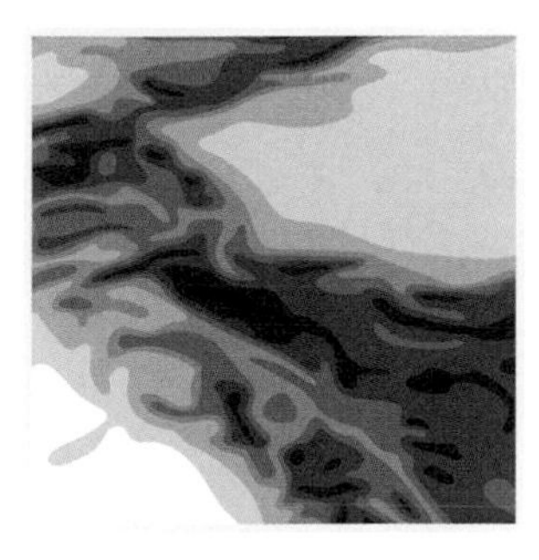

surface

surface fitting [STATISTICS] Generating a statistical surface that approximates the values of a set of known x,y,z points.

surface model A representation of a geographic feature or phenomenon that can be measured continuously across some part of the earth's surface (for example, elevation). A surface model is an approximation of a surface, generalized from sample data. Surface models are stored and displayed as rasters, TINs, or terrains.

surround element *See* map element.

surveying [SURVEYING] Measuring physical or geometric characteristics of the earth. Surveys are often classified by the type of data studied or by the instruments or methods used. Examples include geodetic, geologic, topographic, hydrographic, land, geophysical, soil, mine, and engineering surveys.

survey marker *See* survey monument.

survey monument [SURVEYING] An object, such as a metal disk, permanently mounted in the landscape to denote a survey station.

survey station [SURVEYING] A location on the earth that has been accurately determined by geodetic survey.

symbol [CARTOGRAPHY] A graphic used to represent a geographic feature or class of features. Symbols can look like what they represent (trees, railroads, houses), or they can be abstract shapes (points, lines, polygons), or characters. Symbols are usually explained in a map legend.

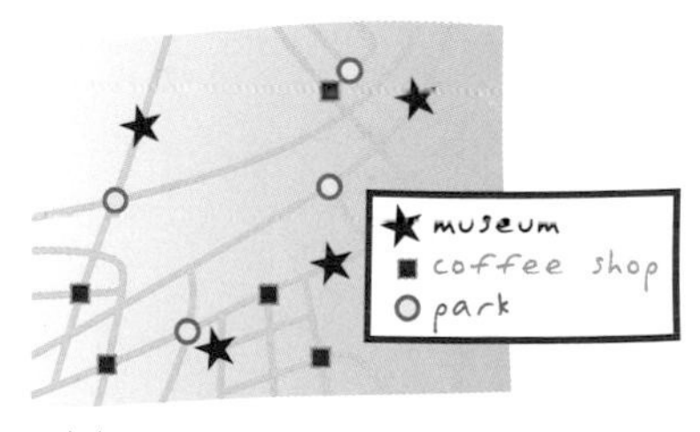

symbol

symbolization [CARTOGRAPHY] The process of devising a set of marks of appropriate size, color, shape, and pattern, and assigning them to map features to convey their characteristics at a given map scale.

symbology [CARTOGRAPHY] The set of conventions, rules, or encoding

S

systems that define how geographic features are represented with symbols on a map. A characteristic of a map feature may influence the size, color, and shape of the symbol used.

synchronous **1.** [PHYSICS] Occurring together, or at the same time. **2.** In data transmission, precisely timed and steady transmission of information that allows for higher rates of data exchange. **3.** [PROGRAMMING] A series of actions or events that must occur in a specified sequence. For example, a program that launches another program and waits for it to finish before continuing is said to be synchronous. *See also* asynchronous.

syntax [PROGRAMMING] The structural rules for using statements in a command or programming language.

T

table A set of data elements arranged in rows and columns. Each row represents a single record. Each column represents a field of the record. Rows and columns intersect to form cells, which contain a specific value for one field in a record.

FID	name	shape
1	road	line
2	market	point
3	lake	polygon

table

table

tabular data Descriptive information, usually alphanumeric, that is stored in rows and columns in a database and can be linked to spatial data. *See also* table.

tangent projection [CARTOGRAPHY] A projection whose surface touches the globe's without piercing it. A tangent planar projection touches the globe at one point, while tangent conic and cylindrical projections touch the globe along a line. At the point or line of tangency, the projection is free from distortion. *See also* secant projection.

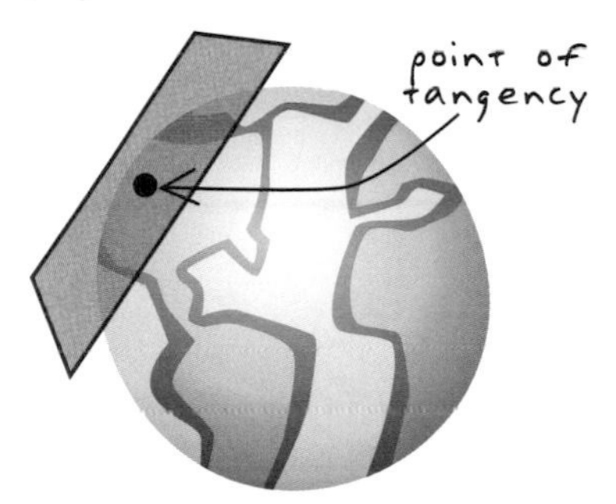

tangent projection

TCP/IP [INTERNET] *Acronym for Transmission Control Protocol/Internet Protocol.* The most common protocol for Internet traffic. The Transmission Control Protocol (TCP) is a communication protocol layered above the Internet Protocol (IP), which is a suite of nonproprietary communication protocols, or sets of rules, that allow computers to send and receive data over networks. *See also* SOAP.

TDOP *See* DOP.

temporal data Data that specifically refers to times or dates. Temporal data may refer to discrete events, such as lightning strikes; moving objects, such as trains; or repeated observations, such as counts from traffic sensors. *See also* thematic data, continuous data, discrete data, spatial data.

temporal GIS An emerging capability in GIS for integrating temporal data with location and attribute data.

tessellation The division of a two-dimensional area into polygonal tiles, or a three-dimensional area into polyhedral blocks, in such a way that no figures overlap and there are no gaps. *See also* Thiessen polygons.

text envelope A rectangle that bounds a text string.

text modifier *See* attribute.

texture A digital representation of the surface of a feature.

thematic data Features of one type that are generally placed together in a single layer.

thematic map [CARTOGRAPHY] A map designed to convey information about a single topic or theme, such as population density or geology. *See also* choropleth map.

theodolite [SURVEYING] An instrument for measuring vertical and horizontal angles, consisting of an alidade, a telescope, and graduated circles mounted vertically and horizontally. *See also* alidade.

Thiessen polygons [MATHEMATICS] Polygons generated from a set of sample points. Each Thiessen polygon defines an area of influence around its sample point, so that any location inside the polygon is closer to that point than any of the other sample points. Thiessen polygons are named for the American meteorologist Alfred H. Thiessen (1872–1931). *See also* Delaunay triangles, Voronoi diagram.

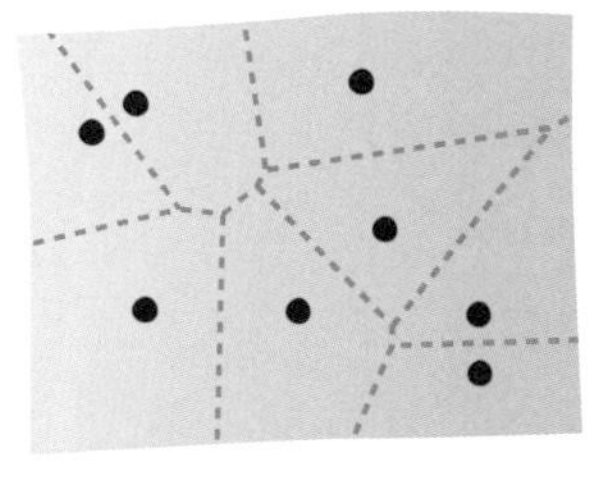

Thiessen polygons

thinning *See* weeding.

third normal form The third level of guidelines for designing table and data structures in a relational database. The third-normal-form guideline incorporates the guidelines of first and second normal form; in addition, it recommends removing from a table those columns that do not depend on the table's primary key. A database that follows these guidelines is said to be in third normal form. *See also* normal form.

three-dimensional shape *See* 3D shape.

T

three-tier configuration [COMPUTING] A software configuration in which three software applications (commonly a client program, application server, and database server) work together to accomplish a task. *See also* two-tier configuration, application server.

threshold ring analysis In business analysis, an operation that creates rings that contain a given population around a store or stores on a map.

thumbnail A miniaturized version of a graphics file. A thumbnail can be used as a visual index for larger data or images.

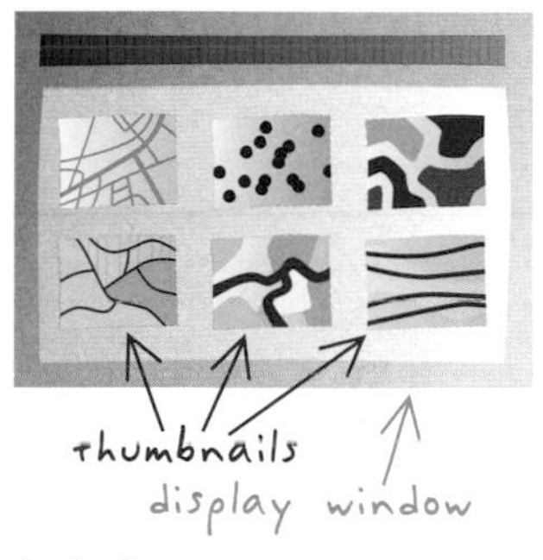

thumbnail

tick marks 1. [CARTOGRAPHY] Graphics that mark divisions of measurement on a scale bar. 2. [CARTOGRAPHY] Short, regularly spaced lines along the edge of an image or neatline that indicate intervals of distance, such as the intersection of longitude and latitude lines to denote the graticule.

tidal datum [GEODESY] A vertical datum in which zero height is defined by a particular tidal surface, often mean sea level. Examples of tidal surfaces include mean sea level, mean low water springs, and mean lower low water. Most traditional vertical geodetic datums are tidal datums. *See also* datum, vertical geodetic datum, local datum.

tie point 1. [SURVEYING] A point whose location is determined by a tie survey. 2. [REMOTE SENSING] A point in a digital image or aerial photograph that represents the same location in an adjacent image or aerial photograph. Usually expressed as a pair, tie points can be used to link images and create mosaics. *See also* tie survey.

tie survey [SURVEYING] A survey that uses a point of known location on the ground to determine the location of a second point.

TIGER *Acronym for Topologically Integrated Geographic Encoding and Referencing.* The nationwide digital database developed for the 1990 census, succeeding the DIME format. TIGER files contain street address ranges, census tracts, and block boundaries. *See also* DIME, GBF/DIME.

TIGER/Line files A digital database of geographic features, covering the entire United States and its territories, that provides a topological description of the geographic structure of these areas. The files are a public product created from the

U.S. Census Bureau Topologically Integrated Geographic Encoding and Referencing (TIGER) database. TIGER/Line files define the locations and spatial relationships of streets, rivers, railroads, and other features to each other and to the numerous geographic entities for which the Census Bureau tabulates data from its censuses and sample surveys. *See also* TIGER.

tight coupling A high or complex degree of interconnections between the components within a program or between programs, that requires substantial overlap between methods, ontologies, class definitions, and so on. *See also* loose coupling.

tiling An internal subsetting of a spatial dataset (commonly raster) into a manageable rectangular set, or rows and columns of pixels, typically used to process or analyze a large raster dataset without consuming vast quantities of computer memory. *See also* tessellation.

TIN *Acronym for triangulated irregular network.* A vector data structure that partitions geographic space into contiguous, nonoverlapping triangles. The vertices of each triangle are sample data points with x-, y-, and z-values. These sample points are connected by lines to form Delaunay triangles. TINs are used to store and display surface models. *See also* Delaunay triangulation.

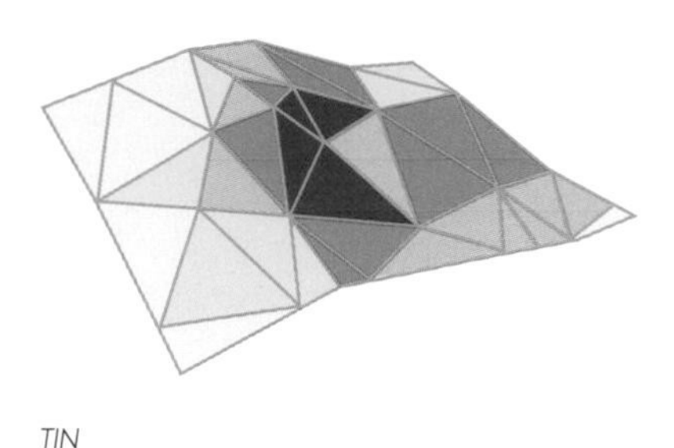

TIN

Tissot's indicatrix [CARTOGRAPHY] A graphical representation of the spatial distortion at a particular map location. The indicatrix is the figure that results when a circle on the earth's surface is plotted to the corresponding point on a map. The shape, size, and orientation of an indicatrix at any given point depend on the map projection used. In conformal (shape-preserving) projections, the indicatrix is a circle; in nonconformal projections, it is an ellipse at most locations. As a visual aid, indicatrices convey a general impression of distortion; as mathematical tools, they can be used to quantify distortion of scale and angle precisely. The indicatrix is named for Nicolas Auguste Tissot, the French mathematician who developed it. *See also* projection, distortion.

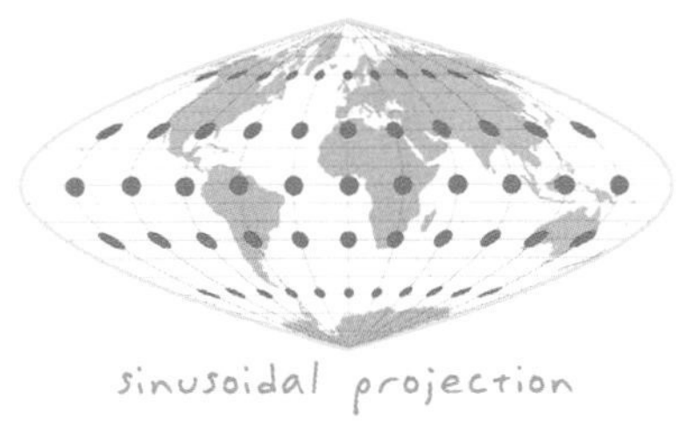

Tissot's indicatrix

Tissot indicatrix *See* Tissot's indicatrix.

TLM [CARTOGRAPHY] *Acronym for topographic line map.* A map that uses line contours to show elevations and depressions of the earth's surface. Topographic line maps may be used to portray topography, elevations, infrastructure, hydrography and vegetation. *See also* topographic map.

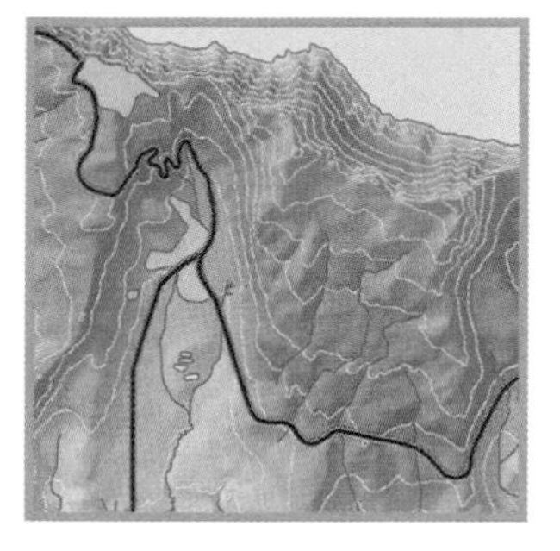

TLM

Tobler's First Law of Geography [GEOGRAPHY] A formulation of the concept of spatial autocorrelation by the geographer Waldo Tobler (1930–), which states "Everything is related to everything else, but near things are more related than distant things." *See also* autocorrelation.

tolerance The minimum or maximum variation allowed when processing or editing a geographic feature's coordinates. For example, during editing, if a second point is placed within the snapping tolerance distance of an existing point, the second point will be snapped to the existing point.

to-node Of an arc's two endpoints, the last one digitized. From- and to-nodes give an arc left and right sides and, therefore, direction. *See also* from-node.

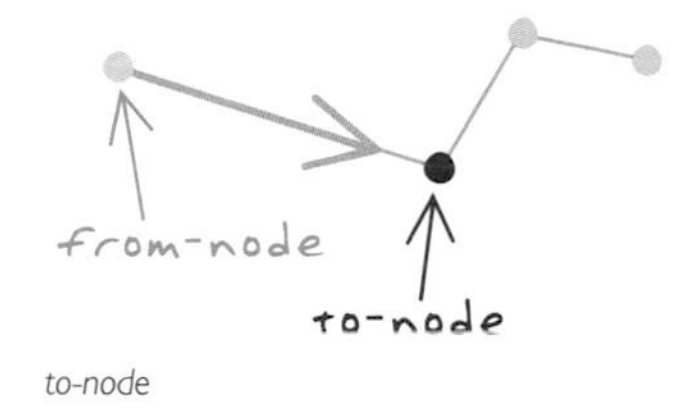

to-node

tool [COMPUTING] A command that requires interaction with the GUI before an action is performed. For example, a zoom tool requires a user to use the mouse to click on or draw a box over a digital map before the tool will cause the map to be redrawn at a larger scale.

toolbar [COMPUTING] A graphical user interface (GUI) with buttons that allow users to execute software commands.

topographic line map *See* TLM.

topographic map [CARTOGRAPHY] A map that represents the vertical and horizontal positions of features, showing relief in some measurable form, such as contour lines, hypsometric tints, and relief shading. *See also* topography, TLM, planimetric map.

topography [CARTOGRAPHY] The study and mapping of land surfaces, including relief (relative positions and elevations) and the position of natural and constructed features. *See also* topographic map.

T

topological overlay *See* overlay.

topology 1. [ESRI SOFTWARE] In geodatabases, the arrangement that constrains how point, line, and polygon features share geometry. For example, street centerlines and census blocks share geometry, and adjacent soil polygons share geometry. Topology defines and enforces data integrity rules (for example, there should be no gaps between polygons). It supports topological relationship queries and navigation (for example, navigating feature adjacency or connectivity), supports sophisticated editing tools, and allows feature construction from unstructured geometry (for example, constructing polygons from lines). 2. [MATHEMATICS] The branch of geometry that deals with the properties of a figure that remain unchanged even when the figure is bent, stretched, or otherwise distorted. 3. [ESRI SOFTWARE] In an ArcInfo coverage, the spatial relationships between connecting or adjacent features in a geographic data layer (for example, arcs, nodes, polygons, and points). Topological relationships are used for spatial modeling operations that do not require coordinate information. **For more information about topology, see* The benefits of using topology in GIS *on page 267.*

topology error *See* error.

toponym [GEOGRAPHY] A place name.

topo sheet *See* quadrangle.

tour A path through a network that visits each stop in the network only once and then returns to its point of origin. The path is determined based on a user-specified set of criteria (such as shortest distance or fastest time).

township 1. [SURVEYING] In the United States, a quadrangle approximately 6 miles on a side, bounded by meridians and parallels and containing 36 sections.

township

2. A governmental subdivision, which may vary from the standard size and shape. *See also* section.

tracing *See* network trace.

tracking data *See* temporal data.

tract *See* census tract.

transaction 1. [COMPUTING] A group of data operations that comprise a complete operational task, such as inserting a row into a table. 2. [INTERNET] An interaction with a Web service. A transaction includes a request to, and a response from, a Web service.

transformation The process of converting the coordinates of a map or an image from one system to another, typically by shifting, rotating, scaling, skewing, or projecting

T

them. *See also* affine transformation, projective transformation, Helmert transformation, Cartesian coordinate system, rectification.

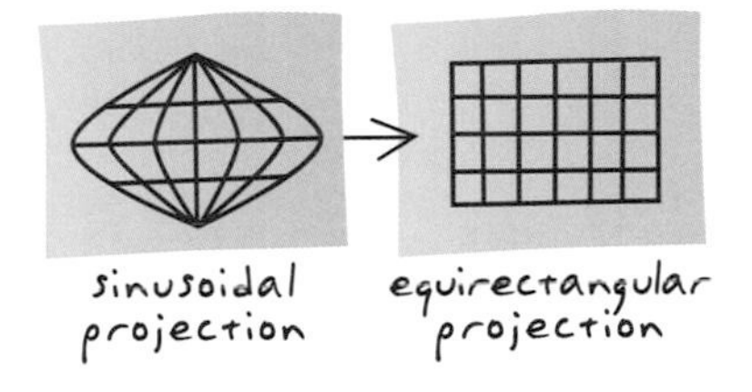

transformation

transit rule [SURVEYING] A rule for adjusting the closure error in a traverse. The transit rule distributes the closure error by changing the northings and eastings of each traverse point in proportion to the northing and easting differences in each course. More specifically, a correction is computed for each northing coordinate as the difference in the course's northings divided by the sum of all the courses' northing differences. Similarly, a correction is computed for each easting coordinate using the easting coordinate differences. The corrections are applied additively to each successive coordinate pair, until the final coordinate pair is adjusted by the whole closure error amount. The transit rule assumes that course directions are measured with a higher degree of precision than the distances. Usually, observed angles are balanced for angular misclosure prior to applying a transit rule adjustment, and corrections are proportional to the x and y components of the measured line. The transit rule is used infrequently since it is only valid in cases in which the measured lines are approximately parallel to the grid of the coordinate system in which the traverse is computed. *See also* closure error.

translation **1.** Adding a constant value to a coordinate. **2.** Converting data from one format to another, usually in order to move it from one system to another. *See also* transformation.

transverse aspect [CARTOGRAPHY] A map projection whose line of tangency is oriented along a meridian rather than along the equator.

traveling salesperson problem A Hamiltonian circuit problem in which a salesperson must find the most efficient way to visit a series of stops, then return to the starting location. In the original version of the problem, each stop may be visited only once. *See also* Hamiltonian circuit.

traverse **1.** [SURVEYING] A predefined path or route across or over a set of geometric coordinates. **2.** [SURVEYING] A method of surveying in which lengths and directions of lines between points on the earth are obtained by or from field measurements across terrain or a digital elevation model. *See also* open traverse, closed loop traverse.

tree *See* hierarchical database.

tree data structure A common data structure consisting of a set of nodes—basic units of data—linked hierarchically. Each node can contain one or more subordinate nodes within it, in which case it is called a parent node. The subordinate nodes are called child nodes. A node without a parent node is the root node; a node without one or more child nodes is called a leaf node. A tree data structure is used to manipulate hierarchical data and make it easily searchable. *See also* hierarchical database.

trend [STATISTICS] In a spatial model, nonrandom variation in the value of a variable that can be described by a mathematical function such as a polynomial. *See also* short-range variation.

trend surface analysis [MATHEMATICS] A surface interpolation method that fits a polynomial surface by least-squares regression through the sample data points. This method results in a surface that minimizes the variance of the surface in relation to the input values. The resulting surface rarely goes through the sample data points. This is the simplest method for describing large variations, but the trend surface is susceptible to outliers in the data. Trend surface analysis is used to find general tendencies of the sample data, rather than to model a surface precisely.

triangle 1. [MATHEMATICS] Any closed, three-sided, two-dimensional polygon. 2. [MATHEMATICS] A face on a TIN surface. Each triangle on a TIN surface is defined by three edges and three nodes and is adjacent to one to three other triangles on the surface. TIN triangles can be used to derive aspect and slope information and may be attributed with tag values.

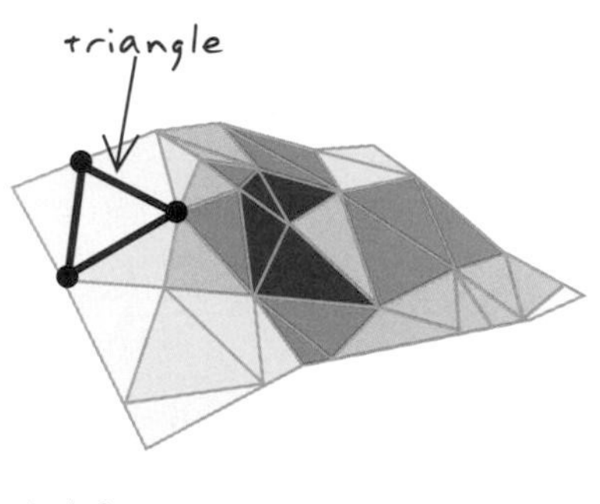

triangle 2

triangulated irregular network *See* TIN.

triangulation [SURVEYING] Locating positions on the earth's surface using the principle that if the measures of one side and the two adjacent angles of a triangle are known, the other dimensions of the triangle can be determined. Surveyors begin with a known length, or baseline, and from each end use a theodolite to measure the angle to a distant point, forming a triangle. Once the lengths of the two sides and the other angle are known, a network of triangles can be extended from the first. *See also* trilateration.

trilateration [SURVEYING] Determining the position of a point on the earth's surface with respect to two other points by measuring the distances between all three points. *See also* triangulation.

T

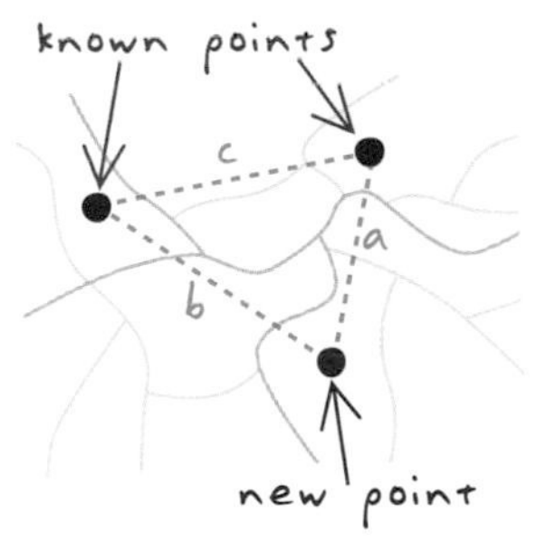

trilateration

true bearing [NAVIGATION] A bearing measured relative to true north. *See also* bearing.

true curve *See* parametric curve.

true-direction projection *See* azimuthal projection.

true north [GEOGRAPHY] The direction from any point on the earth's surface to the geographic north pole. *See also* grid north, magnetic north.

TSP *See* traveling salesperson problem.

tuple An individual row or record in a database table. Each tuple records the values for the columns defined in the table. *See also* record.

turn In network analysis, a movement that explicitly models transitions between edge elements during navigation.

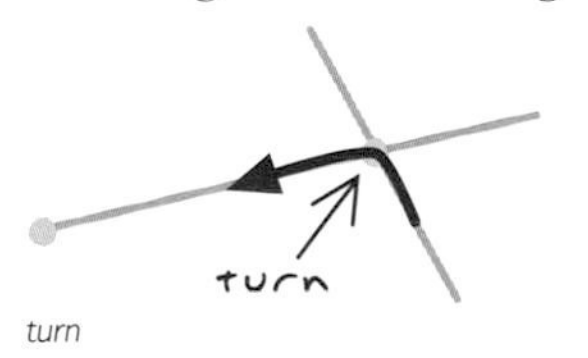

turn

turn-by-turn maps [CARTOGRAPHY] A series of small maps detailing where route segments meet.

turn impedance In network analysis, the cost of making a turn at a network node. The impedance for making a left turn, for example, can be different from the impedance for making a right turn or a U-turn at the same place.

two-tier configuration [PROGRAMMING] A software configuration in which two software applications (commonly a client and a server) work together to accomplish a task. *See also* three-tier configuration.

type library [PROGRAMMING] A collection of reusable classes, interfaces, enumerations, and so on that can be included in programs. Type libraries usually have the extension .olb.

T

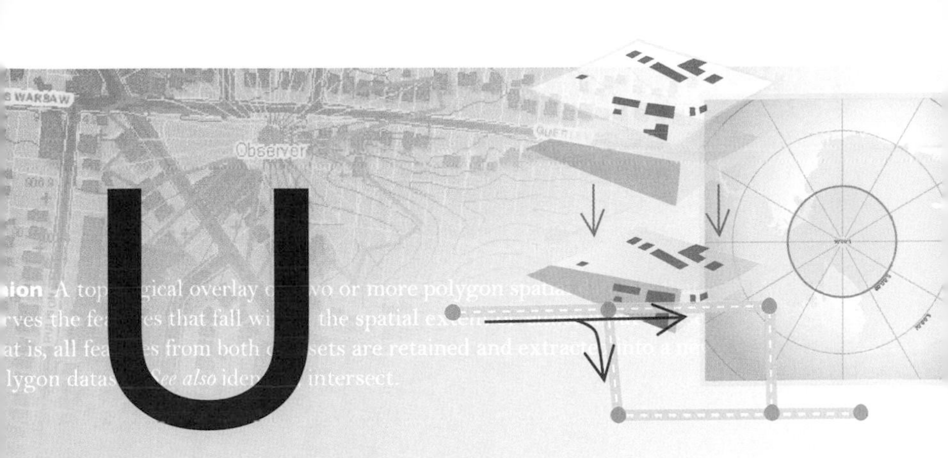

UDDI *Acronym for Universal Description, Discovery, and Integration.* An XML-based standard for creating online directories of Web services.

UI [PROGRAMMING] *Acronym for User Interface.* The portion of a computer's hardware and software that facilitates human interaction. The UI includes items that can be displayed on screen, and interacted with by using the keyboard, mouse, video, printer, and data capture.

UML [PROGRAMMING] *Acronym for Unified Modeling Language.* A modeling language that uses a series of diagrams to model the objects in a system.

uncertainty The degree to which the measured value of some quantity is estimated to vary from the true value. Uncertainty can arise from a variety of sources, including limitations on the precision or accuracy of a measuring instrument or system; measurement error; the integration of data that uses different scales or that describe phenomena differently; conflicting representations of the same phenomena; the variable, unquantifiable, or indefinite nature of the phenomena being measured; or the limits of human knowledge. Uncertainty is often used to describe the degree of accuracy of a measurement. *See also* accuracy, error, vagueness.

unclosed parcel [SURVEYING] A parcel that is only partially defined or that is missing a sequence of one or more lines that would otherwise close the parcel back onto its point of beginning.

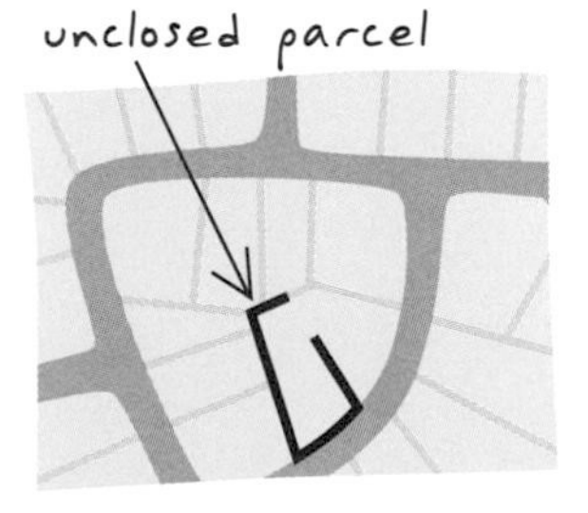

unclosed parcel

undershoot A line that falls short of another line that it should intersect. *See also* dangling arc.

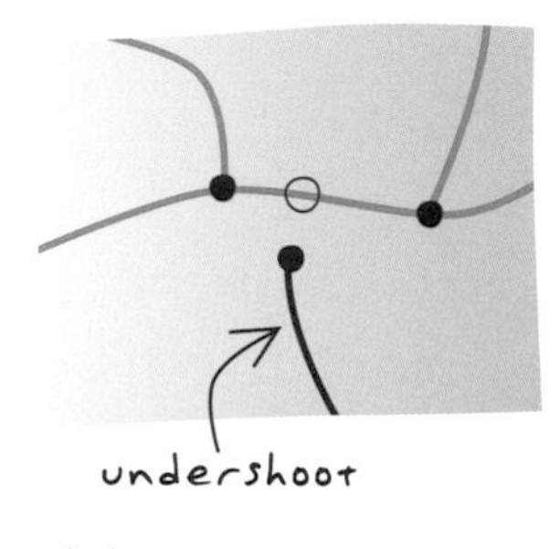

undershoot

undevelopable surface [CARTOGRAPHY] A surface, such as the earth's, that cannot be flattened into a map without stretching, tearing, or squeezing it. To produce a flat map of the round earth, its three-dimensional surface must be projected onto a developable shape such as a plane, cone, or cylinder. *See also* projection.

undirected network flow A network state in which each edge may or may not have an associated direction of flow. In an undirected network flow, the resource that traverses a network's components can decide which direction to take, such as traffic in transportation systems. *See also* directed network flow.

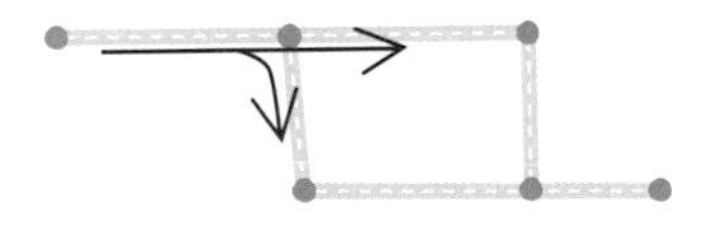

undirected network flow

uninitialized flow direction [ESRI SOFTWARE] A condition that occurs in a network when an edge feature is not connected through the network to sources and sinks or if the edge feature is only connected to sources and sinks through disabled features.

union A topological overlay of two or more polygon spatial datasets that preserves the features that fall within the spatial extent of either input dataset; that is, all features from both datasets are retained and extracted into a new polygon dataset. *See also* identity, intersect.

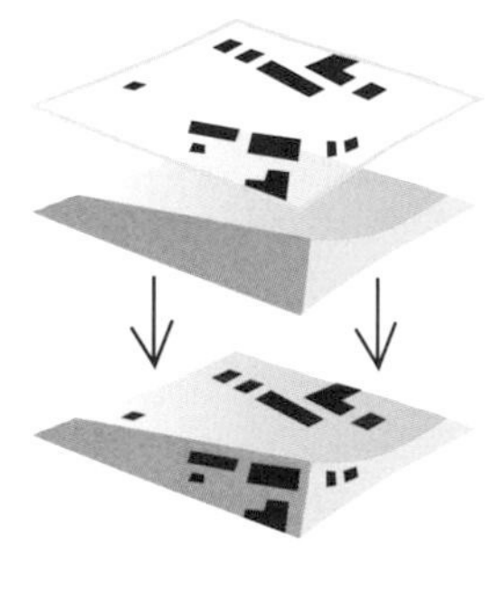

union

United States Geological Survey *See* U.S. Geological Survey.

unit of measure A standard quantity used for measurements such as length, area, and height. *See also* angular unit, linear unit, map unit, page unit, display unit, distance unit.

univariate analysis [STATISTICS] Any statistical method for evaluating a single variable, rather than the relationship between two or more variables.

univariate distribution [STATISTICS] A function for a single variable that gives the probabilities that the variable will take a given value.

Universal Description, Discovery, and Integration *See* UDDI.

universal kriging [STATISTICS] A kriging method often used on data with a significant spatial trend, such as a sloping surface. In universal kriging, the expected values of the sampled points are modeled as a polynomial trend. Kriging is carried out on the difference between this trend and the values of the sampled points. *See also* kriging.

universal polar stereographic [CARTOGRAPHY] A projected coordinate system that covers all regions not included in the UTM coordinate system; that is, regions above 84 degrees north and below 80 degrees south. Its central point is either the north or south pole. *See also* UTM.

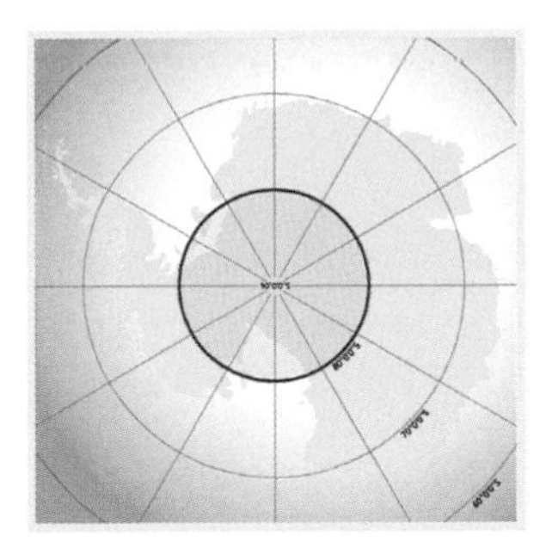

universal polar stereographic

Universal Soil Loss Equation An erosion model developed by the Agricultural Research Service of the United States Department of Agriculture that computes average annual soil loss caused by rainfall and associated overland flow. Factors used in the equation include rainfall, soil characteristics, topography, and land use and land cover. Each major factor is divided into numerous subfactors.

universal time [ASTRONOMY] A time-keeping system that defines local time throughout the world by relating it to time at the prime meridian. Universal time is based on the average speed at which the earth rotates on its axis. For official purposes, universal time has been replaced by coordinated universal time; universal time is, however, still used in navigation and astronomy. Different versions of universal time correct for irregularities in the earth's rotation and orbit. *See also* coordinated universal time.

universal transverse Mercator *See* UTM.

UNIX time [ASTRONOMY] The number of seconds, in coordinated universal time format, since January 1, 1970 (the start of the UNIX system). *See also* coordinated universal time.

unprojected coordinates *See* geographic coordinates.

UPS *See* universal polar stereographic.

upstream In network tracing, the direction along a line or edge that opposes the direction of flow. *See also* downstream, directed network flow.

urban geography [GEOGRAPHY] The field of geography concerning the spatial and cultural patterns and processes of cities and neighborhoods. *See also* geography.

Urban Vector Map *See* UVMap.

URL [INTERNET] *Acronym for Uniform Resource Locator.* A standard format for the addresses of Web sites. A URL may look like this: http://www.esri.com. The first part of the address indicates what protocol to use (such as http: or ftp:), while the second part specifies the IP address or the host name (including the domain name) where the Web site is located. An optional third part may specify the path to a specific file or resource (http://www.esri.com/products.html).

user interface [COMPUTING] The aspects of a computer system or program with which a software user can interact, and the commands and mechanisms used to control its operation and input data. *See also* GUI.

U.S. Geological Survey *Acronym for United States Geological Survey.* A scientific agency of the U.S. government, part of the Department of the Interior. The U.S. Geological Survey is a fact-finding research agency that monitors, analyzes, and provides scientific understanding about natural resource issues and conditions, the environment, and natural hazards. The U.S. Geological Survey is the primary civilian mapping agency in the United States. It produces digital and paper map products; aerial photography; and remotely sensed data on land cover, hydrology, geology, biology, and geography.

USGS *See* U.S. Geological Survey.

USLE *See* Universal Soil Loss Equation.

U.S. National Geodetic Survey The U.S. government agency responsible for maintaining the National Spatial Reference System (NSRS), the national coordinate system of the United States.

UT *See* universal time.

UTC *See* coordinated universal time.

UTM [CARTOGRAPHY] *Acronym for universal transverse Mercator.* A projected coordinate system that divides the world into 60 north and south zones, 6 degrees wide.

UVMap [CARTOGRAPHY] *Acronym for Urban Vector Map.* A vector-based data product in vector product format (VPF), typically at larger scales ranging from 1:2,000 to 1:25,000. UVMap data is typically collected over densely populated urban areas.

UVMap

U

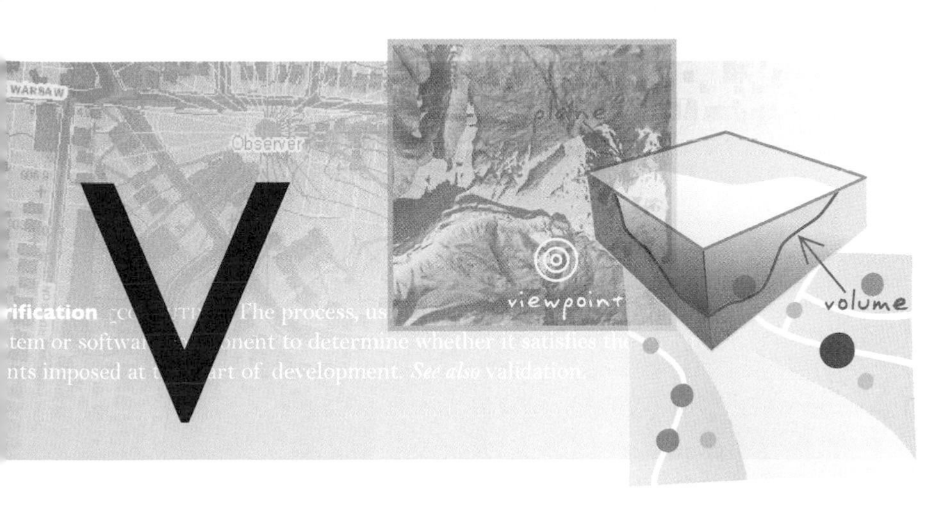

vagueness A state of uncertainty in data classification that exists when an attribute applies to an indeterminate quality of an object or describes an indefinite quantity. For example, the classification of an area of land as the range of golden-winged warblers (a rare species of bird) is vague for two reasons. The area populated by the birds is indefinite: it is changing constantly and can never be precisely defined. The term "range" is also somewhat vague since the birds migrate and occupy the territory for only part of the year. *See also* ambiguity, uncertainty.

valency [ESRI SOFTWARE] In coverages, the number of arcs that begin or end at a node.

validation **1.** [COMPUTING] The process, using formal methods, of evaluating a system or software component to determine whether it functions as expected and achieves the intended results. **2.** The process, using formal methods, of evaluating the integrity and correctness of data or a measurement. **3.** In modeling, the evaluation of a method to show whether it is assessing the parameter of interest rather than something else. *See also* verification, cross validation.

value **1.** [MATHEMATICS] A measurable quantity that may be passed to a function. Values are either assigned or determined by calculation. **2.** The lightness or darkness of a color. **3.** [PHYSICS] The brightness of a color or how much light it reflects; for instance, blue, light blue, dark blue. *See also* saturation, hue.

variable **1.** [MATHEMATICS] A symbol or placeholder that represents a changeable value or a value that has not yet been assigned.

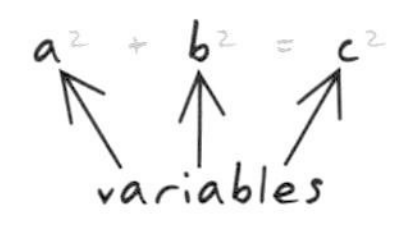

variable 1

2. [COMPUTING] A symbol or quantity that can represent any value or set of values, such as a text string or number. Variables may change depending on how they are used and applied.

variance [STATISTICS] A numeric description of how values in a distribution vary or deviate from the mean. The larger the variance, the greater the dispersion of values around the mean. The standard deviation for a distribution is the square root of the variance. *See also* standard deviation.

variance-covariance matrix [SURVEYING] The symmetric 3x3 matrix that mathematically expresses the correlation between errors in coordinates x, y, and z.

variogram [STATISTICS] A function of the distance and direction separating two locations that is used to quantify dependence. The variogram is defined as the variance of the difference between two variables at two locations. The variogram generally increases with distance and is described by nugget, sill, and range parameters. If the data is stationary, then the variogram and the covariance are theoretically related to each other. *See also* covariance, nugget, range, sill, semivariogram.

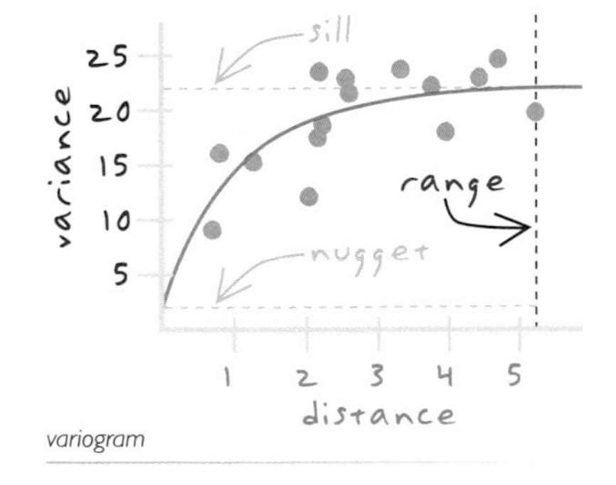

variogram

variography [STATISTICS] The process of examining spatial dependence using a variogram; a set of procedures (as much art as science) for interpreting variograms.

VDOP *See* DOP.

vector 1. A coordinate-based data model that represents geographic features as points, lines, and polygons. Each point feature is represented as a single coordinate pair, while line and polygon features are represented as ordered lists of vertices. Attributes are associated with each vector feature, as opposed to a raster data model, which associates attributes with grid cells.

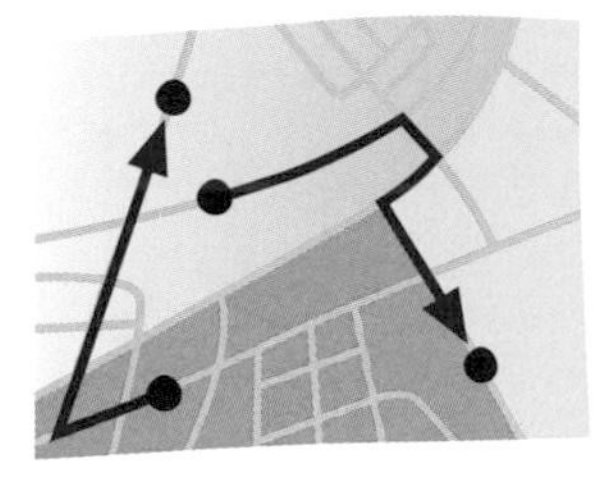

vector

2. Any quantity that has both magnitude and direction. *See also* raster.

vector data model A representation of the world using points, lines, and polygons. Vector models are useful for storing data that has discrete boundaries, such as country borders, land parcels, and streets. *See also* raster data model.

vectorization The conversion of raster data (an array of cell values)

V

to vector data (a series of points, lines, and polygons). *See also* rasterization, batch vectorization, centerline vectorization, interactive vectorization.

vector model *See* vector data model.

Vector Product Format *See* VPF.

verbal scale [CARTOGRAPHY] A map scale that expresses the relationship between distance on the map and distance on the ground in words; for example, "One inch represents 20 miles." *See also* scale bar, representative fraction.

verification [COMPUTING] The process, using formal methods, of evaluating a system or software component to determine whether it satisfies the requirements imposed at the start of development. *See also* validation.

version 1. In databases, an alternative state of the database that has an owner, a description, a permission (private, protected, or public), and a parent version. Versions are not affected by changes occurring in other versions of the database. **2.** [COMPUTING] An edition of a software product that incorporates major changes to the software from the previous version. A version is often called a release.

version merging The process of reconciling two versions of a dataset into a common version. If conflicting edits have been made in either of the merged versions, these conflicts are resolved, either automatically or by an interactive process.

vertex 1. [MATHEMATICS] One of a set of ordered x,y coordinate pairs that defines the shape of a line or polygon feature.

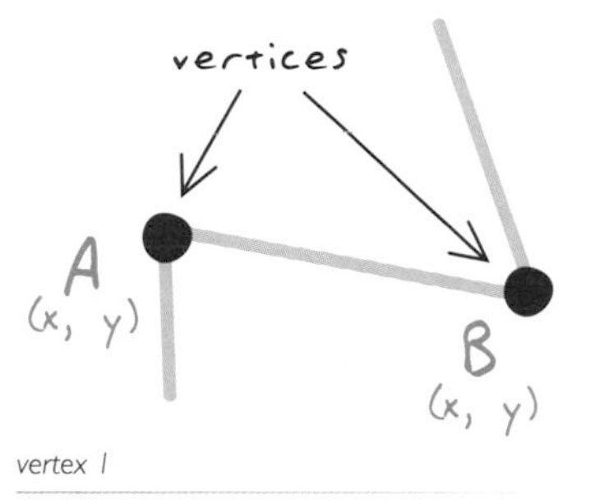

vertex 1

2. [MATHEMATICS] The junction of lines that form an angle. **3.** [GEODESY] The highest point of a feature. *See also* node.

vertical control *See* control.

vertical control datum *See* vertical geodetic datum.

vertical coordinate system [CARTOGRAPHY] A reference system that defines the location of z-values relative to a surface. The surface may be gravity related, such as a geoid, or a more regular surface like a spheroid or sphere. *See also* coordinate system.

V

vertical exaggeration [CARTOGRAPHY] A multiplier applied uniformly to the z-values of a three-dimensional model to enhance the natural variations of its surface. Scenes may appear too flat when the range of x- and y-values is much larger than the z-values. Setting vertical exaggeration can compensate for this apparent flattening by increasing relief.

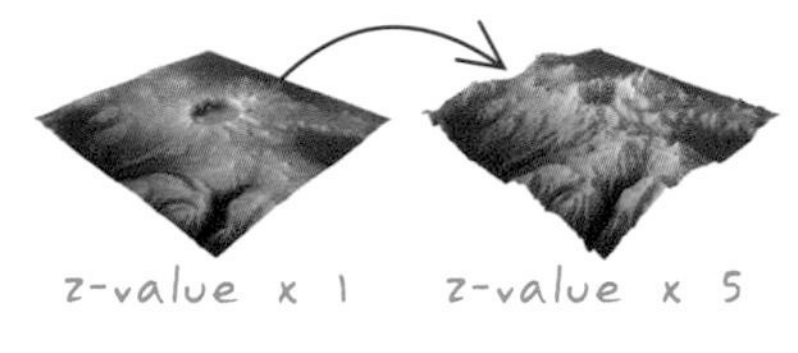

vertical exaggeration

vertical geodetic datum [GEODESY] A geodetic datum for any extensive measurement system of heights on, above, or below the earth's surface. Traditionally, a vertical geodetic datum defines zero height as the mean sea level at a particular location or set of locations; other heights are measured relative to a level surface passing through this point. Examples include the North American Vertical Datum of 1988; the Ordnance Datum Newlyn (used in Great Britain); and the Australian Height Datum. *See also* datum, geodetic datum, elevation.

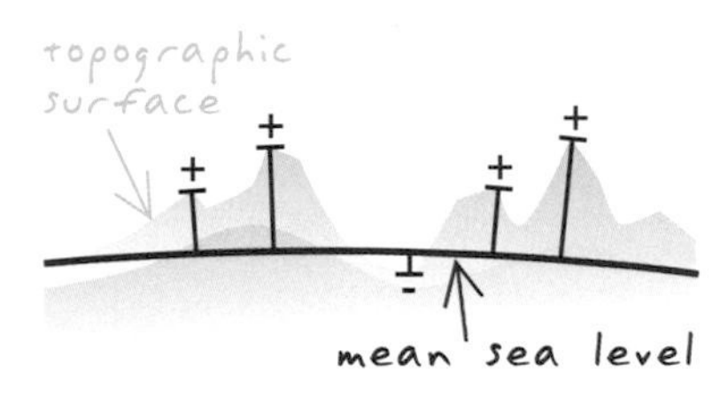

vertical geodetic datum

vertical line *See* plumb line.

vertical photograph [REMOTE SENSING] An aerial photograph taken with the camera lens pointed straight down. *See also* aerial photograph.

vertical photograph

viewshed The locations visible from one or more specified points or lines. Viewshed maps are useful for such applications as finding well-exposed places for communication towers, or hidden places for parking lots.

viewshed

virtual directory [INTERNET] A directory name, used in a URL, that corresponds to a real or actual directory on a Web server. *See also* directory.

virtual table [COMPUTING] A logical table in a database that stores a pointer to the data rather than the data itself.

V

visible scale range A minimum and maximum value that a map scale must fall between in order for the map layers to be displayed.

visual center [CARTOGRAPHY] The point on a rectangular map or image to which the eye is drawn. The visual center lies slightly (about 5 percent of the total height) above the geometric center of the page.

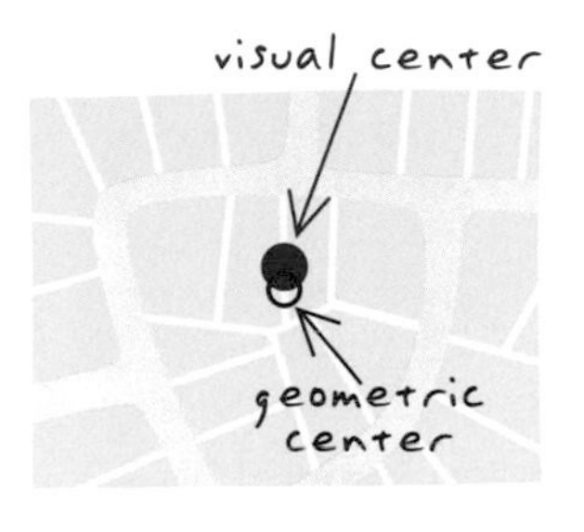

visual center

visual hierarchy [CARTOGRAPHY] The presentation of features on a map in a way that implies relative importance, usually achieved with visual contrast.

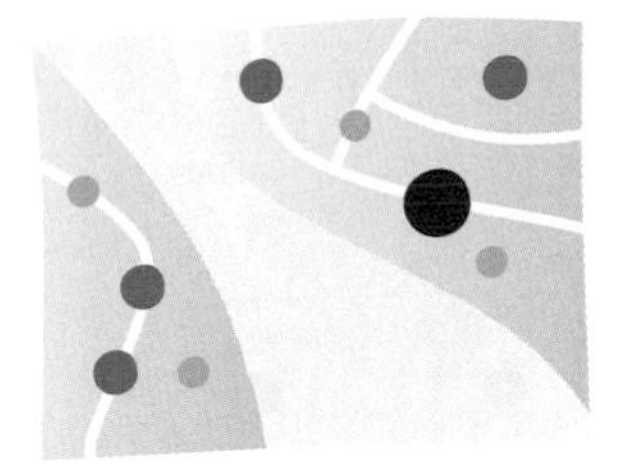

visual hierarchy

visualization The representation of data in a viewable medium or format. In GIS, visualization is used to organize spatial data and related information into layers that can be analyzed or displayed as maps, three-dimensional scenes, summary charts, tables, time-based views, and schematics.

VMap *Acronym for Vector Map.* A vector-based data product in vector product format (VPF) at several scales divided into groups, referred to as levels. For example, VMap Level 1 includes vector maps at a scale of 1:250,000, and VMap Level 2 includes vector maps at a scale of 1:50,000. *See also* VPF.

volume 1. In a TIN, the space (measured in cubic units) between a TIN surface and a plane at a specified elevation. Volume may be calculated above or below the plane. **2.** [MATHEMATICS] The space contained within any geometric solid, usually expressed in cubic units.

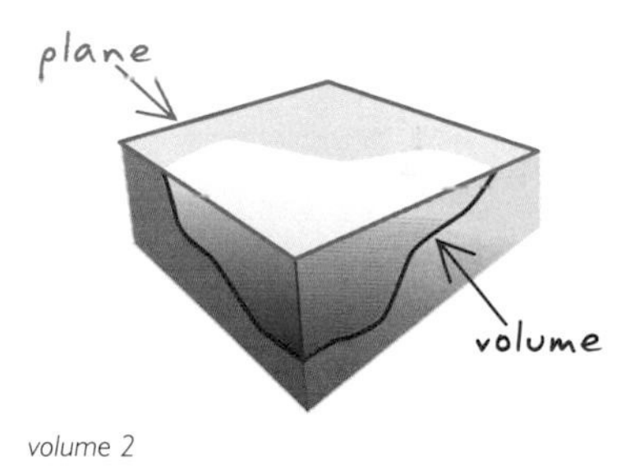

volume 2

Voronoi diagram [MATHEMATICS] A partition of space into areas, or cells, that surround a set of geometric objects (usually points). These cells, or polygons, must satisfy the criteria for Delaunay triangles. All locations within an area are closer to the object

V

it surrounds than to any other object in the set. Voronoi diagrams are often used to delineate areas of influence around geographic features. Voronoi diagrams are named for the Ukrainian mathematician Georgy Fedoseevich Voronoi (1868–1908). *See also* Delaunay triangulation, Thiessen polygons, Delaunay triangles.

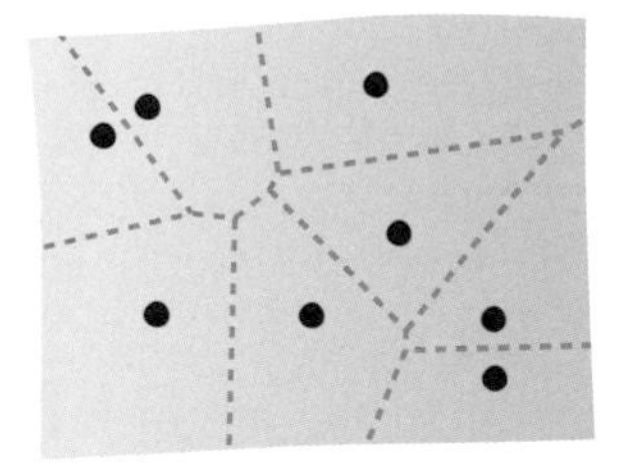

Voronoi diagram

voxel A three-dimensional pixel used to display and rotate three-dimensional images. *See also* pixel.

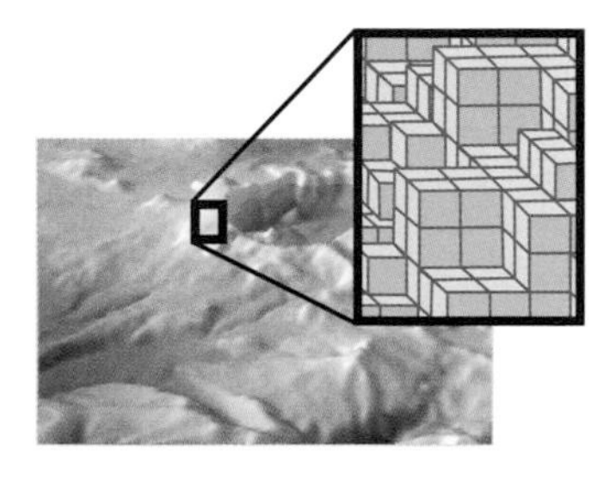

voxel

V

VPF *Acronym for Vector Product Format.* A vendor-neutral data format used to structure, store, and access geographic data according to a defined standard.

VPF dataset *See* VPF.

VPF feature class *See* feature class.

W3C [INTERNET] *Acronym for World Wide Web Consortium.* An organization that develops standards for the World Wide Web and promotes interoperability between Web technologies, such as browsers, programming languages, and devices. Members from around the world contribute to standards for XML, SOAP, HTML, and many other Web-based protocols.

wait time [COMPUTING] The amount of time between the time that a client requests an object from a server and the time the client receives that object.

WAN [COMPUTING] *Acronym for wide area network.* A computer network that connects computers in a large area, such as in different cities or countries. The Internet is the most well-known example of a WAN. *See also* LAN.

warping *See* rubber sheeting.

waterfall model [PROGRAMMING] A software design methodology in which development proceeds through a top-down process of overlapping stages. First proposed in 1970, the waterfall model is a highly structured approach to a project life cycle that cascades linearly through the developmental phases of requirements analysis, design, implementation, testing, integration, and maintenance.

watershed A basin-like terrestrial region consisting of all the land that drains water into a common terminus. *See also* source, sink.

watershed

wavelength [PHYSICS] The distance between two successive crests on a wave, calculated as the velocity of the wave divided by its frequency. ▸

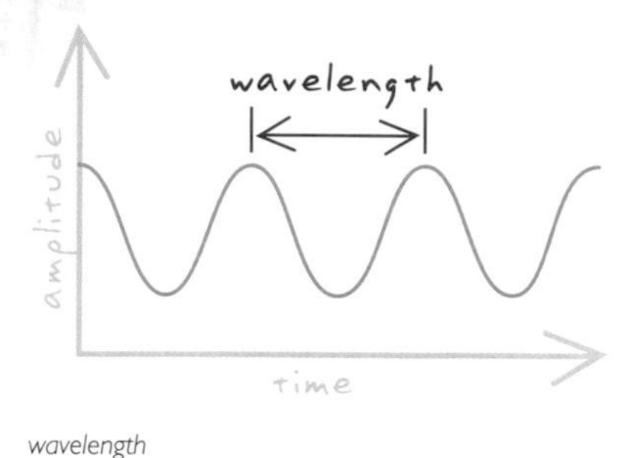

wavelength

wavelet compression A lossy method of data compression that uses mathematical functions and is best used in image or sound compression.

wayfinding 1. [GEOGRAPHY] The mental activities engaged in by a person trying to reach a destination, usually an unfamiliar one, in real or virtual space. Wayfinding consists of acquiring information that is relevant to choosing a route, or a segment of a route, and of evaluating that information in the course of travel so the route can be changed as needed. Wayfinding is the cognitive component of navigation. 2. [GEOGRAPHY] The academic study of wayfinding behavior; also, the scientific art of designing real or virtual environments to make wayfinding easier. 3. [NAVIGATION] Long-distance, open-sea navigation without instruments, as traditionally practiced by Pacific Islanders. *See also* navigation, locomotion.

waypoint [GPS] A location of interest, or a reference point on a route, stored as latitude-longitude coordinates and often captured by a GPS receiver.

Web application 1. [INTERNET] A software program that communicates via the World Wide Web and delivers Web-based information to the user in HTML format. Web applications are typically used to add customization and interactivity to Web pages. Web applications may also be called Web-based applications. 2. [INTERNET] A Web-based program that uses a Web site as the front end of a software application. Web applications allow end users to modify and pass data between a server and a client. Web applications are typically used to provide Web site search capabilities, retrieve and display user information from a database, and provide the ability to purchase items from a Web site.

Web browser [INTERNET] An application that allows users to access and view Web pages on their computer screens. Web browsers enable users to view HTML documents on the World Wide Web. *See also* Web site, HTTP.

Web Feature Server specification [INTERNET] A set of interface specifications that standardizes data manipulation and map display on the Internet. The Web Feature Server (WFS) specification is the result of a collaborative effort assembled by the Open Geospatial Consortium, Inc. (OGC).

Web Map Server specification [INTERNET] A set of interface specifications that provides uniform access by Web clients to maps rendered by map servers on the Internet. The Web Map

Server (WMS) is the result of a collaborative effort assembled by the Open Geospatial Consortium, Inc. (OGC).

Web page [INTERNET] A page of information stored on a Web site and viewed in a Web browser. Web pages may contain text, graphics, animations, forms for data entry, and links to other Web pages.

Web portal *See* portal.

Web server [COMPUTING] A computer that manages Web documents, Web applications, and Web services and makes them available to the rest of the world.

Web service [INTERNET] A software component accessible over the World Wide Web for use in other applications. Web services are built using industry standards such as XML and SOAP and thus are not dependent on any particular operating system or programming language, allowing access to them through a wide range of applications.

Web Service Description Language *See* WSDL.

Web site [INTERNET] A collection of Web pages (HTML files) that are interconnected with hyperlinks and published on the World Wide Web.

weeding Reducing the number of points that define a line while preserving its essential shape. *See also* line smoothing.

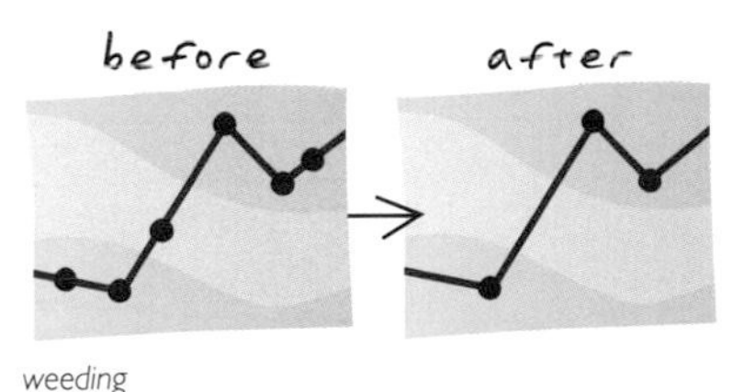

weeding

weed tolerance The minimum distance allowed between any two vertices along a line, set before digitizing. When new lines are added, vertices that fall within that distance of the last vertex are ignored. Weed tolerance applies only to vertices, not to nodes. *See also* weeding.

weight **1.** [MATHEMATICS] A number that indicates the importance of a variable for a particular calculation. The larger the weight assigned to the variable, the more that variable will influence the outcome of the operation. **2.** [ESRI SOFTWARE] A property of a network element typically used to describe the element or to assign a cost for traversing the element. For example, this value may represent the phase or the length of a primary conductor in an electrical distribution system. Weights are calculated based on an attribute of each network feature.

weighted mean center [STATISTICS] The geographic center of a set of points as adjusted for the influence of a value associated with each point. For example, while the mean center of a group of grocery stores would be the location obtained by averaging the stores' x,y coordinates, the weighted

W

mean center would be shifted closer to stores with higher sales, more square footage, or a greater quantity of some other specified attribute.

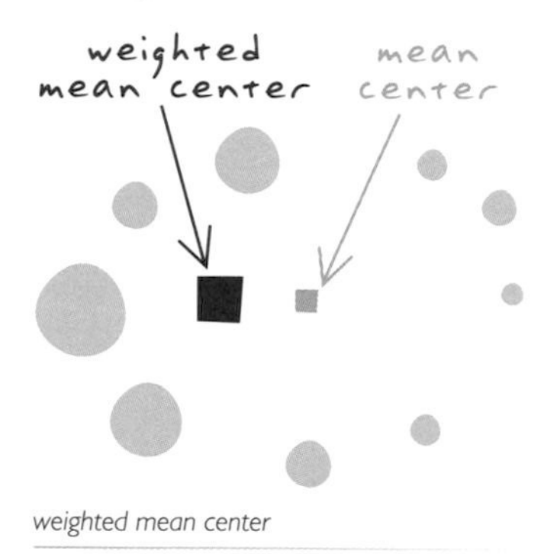

weighted mean center

weighted moving average [STATISTICS] The value of a point's attribute computed by averaging the values of its surrounding points, taking into account their importance or their distance from the point.

weighted overlay A technique for combining multiple rasters by applying a common measurement scale of values to each raster, weighting each according to its importance, and adding them together to create an integrated analysis. *See also* weight.

weird polygon *See* nonsimple polygon.

W

WFS *See* Web Feature Server specification.

WGS 1972 *See* WGS72.

WGS 1984 *See* WGS84.

WGS72 [GEODESY] *Acronym for World Geodetic System 1972.* A geocentric datum and coordinate system designed by the U.S. Department of Defense, no longer in use. *See also* WGS84.

WGS84 [GEODESY] *Acronym for World Geodetic System 1984.* The most widely used geocentric datum and geographic coordinate system today, designed by the U.S. Department of Defense to replace WGS72. GPS measurements are based on WGS84. *See also* geocentric datum.

whisk broom scanner *See* across-track scanner.

wide area network *See* WAN.

windowing [CARTOGRAPHY] The process of limiting the viewable extent of a map or data by panning and zooming.

wind rose A diagram showing, for a given place and time period, how much of the time the wind blows from each direction. Wind roses have many variations, but in the typical pattern, a number of wedges (usually eight, twelve, or sixteen) radiate from the center of a circle. The width and orientation of a wedge represent the direction from which the wind blows; the length of a wedge represents the percentage of time the wind blows from that direction. More complex wind roses use color schemes and other graphic devices to represent wind speed and related information.

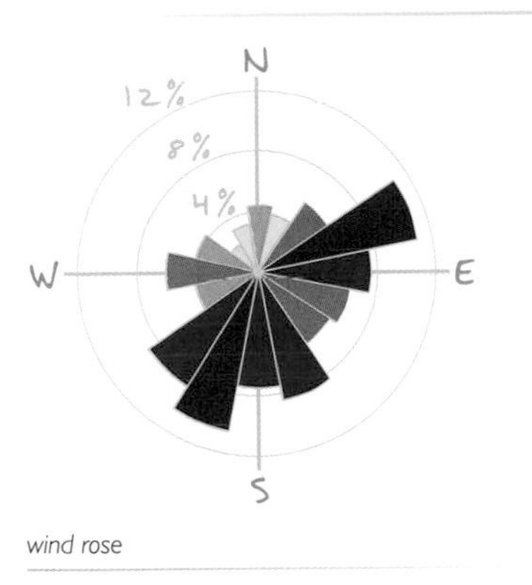

wind rose

wireframe A three-dimensional picture of an object, composed entirely of lines (wires). The lines represent the edges or surface contours, including those that would otherwise be hidden by a solid view. Wireframes make editing easier, since the screen redraws much more quickly.

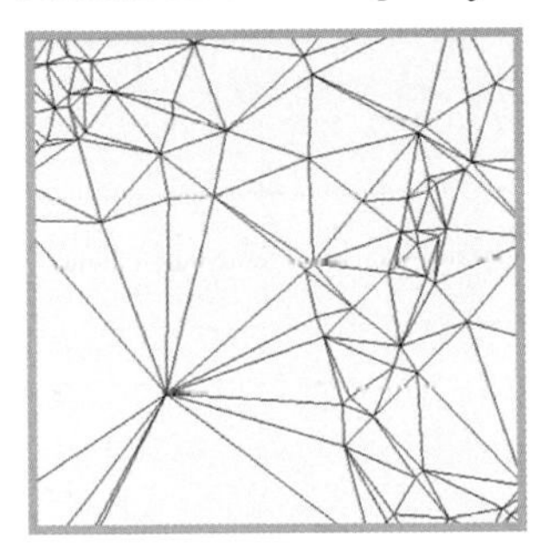

wireframe

wireless application [INTERNET] A ready-to-use Web application for a wireless client such as a handheld device, PDA, or cellular phone, designed for a specific purpose such as maps and routing for travel assistance.

wizard [COMPUTING] An interactive user interface that helps a software user complete a task one step at a time. Wizards are often implemented as a sequence of dialog boxes that the user can move through, filling in required details. Wizards are usually used to simplify long, difficult, or complex tasks.

WMS *See* Web Map Server specification.

word abbreviation *See* abbreviation.

workflow 1. An organization's established processes for design, construction, and maintenance of programs, products, and business objectives. 2. A set of tasks carried out in a certain order to achieve a goal.

working directory [COMPUTING] A directory that indicates the appropriate location on disk to place results from analysis. *See also* workspace, directory.

workspace [ESRI SOFTWARE] A container for geographic data. A workspace can be a folder that contains shapefiles, a geodatabase, a feature dataset, or an ArcInfo workspace. Other multidimensional data formats such as netCDF or HDF can also be considered workspaces, and are often treated in this manner within GIS software packages.

world file A text file containing information about where an image should be displayed in real-world

W

coordinates. When an image has a properly configured world file, GIS software can use the information (a total of six values, including the starting coordinates, the cell size in both x- and y-dimensions, and any rotation and scaling information) to accurately overlay the image with any other data already in a projected or geographic coordinate system.

World Geodetic System 1972
See WGS72.

World Geodetic System 1984
See WGS84.

World Wide Web [INTERNET] A worldwide, decentralized, public information space for sharing documents and conducting business on the Internet. Components of the World Wide Web include information in the form of HTML documents; identification tags (URLs) for the millions of computers that host this information; a set of technical specifications, called HTTP, for sending information from one computer to another; and Web browser software for accessing and displaying information.

World Wide Web Consortium
See W3C.

WSDL [INTERNET] *Acronym for Web Service Description Language.* An XML format for describing the methods, types, and connection point of a SOAP Web service. *See also* SOAP.

W-test [SURVEYING] A type of statistical test used in surveying to detect blunders in a measurement network. The W-test is based on the assumption that the null hypothesis is rejected due to a gross error in one of the measurements and uses an alternative hypothesis to identify the erroneous measurement.

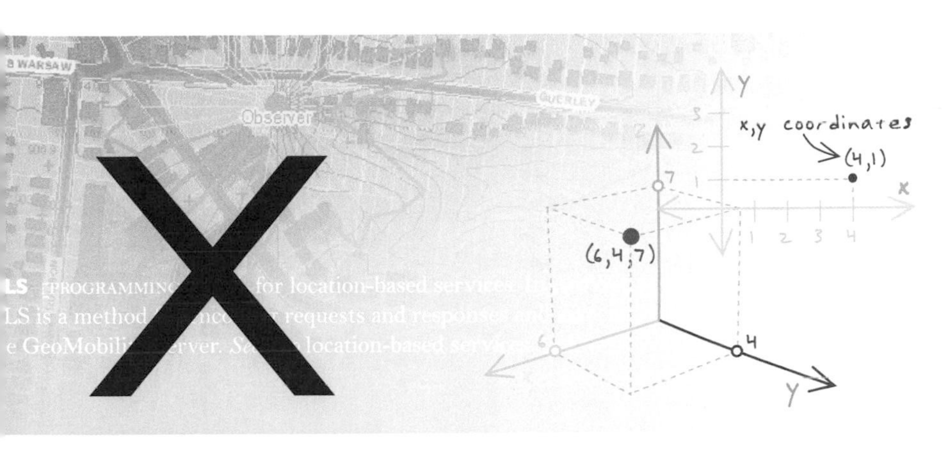

x,y coordinates [CARTOGRAPHY] A pair of values that represents the distance from an origin (0,0) along two axes, a horizontal axis (x), and a vertical axis (y). On a map, x,y coordinates are used to represent features at the location they are found on the earth's spherical surface. *See also* x-axis, y-axis.

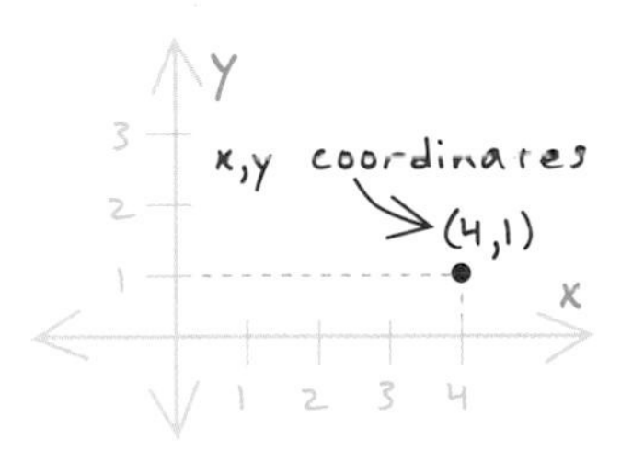

x,y coordinates

x,y values *See* x,y coordinates.

x,y,z coordinates [CARTOGRAPHY] In a planar coordinate system, three coordinates that locate a point by its distance from an origin (0,0,0) where three orthogonal axes cross. Usually, the x-coordinate is measured along the east–west axis, the y-coordinate is measured along the north–south axis, and the z-coordinate measures height or elevation. *See also* x-axis, y-axis, z-axis.

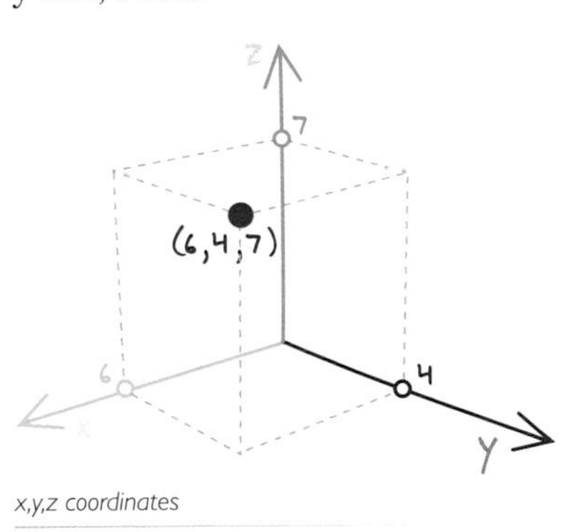

x,y,z coordinates

x-axis 1. [CARTOGRAPHY] In a planar coordinate system, the horizontal line that runs right and left (east and west of) the origin (0,0). ▸

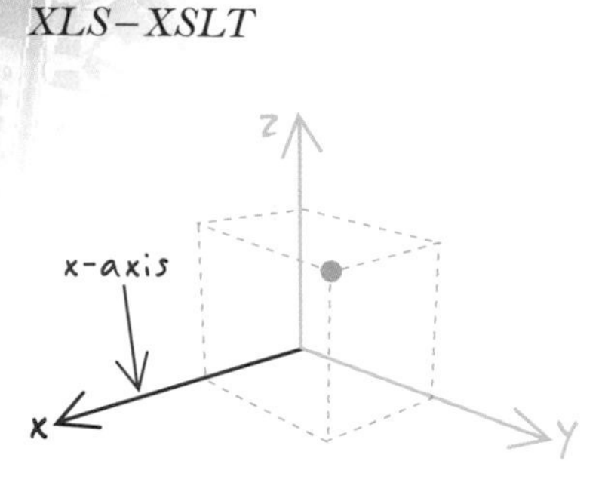

x-axis 1

2. [CARTOGRAPHY] In a spherical coordinate system, a line in the equatorial plane that passes through 0 degrees longitude. **3.** [MATHEMATICS] On a chart, the horizontal axis. *See also* y-axis, z-axis, Cartesian coordinate system.

XLS [PROGRAMMING] XML for location-based services. *See also* location-based services.

XMI [PROGRAMMING] *Acronym for XML Metadata Interchange.* A standard that specifies how to store a UML model in an XML file.

XML [PROGRAMMING] *Acronym for Extensible Markup Language.* Developed by the W3C, a standardized general purpose markup language for designing text formats that facilitates the interchange of data between computer applications. XML is a set of rules for creating standard information formats using customized tags and sharing both the format and the data across applications.

XML Metadata Interchange *See* XMI.

XSL [PROGRAMMING] *Acronym for extensible style language.* A set of standards for defining XML document presentation and transformation. An XSL style sheet may contain information about how to display tagged content in an XML document, such as font size, background color, and text alignment. An XSL style sheet may also contain XSLT code that describes how to transform the tagged content in an XML document into an output document with another format. The W3C maintains the XSL standards. *See also* style sheet, XML, W3C.

XSLT [PROGRAMMING] *Acronym for Extensible Style Language Transformations.* A language for transforming the tagged content in an XML document into an output document with another format. An XSL style sheet contains the XSLT code that defines each transformation to be applied. Transforming a document requires the original XML document, an XSL document containing XSLT code, and an XSLT processor (parser) to execute the transformations. The W3C maintains the XSLT standard. *See also* style sheet, XML, XSL, W3C.

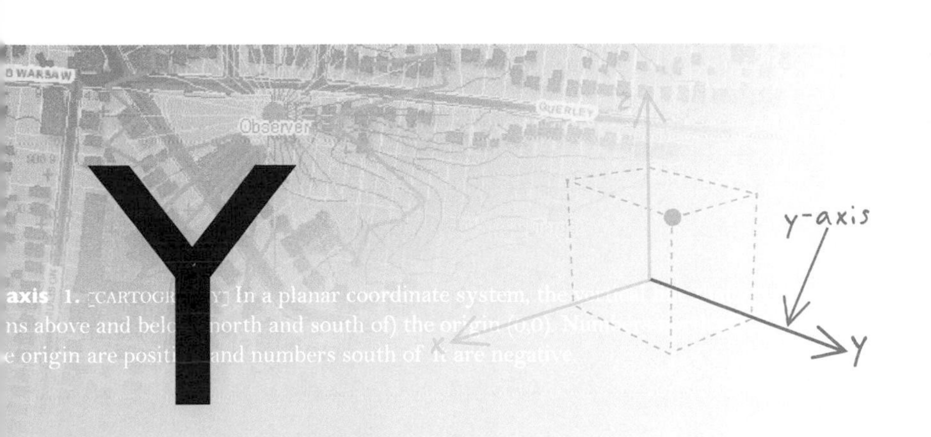

y-axis **1.** [CARTOGRAPHY] In a planar coordinate system, the vertical line that runs above and below (north and south of) the origin (0,0). Numbers north of the origin are positive, and numbers south of it are negative.

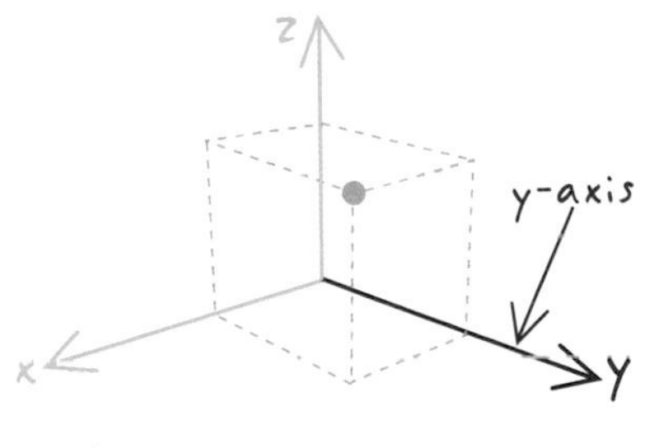

y-axis I

2. [CARTOGRAPHY] In a spherical coordinate system, a line in the equatorial plane that passes through 90 degrees east longitude. **3.** [MATHEMATICS] On a chart, the vertical axis. *See also* x-axis, z-axis, Cartesian coordinate system.

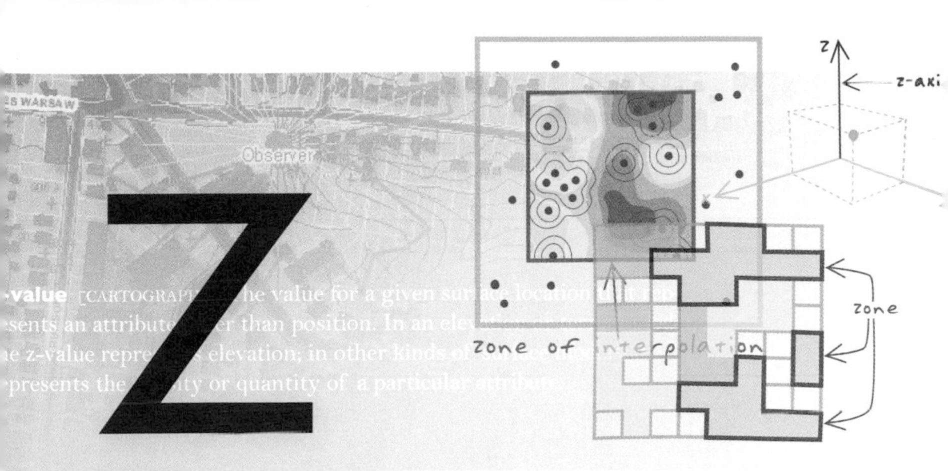

z-axis [CARTOGRAPHY] In a spherical coordinate system, the vertical line that runs parallel to the earth's rotation, passing through 90 degrees north latitude, and perpendicular to the equatorial plane, where it crosses the x- and y-axes at the origin (0,0,0). *See also* x-axis, y-axis.

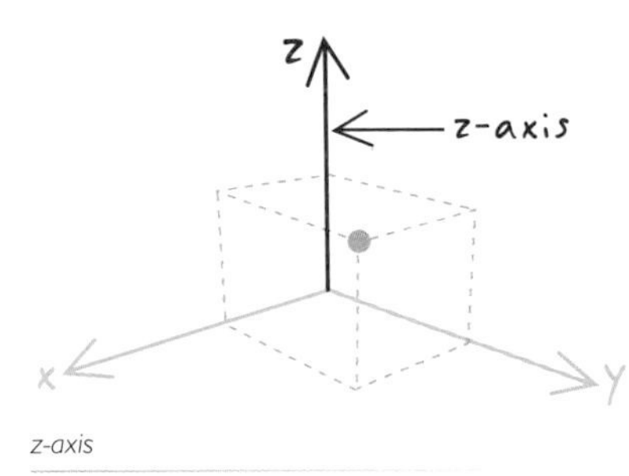

z-axis

z-coordinate *See* z-value.

zenith [ASTRONOMY] The point on the celestial sphere directly above an observer. Both the zenith and nadir lie on the observer's meridian; the zenith lies 180 degrees from the nadir, and is observable. *See also* nadir.

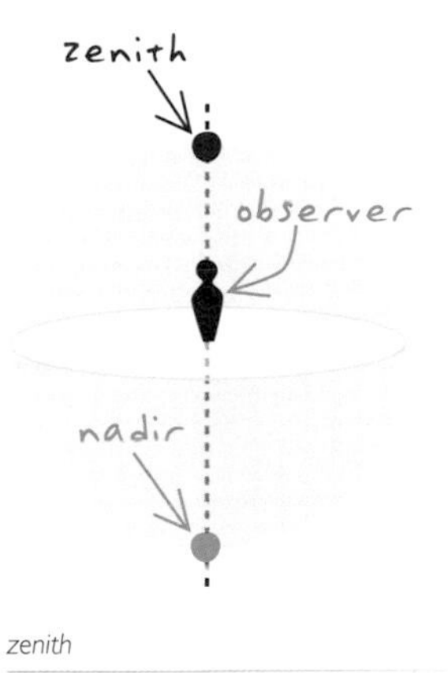

zenith

zenithal projection *See* azimuthal projection.

z-factor [CARTOGRAPHY] A conversion factor used to adjust vertical and horizontal measurements into the same unit of measure. Specifically, the number of vertical units (z-units) in each horizontal unit. For example, if a surface's horizontal units are meters and its elevation (z) is measured in feet, the z-factor is 0.3048 (the number of meters in a foot).

ZIP+4 Code An enhanced ZIP Code that consists of the five-digit ZIP Code plus four additional digits that identify a specific geographic segment within the five-digit delivery area, such as a city block, office building, or other unit. *See also* ZIP Code.

ZIP Code *Acronym for zone improvement plan code.* A five-digit code, developed by the U.S. Postal Service, that identifies the geographic delivery area served by an individual post office or metropolitan area delivery station. *See also* ZIP+4 Code.

zonal analysis The creation of an output raster in which the desired function is computed on the cell values from the input value raster that intersect or fall within each zone of a specified input zone dataset. The input zone dataset is only used to define the size, shape, and location of each zone, while the value raster identifies the values to be used in the evaluations within the zones.

zonal functions *See* zonal analysis.

zone 1. All cells in a raster with the same value, regardless of whether or not they are contiguous.

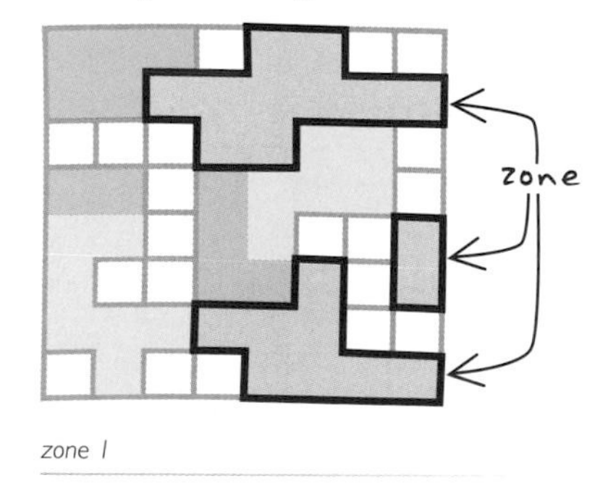

zone 1

2. Additional information about a location or address, used to narrow a geocoding search and increase search speed. Address elements and their related locations such as city, postal code, or country all can act as a zone.

zone of interpolation [SURVEYING] The area in a TIN layer for which values (elevation, slope, and aspect) are calculated. When a TIN layer is clipped to a smaller size to create a more focused study area, the parts that lie outside the study area remain triangulated and are represented as outside lines, but they have no values. These parts are said to be outside the zone of interpolation.

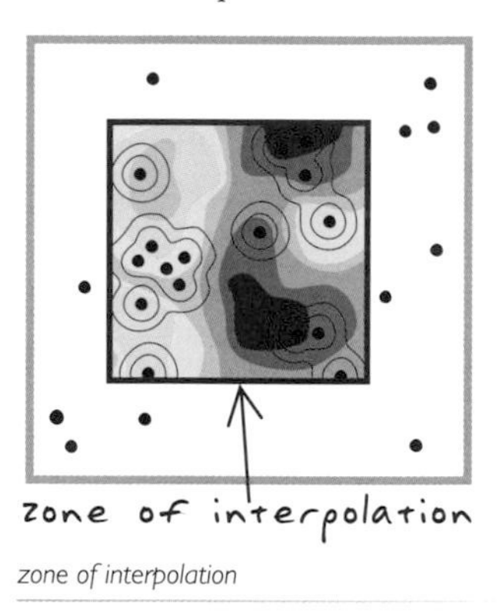

zone of interpolation

zoning The application of local government regulations that permit certain land uses within geographic areas under the government's jurisdiction. Zoning regulations typically set a broad category of land use permissible in an area, such as residential, commercial, agricultural, or industrial. Zoning regulations can also set constraints on building construction within areas, which may affect factors such as the maximum height of structures, minimum setbacks from property lines, the amount of parking that must be provided, or the density of housing.

zoom [COMPUTING] To display a larger or smaller region of an on-screen map or image. *See also* pan.

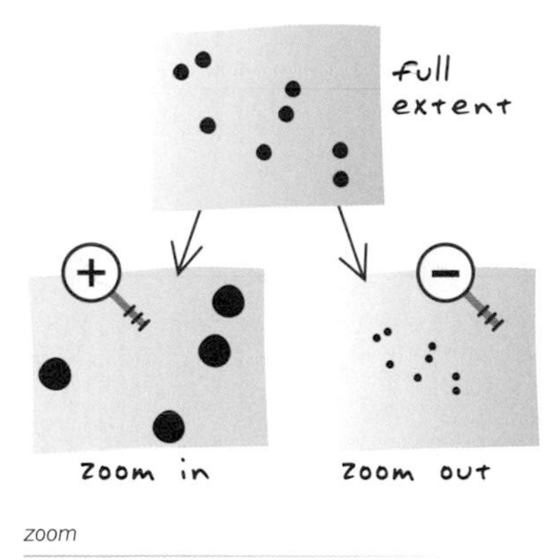

zoom

z-score [STATISTICS] A statistical measure of the spread of values from their mean, expressed in standard deviation units, where the z-score of the mean value is zero and the standard deviation is one. In a normal distribution, 68 percent of the values have a z-score of plus or minus 1, meaning they lie within one standard deviation of the mean. Ninety-five percent of the values have a z-score of plus or minus 1.96, meaning they lie within two standard deviations of the mean; 99 percent of the values have a z-score of plus or minus 2.58. Z-scores are a common scale on which different distributions, with different means and standard deviations, can be compared. *See also* standard deviation, normal distribution.

z-tolerance [CARTOGRAPHY] In raster-to-TIN conversion, the maximum allowed difference between the z-value of the input raster cell and the z-value of the output TIN at the location corresponding to the raster cell center.

z-value [CARTOGRAPHY] The value for a given surface location that represents an attribute other than position. In an elevation or terrain model, the z-value represents elevation; in other kinds of surface models, it represents the density or quantity of a particular attribute.

z-value

Z

1NF *See* first normal form.

2NF *See* second normal form.

3D feature A representation of a three-dimensional, real-world object in a map or scene, with elevation values (z-values) stored within the feature's geometry. Besides geometry, 3D features may have attributes stored in a feature table. In applications such as CAD, 3D features are often referred to as 3D models.

3D graphic A representation of a three-dimensional, real-world object in a map or scene, with elevation values (z-values) stored within the feature's geometry. Unlike 3D features, 3D graphics do not have attributes.

3D model A construct used to portray an object in three dimensions. In GIS, 3D models are often referred to as 3D features.

3D polygon *See* polyhedron.

3D shape A point, line, or polygon that stores x-, y-, and z-coordinates as part of its geometry. A point has one set of z-coordinates; lines and polygons have z-coordinates for each vertex in a shape.

3D symbol [CARTOGRAPHY] A symbol with properties that allow it to be rendered in three dimensions.

3NF *See* third normal form.

Appendix

Cartographic text in ArcGIS software: Annotation, labeling, and graphic text

By Craig Williams, ESRI product engineer

Text can be used in a variety of ways to enhance the information on a map. Maps convey information about geographic features, yet displaying only features on a map—even with symbols that convey their meaning—isn't always enough to make your point. Adding text to your map improves the visualization of geographic information.

There are various uses for text on a map. Descriptive text can be used to help identify individual map features. For example, you might add the name of each major city in Africa to your map. You can also add text to a map to draw attention to an area, such as the general location of the Sahara Desert. Text can also provide information and improve the presentation of your map. For example, you might consider including information such as the map's author, data source, or date, or a map title to provide context.

Using different kinds of text

Because text serves so many different mapping purposes, ArcGIS software offers several different types. The main types of text are labels, annotation, and graphic text. In ArcGIS, these terms have more specific meanings than they have in the general cartographic world. ArcGIS differentiates these terms by their function and storage models. All types of text are based on the same text symbol; therefore, they support the same graphic abilities, but they are each tailored to different situations. It is important to know the differences between these types of text when considering adding text to a map. It is also important to realize that in ArcGIS annotation and labeling mean something different from what they might mean in general cartographic practice.

Labeling

In ArcGIS, a label is a piece of text that is automatically positioned and whose text string is based on feature attributes. Labels offer the fastest and easiest way to add descriptive text for individual features to your map. For example, you can turn on dynamic labeling for a layer of major cities to quickly add all the city names to your map. Because labels are based on attribute fields, they can only be used to add descriptive text to features. Automatic label positioning is performed by a label engine. The ESRI Standard Label Engine is provided with all license levels of ArcGIS. The extension Maplex for ArcGIS provides the ESRI Maplex Label Engine for advanced cartographic labeling.

Annotation

Annotation can be used to describe particular features or add general information to the map. You can use annotation like labels, to add descriptive text for map features, or just add a few pieces of text manually to describe an area on your map. Unlike labels, each piece of annotation stores its own position, text string, and display properties. Compared to labels, annotation provides more flexibility over the appearance and placement of your text because you can select individual pieces of text and edit their position and appearance. You can use ArcMap software to convert labels to annotation. Annotation can be further divided based on where it is stored: in a geodatabase, in a map document, or in one of the read-only formats that ArcGIS supports.

Geodatabase annotation is the native annotation type of ArcGIS. With this type, pieces of text or graphics are stored as features in a geodatabase annotation feature class. The text positions are stored in spatially referenced coordinates and are sized graphically according to the reference scale. Geodatabase annotation is fully editable using the ArcGIS editing tools. There are two types of geodatabase annotation. Standard annotation is geodatabase annotation that does not have a maintained link to any features. Feature-linked annotation is a special type of geodatabase annotation that is directly linked to features and has special updating behavior when the linked features are updated.

Map annotation consists of text or graphics stored in the map document. With this type of annotation, pieces of text or graphics are stored as elements in a map's annotation group. A map can have multiple annotation groups to aid with organization. The text positions are stored in spatially referenced coordinates and are sized graphically according to the map scale or, optionally, a reference scale. Map annotation is fully editable using the ArcGIS graphic tools.

ArcGIS supports a variety of other annotation types as read-only data sources, including ArcInfo Workstation coverage, PC ARC/INFO coverage, Spatial Database Engine (SDE) 3.x, computer-aided design (CAD), and Vector Product Format (VPF) annotation. ArcGIS provides tools to convert these types of annotation to geodatabase annotation or map document annotation, both of which are editable formats.

Graphic text

Text is also a fundamental part of page design in the page layout view in ArcGIS. This type of text is called graphic text, and is useful for adding information on and around your map that exists in page space—as opposed to annotation, which is stored in geographic space. Graphic text editing is performed using the graphic editing tools in ArcGIS.

Geometry: Creating map features with points, lines, and polygons

By Rob Burke, ESRI instructor

When you make a map, even a simple hand-drawn one like the one shown below, you draw a few points, lines, and polygons and assign each one a meaning. You might also include directions that follow the geometric shapes in the map, such as the following: start at your house (point); turn left on 1st Street (line); drive past the mall (polygon), and continue until you come to Pine Street (line); turn right on Pine Street (line), and stop at the third house (point) on the left. Points, lines, and polygons are the geometry figures we use to represent real things and places.

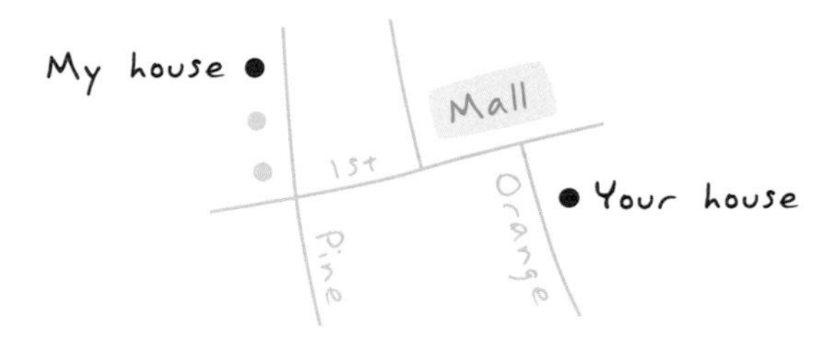

This hand-drawn map would be your graphic version of reality, but it wouldn't be drawn to scale with true distances and direction. If you wanted to be able to take measurements and perform analysis on the data, it would need coordinates that represent the real locations on earth.

GIS technicians work painstakingly to make sure their geometry has precise and accurate coordinate information, using heavy-duty math, satellites, and expensive survey equipment to make sure the x- and y-values (or longitude and latitude values) are as close to true values as possible.

Points in ArcGIS software have one x-value and one y-value to define their location. For example, the point in the illustration below is identified with two numbers, −117.3 and 34.4. These values represent the longitude and latitude of Redlands, California. A line is a pair of points where one defines its start and the other its end. An arc, which also has starting and ending points, is defined as part of a circle using geometric values like angles, chord length, radius, and tangency to precisely describe the arc's location on the circle.

Lines and arcs are called segments. One or more segments are used to construct a polyline. A polyline can be made up of any number of segments, and the segments can be a mixed set of lines and arcs.

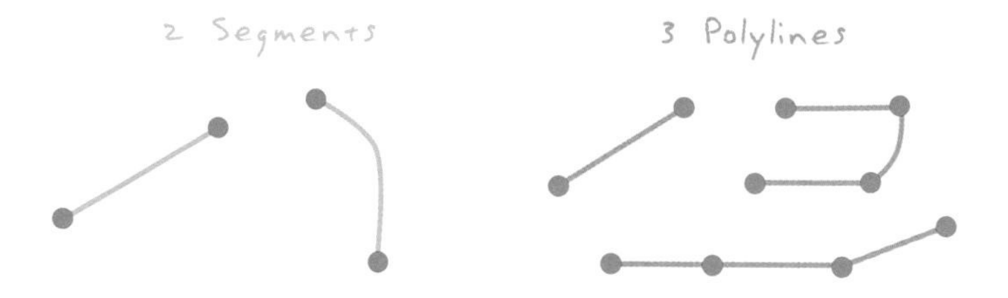

In casual GIS talk, we interchange the words line and polyline, but in GIS software, a polyline has a distinct meaning. A polyline is really a collection of lines (segments). Non-GIS people look at a polyline and call it a line. Since line is the more commonly used vernacular, the GIS-savvy usually just follow along and call polylines lines to avoid argument.

The points that make up a polyline are called vertices. In casual conversation these points are sometimes called shape points, probably because they give a polyline its shape. During an editing session, you can view and edit a polyline's vertices and their corresponding coordinates. Not only can you add a vertex to a polyline, but you can move or remove them, too.

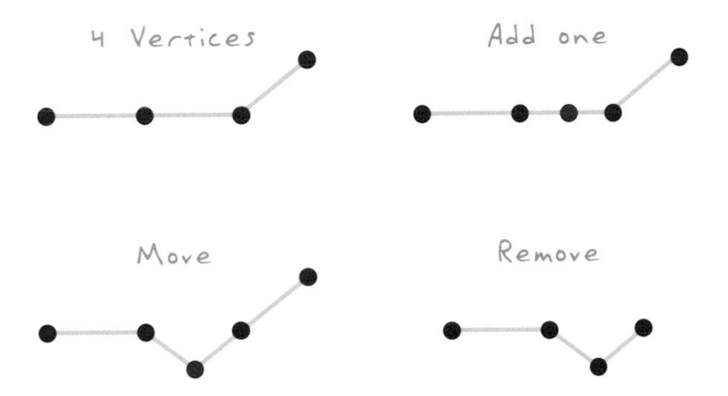

A polyline whose last vertex is the same as its first forms an enclosed area called a ring. In the most typical mapping situations, a ring and a polygon are exactly the same thing. To allow for the complex structure of reality, a specialized polygon called a multipart polygon can be made up of many rings. Multipart polygons, for example, can have holes in them or be noncontiguous. Either way, they are considered a single polygon. For example, sometimes within a city boundary there is an unincorporated zone that is not an actual part of the city. The city boundary could be one ring and the unincorporated zone could be a second ring inside the first. Both rings work together to form a single multipart polygon that represents the city.

A multipart polygon can also be composed of a collection of noncontiguous rings. For example, the state of Hawaii is made up of seven primary islands, which could be stored as seven separate single-ring (single-part) polygons. In some situations, however, you could combine the seven rings into a single multipart polygon. If you need to store attributes about each island, for example, you would use seven single polygons. If you only need to store attributes about the state, however, one multipart polygon will do.

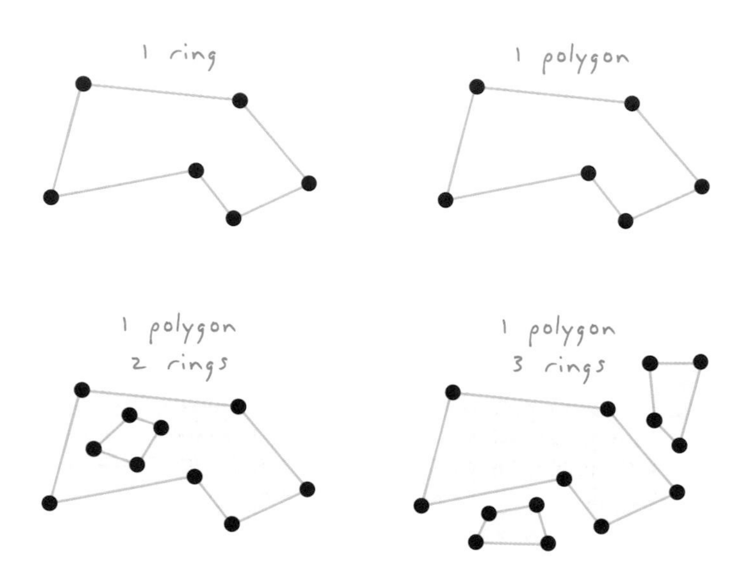

In review, a point is a unique location; a line is made up of two points; an arc is part of a circle also defined with two points, plus some geometric descriptors; lines and arcs are called segments; segments are grouped together to form polylines; a polyline that starts and ends at the same location and forms an enclosed area is called a ring; and a polygon can be composed of one or more rings.

How features and feature attributes are stored

By Rob Burke, ESRI instructor

Geometries like points, polylines, and polygons are used to represent the location of a place or thing on earth. Features add to those geometric descriptions by also storing a location's characteristics.

Features are stored as records in a database table. Fields in the table represent the possible characteristics that each feature might have. A value represents one characteristic of a feature. For example, in table A below, the feature with FID (feature identification number) 202 represents the country San Marino. Its population value of 23,758 is slightly smaller than Monaco's population of 27,409.

Table A

Attributes of World Countries

FID	Shape	Name	Population	SquareKM	
230	Polygon	United Kingdom	56420180	243137.203	Field (arrow to SquareKM header)
213	Polygon	Switzerland	6713839	41178.398	
205	Polygon	Spain	39267780	505674.406	
202	Polygon	San Marino	23758	98.171	Value (arrow to 23758)
181	Polygon	Portugal	9625516	92098.273	
165	Polygon	Netherlands	15447470	35492.691	
147	Polygon	Monaco	27409	11.988	Record (arrow to this row)

Tables can be spatial or nonspatial. A nonspatial table might be a table of world countries, like table B on the following page, in which each record represents a country, and each field indicates a factor about each country like its name or the name of its currency. Nonspatial tables do not contain any geometry information. For example, there aren't any attributes in table B that would help you determine where San Marino is located, or how its population compares with that of Monaco.

Table B

Attributes of World Countries

Name	Currency
United Kingdom	Pound Sterling
Switzerland	Franc
Spain	Peseta
San Marino	Lira
Portugal	Escudo
Netherlands	Guilder
Monaco	Franc

The two tables above both represent information about countries. Table A, however, has an extra field that makes it spatial. The Shape field, sometimes referred to as the spatial column or the geometry field, contains the geometric description of each feature. Visually, the words point, line, or polygon appear in the Shape field. Technically, behind the scenes, the values that get stored are coordinate pairs. For point features, one pair of coordinates is stored; for line and polygon features, a list of coordinate pairs is stored.

An overview of layers in ArcGIS

By Jeff Shaner and Anthony Burgon, ESRI product engineers

There are four types of geographic information in ArcGIS: features, images, surfaces, and tables. Each of these types of geographic information is organized into GIS datasets based upon the information or entity it portrays. These datasets can be displayed in ArcGIS as a layer that, when combined with other layers in a data frame, form a map.

Properties of a layer

Regardless of the type of geographic information a layer represents, it will include a set of properties common to all layers. These properties include the following:

- **Name.** Each layer has a name. By default, a layer will use the name of the data source it is referencing. If that data source contains an alias name, that name will be used.
- **Description.** You can add descriptive text to a layer. The description is displayed and can be edited from the properties of a layer.
- **Data source.** Each layer references some type of geographic information. It could be a geodatabase feature class, a CAD file, a TIFF image, or even a surface.
- **Spatial reference.** Each layer has a spatial reference. The spatial reference is defined by the data source and is projected to the coordinate system of the data frame.
- **Scale.** You can define the minimum and maximum scale at which the contents of the layer are drawn on the map.

In addition to these properties that all layers have in common, there are several key types of layers that comprise a map:

- **Feature layers** display vector data sources (points, multipoints, lines, and polygons) that are stored in coverages, shapefiles, and geodatabases.
- **Raster layers** display raster image files and raster datasets.
- **Graphics layers** display map graphics, text, and geodatabase annotation.

- **Web service layers** display ArcIMS, ArcGIS Server, and WMS data.
- **Data layers** display the contents of a topology, network, terrain, and so on.

Types of layers

There are several different types of layers in ArcGIS. Each type of layer has specific properties based on the characteristics of the geographic information it represents. The following is a list of all major layer types in ArcGIS:

Layer type	Description
Annotation layer	Displays annotation stored in a geodatabase
CAD annotation layer	Displays CAD text
CAD feature layer	Displays CAD entities
Coverage annotation layer	Displays annotation stored in a coverage
Dimension layer	Displays dimensions stored in a geodatabase
Feature layer	Displays vector features (points, lines, polygons)
Geoprocessing layer	Displays the output of a geoprocessing tool
Graphics layer	Displays graphics on a map or page
Group layer	A collection of layers with the properties of a single layer

Layer type	Description
IMS layer	Displays IMS data
Map server layer	Displays ArcGIS Server map server data
Raster catalog layer	Displays a raster catalog
Raster layer	Displays raster data
Terrain layer	Displays terrain information
Tin layer	Displays tin data
Topology layer	Displays geodatabase topology errors, dirty areas
Tracking layer	Displays tracking information
WMS layer	Displays WMS data

How layers display geographic information

Layers use symbology to draw geographic information onto the map display. Symbology defines the appearance of the geographic information in a layer. By default, when you add a GIS dataset to a map, symbology is assigned to the layer that gets created for it. Depending upon the type of layer created, you will find different ways to symbolize the data contained within the layer. Usually, information contained within the dataset itself is used to determine how the data is symbolized.

Projected and geographic coordinate systems: What is the difference?

By Melita Kennedy, ESRI product engineer

A coordinate system provides a regular and consistent method of quantifying the locations of objects and their spatial relationships. By locating objects within a coordinate system, we can determine whether two objects overlap, are neighbors, or are far apart. Geographic and projected coordinate systems are often used in GIS.

Geographic and projected coordinate systems represent the world in very different ways. A geographic coordinate system maps locations on a spherical surface that is a simplification of the actual earth. A projected coordinate system further simplifies the representation of the earth by mapping locations on a flat surface with a map projection.

Geographic coordinate systems

In geographic coordinate systems, the three-dimensional spherical surface used to define locations on the earth is itself defined by an angular unit of measure like degrees, a prime meridian, and a datum. Points are referenced by their longitude and latitude values. Longitude values are measured east-west, while latitude values are measured north-south. The linear distance may differ between two points separated by the same angular distance, depending on their locations. This is mainly because the reference surface is curved, and because lines of longitude converge at the poles.

Projected coordinate systems

For smaller areas, and to simplify calculations, GIS data is often georeferenced to a projected coordinate system. A projected coordinate system is a planar, two dimensional coordinate system. The connection between geographic and projected coordinate systems involves mathematical formulas collectively called a map projection, or simply a projection.

Representing data that was originally curved on a two-dimensional surface will cause distortions in one or more of the following spatial properties: distance, area, shape, or direction. No map projection can preserve all these properties; as a result, all flat maps contain distortions. Each map projection is, however, distinguished by its suitability for representing a particular portion and amount of the earth's surface, or by its ability to preserve distance, area, shape, or direction. Some map projections minimize distortion in one property at the expense of another, while others strive to balance the overall distortion.

A projected coordinate system is always based on a geographic coordinate system. The information contained within a geographic coordinate system is needed to properly convert locations to a projected coordinate system.

Each type of coordinate system has its advantages and disadvantages. Geographic coordinate systems make it easy to identify locations on a globe, but it is much easier to calculate spatial locations and relationships when you are working in a projected coordinate system. Coordinates based on a geographic coordinate system contain less distortion, but are more difficult to work with. Projected coordinate system coordinates are easier to work with, but quantities like distances and angles are often distorted due to the map projection.

Remote sensing

By Timothy Kearns, ESRI product engineer

Most textbook definitions describe remote sensing as the art and science of observing phenomena and collecting information about those phenomena without coming into physical contact with them. It is a science because the discipline employs theories and methodologies that rely on extracting information about surfaces through the interaction of energy and matter. It is also an art because it requires the knowledge and skills used to produce visual products such as images, elevation surfaces, and three-dimensional models that are generated for qualitative and quantitative interpretation and analysis.

The value of remote sensing to geographic information systems (GIS) is immeasurable. Most contour maps, topographic maps, and base data for everyday map use have their origins in remotely sensed data. Aerial photography and its complementary discipline, photogrammetry, have long provided the GIS industry with high-resolution images of the earth that are used to extract spot elevations for contours, digital elevation models (DEM), and basemap vectors such as roads, coastlines, power lines, and so on. Data collected from space and the sea floor allows GIS analysts to generate and archive information that includes surficial geology, land use, sea surface temperature, forestry, agriculture, urban sprawl, and much more.

The science

The essential process of a remote-sensing system includes an energy transmission source, a medium for that energy to pass through, matter for the energy to interact with, a reflection or retransmission of energy from matter, and a sensor or receiver to measure the reflected or retransmitted energy. These sensors are calibrated to measure particular wavelengths reflected and transmitted from surface material and convert the energy received into digital numbers that are subsequently used to create images of the surface.

All remote-sensing systems are dependent on a common element: energy. Energy comes in many forms, yet most traditional remote-sensing systems measure wavelengths of electromagnetic radiation from within the electromagnetic spectrum. Humans can only see a small portion of the electromagnetic spectrum, called visible light, which ranges from 0.4μm to 0.7μm. Remote sensing is about detecting and measuring not only visible light, but energy that we cannot see.

The most common energy source for remote sensing systems is the sun, which provides most of the visible light and thermal and infrared energy found around us. However, some remote-sensing systems generate their own energy, such as radar satellites, laser light systems, and acoustic transmitters. Energy transmitted from the source is affected by the medium through which it travels. Atmospheric windows allow the transmission of radiation; conversely, atmospheric particles can attenuate energy. The matter that energy comes in contact with plays a significant role in the process of remote sensing. Energy can be transmitted, absorbed, emitted, scattered, and reflected by the matter that it encounters. The composition, texture, and shape of matter affect the energy interaction. For example, water reflects short-wavelength energy, but absorbs longer wavelengths, whereas healthy vegetation strongly reflects infrared energy.

The technology

Modern remote sensing is derived from aerial photography, which was developed during World War II. In the 1960s, scientists wanted more out of earth observation than just looking at photographs captured with visible light. First, infrared aerial photography was developed, which led to the deployment of multispectral scanners on satellites in the early 1970s. The sensors on satellites and airplanes could detect wavelengths outside of the visible spectrum and record the energy associated with the ground beneath. As the technology evolved, remote-sensing platforms achieved higher spatial resolution and finer spectral resolution. They began to fly in lower orbits, use laser light to measure elevation, and use directional imaging from a satellite. All of these changes have led to a robust remote-sensing industry that now consists of more than 30 commercial earth-observation satellites operating and more than 175,000 people employed in remote-sensing related jobs in the United States alone.

There are two categories of sensors used in remote sensing: active and passive. Examples of active sensors include radar, lidar, and sonar. Most traditional remote-sensing platforms, called multispectral scanners, are passive systems. Passive imagers measure incoming energy without emitting any of their own. These systems are generally limited to daytime acquisition, when the light from the sun is available to provide reflected energy. Passive sensors include aerial photography, multispectral satellites, and hyperspectral systems.

The mechanical methods sensors use to acquire information vary from device to device. Image-based sensors are generally categorized into across-track or along-track scanners, whereas lidar employs an oscillating mirror to rotate laser light in a zigzag pattern across the surface beneath. Aerial-based remote sensing can include aerial photography, hyperspectral scanning, radar, or lidar, and generally follow line-based surveys that traverse back and forth over long distances on the ground.

Satellites, however, are placed in orbit above the surface of the earth. Lower-orbit platforms, orbiting between 500 and 800 kilometers above the earth, are called polar-orbiting satellites because they pass near the north and south pole on each revolution around the earth. It takes approximately 99 minutes for these types of satellites to make a single orbit around the planet. Higher-altitude satellites, those that orbit the earth at approximately 36,000 kilometers, are usually positioned so that they hover over the same area. Their orbit is timed to coincide with the rotation of the earth; thus, they only make one complete revolution every 24 hours. These satellites are referred to as geostationary; weather-observing satellites are the best example of these.

Each remote-sensing platform is unique, and the sensors on board vary greatly in their resolutions. There are four main types of resolutions: spatial, spectral, radiometric, and temporal. Commercial satellites with high spatial resolution can discriminate features on the ground approximately 1 meter apart, whereas low-resolution satellites are on the order of 500 to 1,000 meters. The revisit time it takes for a satellite to image the exact same area on the earth is referred to as the temporal resolution and can vary from three to twenty-five days, depending on the characteristics of the satellite and the latitude at which it is imaging. The spectral resolution of a sensor describes the width, in nanometers, of a spectral band that the platform is capable of sensing. Hyperspectral sensors have the ability to sense hundreds of very narrow portions of the electromagnetic spectrum, whereas traditional multispectral scanners capture three to seven spectral bands.

The information associated with each band of energy is stored digitally on board the sensor until it is downloaded using multichannel and multiplexing channel receivers. If it is an airborne platform, the data is retrieved and downloaded at the end of each day. For satellite platforms, however, the data must be downloaded to ground receiving stations placed at strategic locations around the world.

The data

Remote-sensing devices typically capture information that is encoded and organized into images. Some remote-sensing devices collect discrete point information, such as lidar or sonar. This data is usually stored in large binary files that are ingested into applications for postprocessing and surface modeling. Aerial photograph surveys historically collected film photos, but this technology is evolving to collect digital imagery instead.

All satellite earth-observation systems collect digital imagery, as do airborne hyper- and multispectral systems, including radar. Digital images are an array of rows and columns made up of cells, or pixels. Each cell has a single numerical attribute, called a digital number, and represents the amount of energy recorded by the sensor on a binary scale dictated by the bit depth of the image. The collection of all the cells in a single image represents the area covered by one scene of the imaging device, or sensor. For systems that collect multiple portions of the electromagnetic spectrum simultaneously, separate images for each band are acquired. Each image represents a single portion of the electromagnetic spectrum and covers the exact same area on the ground. Combining three images together by projecting them through red, green, and blue light forms color composites.

Most imagery collected by aerial and satellite systems needs to be geometrically corrected to account for distortions, warping, and proper positioning in either a global or projected coordinate system. The use of ephemeris metadata, GPS ground truthing, and ground control points can assist in this process.

How remote sensing is related to GIS

The information content of remote-sensing imagery and discrete data is a fundamental source of base-layer information for GIS. Both qualitative and quantitative analysis, in the form of digital image processing, allow GIS analysts to use the information content of imagery to make decisions and support spatial queries. Digital image processing can include radiometric corrections to remove noise, image enhancement, change detection, and feature extraction.

Remote-sensing imagery is often used by analysts and viewers as backdrop imagery to visually interpret surface features such as roads, buildings, coastlines, and seafloor geomorphology. In this regard, remotely sensed imagery can help analysts gain perspective, or provide them with the visual information necessary for digitizing vector-based features from the image (for example, coastlines, roads, forest stands, and urban sprawl).

The quantitative value of remotely sensed imagery is of paramount importance to GIS. The information in the imagery represents properties of the incident surface. Often, numerical patterns are formed, which can assist algorithms in statistically classifying imagery and aid in decision-making about relationships involving surface features. The purpose of remote sensing is to collect information about some phenomena, and the goal of image processing is to assist in the extraction of that information, either through human interpretation or through automation such as image classification or other statistical means. There are various methods for converting image-based information into GIS vector data, which facilitates geoprocessing, archival in spatial databases, and the creation of maps and visualizations.

Information from space, the air, and ships significantly aids our understanding of the earth. The evolution of remote sensing from aerial photography to include multispectral imaging, lidar, sonar, and radar has exciting implications for the field of GIS. Our abilities to model the environment around us would not be complete without the science and art of remote sensing.

The benefits of using topology in GIS

By Rob Burke, ESRI instructor

Topology is all about spatial relationships between pieces of geometry. It's a branch of mathematics that takes a step beyond geometry to prove relationships like area, connectivity, contiguity, containment, and direction. Topology enhances GIS functionality in spatial analysis, error identification, and error correction.

The human eye can see topologic relationships. For example, you can look at a road map and determine the roads that connect one city to another, the state that contains the cities, the neighboring states, and the direction of a river flowing through one of the cities. For these activities, it might seem foolish to even worry about using a math equation. As the number of features being analyzed increases, however, topology is right there to help.

Let's say that you are a major appliance retailer, and today you are responsible for delivering 300 appliances from four warehouses. You'd have to look at the road map and determine 300 separate routes to travel. Sure, you could do it, but with topology the problem can be solved as fast as a computer can run a math formula.

Spatial analysis, like the connectivity analysis described above, is just one facet of topology in GIS. A particularly laborious task performed by GIS technicians is looking over a paper map for hours trying to identify and mark geometry errors. For example, before getting an accurate tracing of connectivity, someone must confirm that all the lines that are supposed to be connected are really connected. Imagine visually examining this connectivity for half a million street segments. It could take one person several days, and, if the person became fatigued, they might leave some connections unverified. In just a few moments, topology can check all the connections and then place graphical markers where potential violations exist (error identification). Instead of days, the technician's work is reduced to an hour or so.

Another benefit of topology is error correction. After error identification, a set of topology tools can be used to automatically correct the errors. The technician can use nontopological editing tools to correct geometry errors one at a time, but the topological tools in ArcGIS software are automatic operations that can process and correct multiple errors all at once.

Topology adds three major benefits to the GIS industry, all of which are time savers and take advantage of computer speeds to carry out processing that humans perform inefficiently. Spatial analysis, locating potential errors, and fixing them quickly helps GIS technicians become very efficient.

BOCA RATON PUBLIC LIBRARY, FLORIDA
3 3656 0531648 2